Laboratory Manual For
PHYSICAL GEOLOGY

Laboratory Manual For
PHYSICAL GEOLOGY

Second Edition

NORRIS W. JONES
The University of Wisconsin–Oshkosh

Featuring the Photography of Doug Sherman

**WCB
McGraw-Hill**

Boston, Massachusetts   Burr Ridge, Illinois   Dubuque, Iowa
Madision, Wisconsin   New York, New York   San Francisco, California   St. Louis, Missouri

WCB/*McGraw-Hill*
A division of the **McGraw-Hill Companies**

LAB MANUAL FOR PHYSICAL GEOLOGY

 This book is printed on recycled, acid-free paper containing 10% postconsumer waste.

1234567890 QPD/QPD 0987

ISBN: 0-697-37451-3

Publisher: *Ed Bartell*
Sponsoring editor: *Lynne Meyers*
Developmental editor: *Daryl Bruflodt*
Marketing manager: *Lisa L. Gottschalk*
Project manager: *Marilyn M. Sulzer*
Production supervsor: *Deborah Donner*
Cover design: *Kim Loscher/Rokusek Design*
Cover photograph:© *Doug Sherman/Geofile*
Photo research coordinator: *Carrie Burger*
Art editor: *Renee A. Grevas*
Compositor: *Carlisle Communications, Ltd.*
Typeface: *10/12 Times Roman*
Printer: *Quebecor, Inc.*

http://www.mhhe.com

TABLE OF CONTENTS

LIST OF TABLES AND MAPS

PREFACE

TO THE STUDENT

Geology and other sciences are major parts of all our lives, and many of the daily decisions we make are affected in some way by science. Because of this, all educated people (scientists or not) should understand, in a general way, how science is done, and should be able to apply the basic ideas of the scientific method. Thus, a principal objective of this laboratory manual is to show you how geologists use the scientific method by having you do it yourselves. You will find that geologists, like other scientists, operate in a very logical and straightforward manner, which differs little from the way you normally do things. You will learn how to gather and use data (for example, making measurements to use in calculations), establish relations based on these data, classify information, develop and test hypotheses or models to explain observations and relations, and use this information to draw conclusions and make predictions. This may sound a bit intimidating, but don't panic; it's not difficult. You'll be surprised at what you'll be able to do.

Laboratory sessions will be most efficient if you *read the entire chapter before coming to class.* If allowed by the instructor, students may work in groups. In a group setting, it soon becomes apparent who has prepared and can contribute to the group. As a courtesy to others, and for your own benefit, *prepare for class!* Make sure you bring those things to class that are needed for the lab, as indicated in the **Materials Needed** section. Don't depend on your classmates.

You will find that geology is relevant to our daily lives. Look around you. Most of the manufactured items you see are made of earth materials—minerals, rocks, or petroleum. We live along rivers that flood, drink water pumped from underground, and dispose of our garbage in the ground. We and our politicians have to understand the consequences of our actions. So pay attention, ask questions, and learn to think critically about these geological issues. Make sure that science-related issues are based on good science; don't let charlatans determine our public policy.

TO THE INSTRUCTOR

This lab manual is for a college-level, introductory course in physical geology. It is based on the following assumptions: (1) A typical student is a freshman or sophomore with little background in science and math. He or she may have had a unit on Earth science in seventh, eighth, or ninth grade, but most likely has had little or no exposure to geology. (2) The course has no prerequisites. (3) Although a few students taking the class plan to major in geology, and more will decide to major once they learn how interesting geology is, most students are taking the class to satisfy a general requirement and will not major in geology or other sciences.

Each chapter contains more exercises than your students may have time to do, so you probably will have to be selective in what you assign.

The first edition of this manual was an outgrowth of 25 years of teaching physical geology and using many different lab exercises. I first wrote a physical geology lab manual in 1975 for use in my own class. That manual went through many revisions, and the first edition of this manual, though much more extensive than the original, benefited much from the comments and criticism of the earlier manual offered by colleagues and students.

The second edition is mostly a refinement of the first edition, with one major exception: new problems based on resources available on the World-Wide Web have been added to all but one chapter. In addition, I now have a home page on the Web with links for all of the Web problems and for *errata;* it also provides an e-mail link for communication with me.

Most photographs in this manual were taken by geology professor Doug Sherman, who also is a professional photographer. Utilizing his expertise in both areas, he prepared the photographs specifically for this manual.

ORGANIZATION AND USE

This manual consists of 17 chapters that cover the usual topics in physical geology. The first five chapters deal with *Earth Materials*—minerals and rocks. Because maps, aerial photographs, and other types of remotely sensed images are used so extensively in geology, two chapters focus on *Maps and Images.* Five chapters then show how *Surface Processes* are studied using maps and images. One laboratory is devoted to *Geologic Time and Sequences,* and in the final chapters, *Internal Processes* and their effects on the surface are studied. Each chapter contains an **Overview,** which summarizes and gives a statement of purpose; a list of materials needed to do the lab (**Materials Needed**), sufficient descriptive text so that all lab problems can be completed without reference to another textbook; a list of **Objectives,** which summarizes the things that a student will be expected to do or know after completing the lab; and laboratory **Problems** or exercises that are required for the lab. The **Objectives** are located immediately before the **Problems,** rather than at the beginning of the chapter, because I want students to relate them to the problems more than the text.

The problems are designed to illustrate basic principles of geology and methods of solving geological problems. A paragraph preceding the objectives in each chapter briefly indicates how the problems relate to the scientific method. As often as possible, problems were chosen to demonstrate practical aspects. Some problems require making measurements and doing simple calculations. Most questions require specific, short answers. A few questions, however, are designed so that several answers may be possible. Students must formulate several hypotheses, then determine what additional information would be needed before a single hypothesis could be chosen. Problems that require more time to complete or are more difficult are preceded by the heading *IN GREATER DEPTH*. It is up to the individual instructor to determine how many problems, or what parts of individual problems, should be assigned in his or her class.

All chapters except Chapter 5 (Metamorphic Rocks) have a **Web Problem,** which can be solved from information available on the World-Wide Web. I have provided specific Web addresses for each problem, so they don't take an unreasonable amount of time. The student can access these Web sites by entering the address given in the individual Web Problem or by using the links provided in my home page. The address for my home page is: **http://www.uwosh.edu/faculty_staff/jonesnw/;** then link to *Problems in Laboratory Manual for Physical Geology.* I encourage you to gain access through my home page, because of the changeable nature of Web sites. If changes occur that affect an individual problem, they will be noted on my home page. I also will maintain a list of corrections for any errors that occur in the text; this list can be accessed by linking to *Errata in Laboratory Manual for Physical Geology.*

The inside of the front cover of the manual contains a geologic time scale. The inside of the back cover lists symbols of chemical elements used in the manual, units of measurement and factors for converting between metric and English units, and a brief explanation of the use of powers of ten to abbreviate large and small numbers. Key terms are in **boldface** in the text. *Italics* are used to call attention to a word or phrase. Answer pages can be torn out and handed in for grading if necessary. Colored, tear-out, structure models at the end of the manual are needed for the problems in Chapter 14.

ACKNOWLEDGEMENTS

I would like to thank my colleagues at the University of Wisconsin Oshkosh, Bill Fetter, Gene LaBerge, Tom Laudon, Jim McKee, Brian McKnight, and Bill Mode, for their comments, suggestions, and moral support over the years. Mary Fahley saved me a bunch of time by finding most of the Web sites used in the manual, and I thank her for that. I also thank the following reviewers: David T. King—Auburn University, Diane I. Doser—The University of Texas—El Paso, Roseann Carlson—Tidewater Community College, John Grover—University of Cincinnati, Michael McKinney—The University of Tennessee—Knoxville, John R. Reid—University of North Dakota, Robert Lawrence—Oregon State University, Charles Thornton—Pennsylvania State University, Gary C. Allen—University of New Orleans, Glen K. Merrill—University of Houston, Bryann Tapp—University of Tulsa, Estella A. Atekwana—Western Michigan University, Bryan Gregor—Wright State University, Dr. Thomas B. Hanley—Columbus State University, Stephen B. Harper—East Carolina University, Gale Martin—Community College of Southern Nevada, John R. Reid—University of North Dakota, Steven V. Stearns—College of Charleston, Christopher A. Suczek—Western Washington University, and Wm. C. Brown Developmental Editor Daryl Bruflodt. This edition of the manual and I still are dedicated to my wife, Judy.

PROLOGUE: *Methods of Science*

Science is a systematized body of knowledge derived mostly from observation and logical inference. The ultimate goal of scientists is to discover the basic principles that determine how things behave. Scientists have the following in mind as they work: (1) natural processes are orderly, consistent, and predictable; (2) there is a cause (or set of causes) for every effect; and (3) the best explanations are usually the simplest (*Doctrine of Simplicity or Parsimony*). Geologists and other scientists who study the past assume that physical and chemical laws, such as the law of gravity, are constant and have not varied over time (*Doctrine of Actualism*).

Scientific advances come in different ways. Some are based on theoretical ideas about things not yet observed (for example, Einstein's theory of relativity). Others result from unplanned, fortuitous events (such as the eruption of Mount St. Helens volcano). But most advances result from a painstaking approach known as the *scientific method.*

The **scientific method** generally involves: (1) formulation of a problem or question; (2) observation and measurement to determine characteristics or properties, to measure variation, classify, and establish relations—in other words, to get a basic understanding of the situation; (3) formulation of one or more explanations, models, or **hypotheses** to explain the observations; (4) experimentation and testing to reject or modify hypotheses; and (5) acceptance of a best hypothesis that withstands all tests and the scrutiny of peers as a **theory.**

The scientific method is *not* a rigid series of steps that must always be followed to do science. For example, most scientists already have a tentative hypothesis in mind when making observations. In fact, as philosopher Karl Popper said, successful scientists often are those with the best "imaginative preconception" of the correct hypothesis. However, to avoid being blinded by such preconception, geologist T. C. Chamberlin recommended maintaining *multiple working hypotheses* until such time as they could be reduced to only one. With this approach, the scientific process involves rejecting those hypotheses that do not withstand testing and experimentation. Commonly, the simplest hypothesis that explains the most observations is the one that survives to become a theory. However, even though a theory is supported by a great deal of evidence and seems to explain all observations, it still is subject to testing, and eventually it may be rejected or modified as a result of new observations. Theories that withstand all testing and are invariable become **scientific laws;** an example is Newton's Law of Gravitation (the law of gravity).

The scientific method may seem rather involved, but it is an approach you probably use in many situations without realizing it. Take a very simple example. Let's say you go to geology lab one day and find nobody in the classroom. That *observation* should raise your curiosity. Why is no one there? Several possible explanations, or *hypotheses,* come to mind: (1) the professor is ill and the class was canceled; (2) no class was scheduled for that day; (3) you are there at the wrong time or the wrong day; (4) the class has gone on a field trip that you forgot; (5) the class is hiding in the room, ready to leap out and yell "Happy birthday"; or (6) everyone has been lifted magically from the room. As a good scientist and rational thinker, you rank them in order of which is the most likely, and quickly eliminate those that are impossible or too unlikely. The last one (6) is not testable and is out of the realm of science, because it calls on a supernatural phenomenon for an explanation. A quick check of the time and date allows you to eliminate number 3. It's not your birthday, so you forget about number 5. Three reasonable hypotheses remain to be tested.

One way to test an hypothesis is to assume it is true, make a prediction based on that assumption, then see if the prediction comes true. If the professor is ill (hypothesis number 1), you might predict that a note to that effect would be posted on the door or the blackboard, but there is none. You might also expect to see other curious students about, but there are none. These negative results strongly suggest that hypothesis 1 is incorrect, although they do not necessarily prove it wrong. After all, it is possible that the professor was not able to have a notice put in the classroom, and it is possible that all the other students decided not to come to class that day. Those are pretty complex and improbable explanations, however. Hypotheses 2 and 4 are simpler and more reasonable. An easy way to test these is to look at the syllabus for the class.

Let's assume that no class is scheduled, according to the syllabus. Then you can establish a *theory.* The theory is that the classroom is empty because there is no class that day. Note that this is a **scientific theory** (although theories in science usually require much more time to develop), one that can be tested in many ways, with the same result. For example, you could check with the Geology Department office or look at another student's syllabus. The public tends to use the word "theory" in quite a different sense, almost disdainful, as something that has little basis in fact. As you can see from the example, a *scientific theory* is firmly established and as close to truth as scientifically possible. However, even a scientific theory is always subject to more testing and possible revision in light of new evidence.

That is how much of science works. There is nothing mysterious about it. It is simply a logical, open-minded approach to a question. Conclusions are not made without considerable thought and testing. It is a rational approach to all kinds of problems or questions, not just scientific ones, and undoubtedly is the approach you have used many times yourself. As you do the exercises in your geology laboratory, you will be using the scientific method as applied by geologists.

Part 1

MINERALS

chapter

1
Properties of Minerals

Overview

To understand most geologic processes, we first must understand minerals, which are the basic chemical entities that make up the Earth. Minerals differ from each other in chemical composition and arrangement of atoms, resulting in distinctive physical properties that enable minerals to be identified. The most useful physical properties for identifying minerals are examined in this chapter. In Chapter 2, these properties are used to identify minerals.

Materials Needed

Set of mineral samples • Equipment for determination of mineral properties: streak plate, copper penny, glass plate or steel knife, magnet, hand lens, dilute hydrochloric acid, contact goniometer (optional)

INTRODUCTION TO MINERALS

A **mineral** is a naturally occurring compound or chemical element made of atoms arranged in an orderly, repetitive pattern. Its chemical composition is expressed with a chemical formula (symbols for chemical elements are given on the back endsheet). Both chemical composition and atomic arrangement characterize a mineral and determine its physical properties. Most minerals form by inorganic processes, but some substances, identical in all respects to inorganically formed minerals, are produced by organic processes (for example, the calcium carbonate in mollusk shells or the silica of certain plankton). A few naturally occurring substances called **mineraloids** have characteristic chemical compositions, but are amorphous; that is, atoms are *not* arranged in regular patterns. Opal is an example.

Because different kinds of minerals have different crystal structures and different chemical compositions, their outward appearance and/or physical properties are different. In most cases, general appearance and a few easily determined physical properties are sufficient to identify the mineral.

PHYSICAL PROPERTIES

Color, luster, streak, hardness, cleavage, fracture, and crystal form are the most useful physical properties for identifying most minerals. Other properties—such as reaction with acid, magnetism, specific gravity, tenacity, taste, odor, feel, and presence of striations—may be helpful for identifying certain minerals.

Color

Color is the most readily apparent property of a mineral, but BE CAREFUL. Slight impurities or defects within the crystal structure determine the color of many minerals. For example, quartz can be colorless, white, pink, purple, green, gray, or black (Fig. 1.1). Color generally is diagnostic for minerals with a metallic luster (defined on next page), but may vary quite a bit in minerals with a non-metallic luster. Check the other properties before making an identification.

FIGURE 1.1
Varieties of quartz with different colors: amethyst (purple); rock crystal (clear); smoky (black).

Luster

Luster describes the appearance of a mineral when light is reflected from its surface. Is it shiny or dull; does it look like a metal or like glass? Most minerals have either a **metallic** or **nonmetallic** luster. As you will see, the first thing you must determine before a mineral can be identified using the tables in Chapter 2 is whether its luster is metallic or nonmetallic.

Minerals with a metallic luster look like a metal, such as steel or copper (Fig. 1.2A, B). They are opaque to light, even if very thin. Many metallic minerals become dull looking when they are exposed to air for a long time. To determine whether or not a mineral has a metallic luster, therefore, you must look at a recently broken part of the mineral.

Minerals with nonmetallic luster can be any color. At first, you may have difficulty determining whether some black minerals have metallic or nonmetallic luster. However, thin pieces or edges of minerals with nonmetallic luster generally are translucent or transparent to light, and even thick pieces give you the sense that the reflected light has entered the mineral a bit before being reflected back. There are several types of nonmetallic lusters. *Vitreous luster* is like that of glass (Fig. 1.2C, D). Remember that glass can be almost any color, including black, so don't be fooled

A.

B.

C.

D.

E.

F.

G.

H.

FIGURE 1.2
Types of luster: Metallic (A. galena (dark mineral) and B. pyrite); Nonmetallic, vitreous (C. muscovite and D. biotite); E. Nonmetallic, dull (kaolinite); F. Nonmetallic, pearly (talc); G. Nonmetallic, waxy (jasper); H. Nonmetallic, resinous (sphalerite).

FIGURE **1.3**
The streak of this dark gray mineral (hematite), obtained by rubbing it on the white streak plate, is reddish brown.

Table 1.1

SCALE OF HARDNESS

MOHS HARDNESS SCALE		MOHS HARDNESS OF COMMON ITEMS	
1	Talc	Fingernail	2 to 2½
2	Gypsum	Copper coin	3½
3	Calcite	Steel knife	5 to 6
4	Fluorite	Glass	5 to 5½
5	Apatite	Streak plate	6½ to 7
6	Orthoclase		
7	Quartz		
8	Topaz or Beryl		
9	Corundum		
10	Diamond		

FIGURE **1.4**
The harder white mineral (calcite) scratches the softer one (gypsum).

by the color. A *dull luster* has an earthy appearance caused by weak or diffuse reflection of light (Fig. 1.2E). Other non-metallic lusters include: *pearly luster,* like a pearl or the inside of a fresh clam shell (Fig. 1.2F); *greasy luster,* as though covered by a coat of oil; *waxy luster,* like paraffin (Fig. 1.2G); and *resinous luster,* like resin or tree sap (Fig. 1.2H).

Streak

Streak is the color of the mineral when finely powdered; it may or may not be the same color as the mineral. Streak is more helpful for identifying minerals with metallic lusters, because those with nonmetallic lusters generally have a colorless or light-colored streak, which is not very diagnostic. Streak is obtained by scratching the mineral on an unpolished piece of white porcelain called a streak plate (Fig. 1.3). Because the streak plate is harder than most minerals, rubbing the mineral across the plate produces a powder of that mineral. When the excess powder is blown away, what remains is the color of the streak. Because the streak of a mineral is usually the same, no matter what the color of the mineral, streak is commonly more reliable than color for identification.

Hardness

Hardness is the resistance of a smooth surface to abrasion or scratching. A harder mineral scratches a softer mineral, but a softer one does not scratch a harder one. To determine the hardness of a mineral, something with a known hardness is used to scratch, or be scratched by, the unknown. The minerals in Mohs Hardness Scale (Table 1.1) are used as standards for comparison. The Mohs hardnesses are also given for some common items, which you can use to determine the hardness of an unknown mineral if you don't happen to have a pocket full of Mohs minerals.

To determine its hardness, run a sharp edge or a point of a mineral with known hardness across a smooth face of the mineral to be tested (Fig. 1.4). Make sure that the contact points of both minerals are the minerals you intend to test, and not impurities. Also, make sure that the mineral has

actually been scratched. Sometimes powder of the softer mineral is left on the harder mineral and gives the appearance of a scratch on the harder one. Brush the tested surface with your finger to see if a groove or scratch remains. You may need to use a hand lens or magnifying glass to see whether a scratch was made. If two minerals have the same hardness, they may be able to scratch each other.

A piece of window glass commonly is used in geology labs as a standard for determining hardness. There are several reasons for this: 1) it's easy to see a scratch on glass; 2) the hardness of glass (5 to 5½) is midway on the Mohs scale; and 3) glass is inexpensive and easily replaced. In fact, you will discover that the hardness of a mineral when compared to glass is one of the principal bases in identifying a mineral using the determinative tables in Chapter 2. Put the piece of glass on a stable, flat surface such as a tabletop. Then rub the mineral on the glass, and check to see if the glass was scratched. DO NOT TRY TO SCRATCH THE MINERAL WITH THE GLASS, because glass chips easily.

More sophisticated methods than the scratch test have been developed to determine hardness; Figure 1.5 illustrates how some of the minerals of the Mohs scale compare with their hardnesses, as determined by another method. As you can see, Friedrich Mohs, who devised the scale in 1822, did a pretty good job of selecting minerals so that the intervals between adjacent minerals on his scale were consistent.

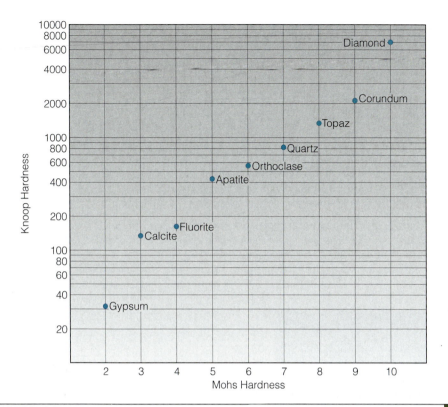

FIGURE 1.5
Comparison between Mohs-scale hardness and "actual" hardness (Knoop hardness). Note that "actual" hardness is shown with a logarithmic scale.

Cleavage and Fracture

The way in which a mineral breaks is determined by the arrangement of its atoms and the strength of the chemical bonds holding them together. Because these properties are unique to the mineral, careful observation of broken surfaces may aid in mineral identification. A mineral that exhibits **cleavage** consistently breaks, or *cleaves*, along parallel flat surfaces called **cleavage planes.** A mineral **fractures** if it breaks along random, irregular surfaces. Some minerals break only by fracturing, while others both cleave and fracture.

The mineral halite (sodium chloride [NaCl], or salt) illustrates how atomic arrangement determines the way a mineral breaks. Figure 1.6 shows the arrangement of sodium and chlorine atoms in halite. Notice that there are planes with atoms and planes without atoms. When halite breaks, it breaks parallel to the planes with atoms but along the planes without atoms. Because there are three *directions* in which atom density is equal, halite has three **directions of cleavage,** each at 90° to each other. The *number of cleavage directions* and the angles between them are important in mineral identification.

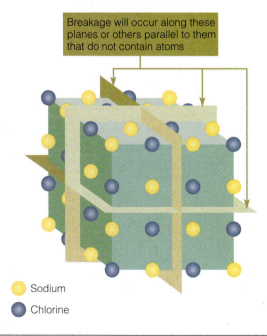

Breakage will occur along these planes or others parallel to them that do not contain atoms

● Sodium
● Chlorine

FIGURE 1.6
The illustration shows that atoms of sodium and chlorine in the mineral halite are parallel to three planes that intersect at 90°. Halite breaks, or *cleaves*, most easily between the three planes of atoms, so has three *directions* of cleavage that intersect at 90°. The photograph of halite illustrates the three directions of cleavage.

FIGURE **1.7**
The parallel cleavage planes in hornblende shine because the light is reflected in the same direction.

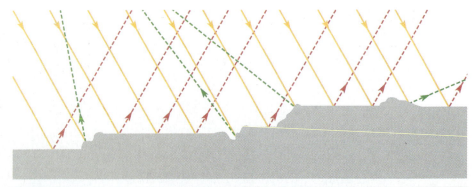

FIGURE **1.8**
When incoming light rays (yellow lines) strike parallel cleavage surfaces, the rays are all reflected in the same direction (pink lines), even though the surfaces are at different elevations. When viewed from that direction, the mineral shines. Rays striking irregular fracture surfaces reflect in many different directions (green lines) causing fracture surfaces to appear duller.

One direction of cleavage

FIGURE **1.9**
Biotite mica has one direction of cleavage.

Cleavage planes, as flat surfaces, reflect more light and appear shinier than other breakage surfaces (Fig. 1.7). A good way to spot a cleavage plane is to turn the sample in your hand and look for positions where you can see good reflections from flat surfaces.

A mineral with *perfect* (excellent) cleavage breaks along flat surfaces that are easy to recognize. In most mineral samples, however, no single cleavage surface extends across the whole sample. In some cases, cleavage may be visible only near a corner where the mineral has been broken recently. Alternatively, cleavage may appear as a series of surfaces that are parallel to each other, but at different elevations. Figure 1.8 is a cross section, or slice, through such a mineral. All rays of light striking the parallel cleavage surfaces in Figure 1.8 are reflected in the same direction; those striking irregular fracture surfaces are reflected in many directions. When viewed at the proper angle, the cleavage surfaces shine. If you turn the sample in your hand, there will be a shiny position where reflections come from many parts of the sample at once.

A mineral may have only *one plane or direction* along which it will break when struck (Fig. 1.9). In such a case, the mineral fragments are likely to be thin plates or sheets (if the cleavage is *perfect*), as in the mica minerals (biotite or muscovite). Although there is only one direction of cleavage, the thin pieces have two parallel surfaces, one on the top and one on the bottom.

When there are *two directions of cleavage*, the cleavage planes may intersect each other at right angles (90°), as in the feldspar minerals, or at other angles, as with the mineral hornblende (56° and 124°) (Fig. 1.10A, B). Typical fragments consist of four cleavage faces, but because a face on one side of the grain is parallel (in the same direction) to one on the other side, there are only two directions of cleavage.

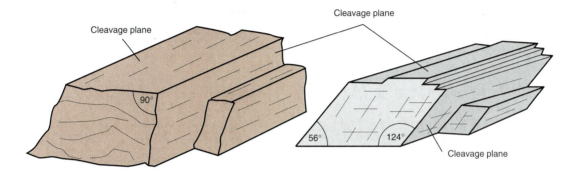

FIGURE **1.10**

Two directions of cleavage: A. Cleavages in potassium feldspar intersect at 90°; B. Cleavages in hornblende intersect at 56° and 124°.

Some minerals have *three directions of cleavage* (Fig. 1.11A, B). With *cubic cleavage*, all three cleavage planes meet at right angles, and it is possible to break the mineral into a six-sided cube (for example, galena, Fig. 1.11A). If the angles are not 90°, the cleavage is *rhombohedral,* and the cleaved fragment is a *rhombohedron* (for example, calcite, Fig. 1.11B).

Four directions of cleavage may yield fragments shaped like octahedra (eight sides, but each side is parallel to one other side) (Fig. 1.12). An example of a mineral with *octahedral cleavage* is fluorite.

Some minerals, such as sphalerite, have *six directions of cleavage* (Fig. 1.13). It is possible, though difficult, to cleave a mineral with *dodecahedral cleavage* into a dodecahedron (a form with 12 diamond-shaped faces).

When determining the number of cleavage directions in a mineral, you must make sure that you are looking at only one crystal at a time (see the discussion under Crystal Form). For example, if the mineral sample consists of a group of crystals, each with a different orientation, you will see many differently oriented cleavages. But each individual crystal will have only the number characteristic of that mineral, so you must concentrate on individual crystals rather than on the whole group.

Fracture surfaces cut the mineral grain in any direction. They generally are rough or irregular, rather than flat, and appear duller than cleavage surfaces (Fig. 1.14). Some minerals fracture in a way that helps to identify them. For example, quartz does not have a cleavage, but like glass, breaks along smooth, curved surfaces. This type of fracture is called *conchoidal fracture*. Other kinds of fracture have descriptive names, such as splintery or irregular.

Both cleavage and fracture are observable on most specimens without actually breaking them, because most of the samples you will see have already been broken from larger pieces. While it is instructive to actually break a mineral specimen yourself, DON'T BREAK SAMPLES WITHOUT YOUR INSTRUCTOR'S APPROVAL!

Crystal Form

A **crystal** is a solid, homogeneous, orderly array of atoms and may be nearly any size. Some crystals have smooth, plane faces and regular geometric shapes (Fig. 1.15); these are what most people think of as crystals. However, a small piece broken from one of these nicely shaped crystals is also a crystal, because the atoms within that small fragment have the same orderly arrangement throughout. When examining minerals, and especially when determining cleavage, you must determine whether you are looking at a single crystal with well-developed crystal faces, a fragment of such a crystal, or a group of small, irregularly shaped fragments of crystals.

The arrangement of atoms in a mineral determines the shape of its crystals. Some minerals commonly occur as well-developed crystals, and their crystal forms are diagnostic. A detailed nomenclature has evolved to describe crystal forms, and some of the common names may be familiar. For example, quartz

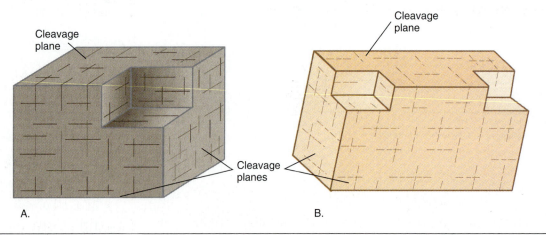

A.

B.

FIGURE **1.11**
Three directions of cleavage: A. Cleavages in galena intersect at 90° (cubic cleavage). B. Cleavages do not intersect at 90° in calcite (rhombohedral cleavage).

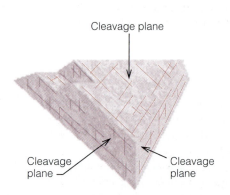

FIGURE **1.12**
Four directions of cleavage (octahedral cleavage) in these samples of fluorite.

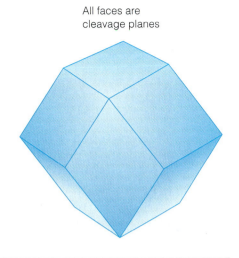

All faces are
cleavage planes

commonly occurs as *hexagonal* (six-sided) *prisms* with pyramid-like faces at the top (Fig. 1.16); pyrite occurs as *cubes* (Fig. 1.17A) or *pyritohedrons* (forms with 12 pentagonal faces, Fig. 1.17B); calcite occurs as *rhombohedrons* (six-sided forms that look like cubes squashed by pushing down on one of the corners; Fig. 1.18A) or more complex, 12-faced forms called *scalenohedrons* (Fig. 1.18B).

Cleavage surfaces may be confused with natural crystal faces; in fact, cleavage planes are parallel to possible (but not always developed) crystal faces. They can be distinguished as follows. (1) Crystal faces are normally smooth, whereas cleavage planes, though also smooth, commonly are broken in a step-like fashion (see Fig. 1.10B or 1.11A). (2) Some crystal faces have fine grooves or ridges on their surfaces (Figs. 1.16, 1.17), whereas cleavage planes do not. Very thin, parallel grooves, or *striations,* are seen on plagioclase cleavage surfaces, but these are features that persist throughout the mineral and are not surficial, as described below. (3) Finally, unless crystal faces happen to coincide with cleavage planes, the mineral will not break parallel to them.

Other Properties

Special properties help identify some minerals. These properties are not distinctive enough in most minerals to help with their identification, or are present only in certain minerals.

Reaction with Acid

Some minerals, especially carbonate minerals, react visibly with acid. (Usually, a dilute hydrochloric acid [HCl] is used.) The acid test is especially useful for distinguishing the two carbonate minerals calcite and dolomite. When a drop of dilute hydrochloric acid is placed on calcite, it readily bubbles or effervesces, releasing carbon dioxide (Fig. 1.19). When a drop of acid is put on dolomite, the reaction is much slower unless the dolomite is powdered first; you may even have to look with a hand lens to see the bubbles. BE CAREFUL when using the acid—even dilute acid can burn your skin or put a hole in your clothing. Only a small drop of acid is needed to see whether

A.

B.

FIGURE **1.17**
Crystals of pyrite in the form of (A.) a cube and (B.) pyritohedron. Note the fine grooves on the faces of the cube.

A.

B.

FIGURE **1.18**
Crystals of (A.) calcite (yellow) in the form of a rhombohedron and (B.) scalenohedron (yellow).

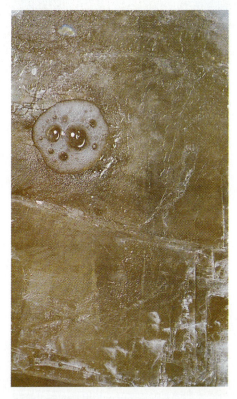

FIGURE **1.19**
Calcite reacting to a drop of acid.

FIGURE **1.20**
Striations are visible on the upper surface of this sample of plagioclase.

FIGURE **1.21**
The thin veinlets seen in some potassium feldspars should not be confused with striations in plagioclase.

or not the mineral bubbles. When you finish making the test, wash the acid off the mineral immediately. Should you get acid on yourself, wash it off right away, or if you get it on your clothing, rinse it out immediately.

Magnetism

Some minerals are attracted to a hand magnet. To test a mineral for magnetism, just put the magnet and mineral together and see if they are attracted. Magnetite is the only common mineral that is magnetic.

Specific Gravity

The **specific gravity** of a mineral is the weight of that mineral divided by the weight of an equal volume of water. The specific gravity of water equals 1.0, by definition. Most of the rock-forming minerals (see Chapter 2) have specific gravities of 2.6 to 3.4; the ore minerals (Chapter 2) are usually heavier, with specific gravities of 5 to 8. If you compare similar-sized samples of two different minerals, the one with the higher specific gravity will feel the heaviest. For

most minerals, specific gravity is not a particularly noteworthy feature, but for some, high specific gravity is distinctive (examples are barite and galena).

Tenacity

The **tenacity,** or toughness, of a mineral describes its resistance to being broken. *Brittle* minerals, such as quartz, shatter when broken; *flexible* minerals, like chlorite, can be bent without breaking, but will not resume their original shape when the pressure is released; *elastic* minerals, such as the micas, can be bent without breaking and will spring back to their original position when the pressure is released; *malleable* minerals can be hammered into thin sheets (examples are gold and copper); *sectile* minerals can be cut with a knife (for example, gypsum).

Taste, Odor, Feel

Some minerals have a distinctive taste (halite is salt, and tastes like it), some a distinctive odor (the powder of some sulfide minerals, such as sphalerite, a zinc sulfide, smells like rotten eggs), and some a distinctive feel (talc feels slippery).

Striations

Plagioclase feldspar can be positively identified and distinguished from potassium feldspar by the presence of *very thin, parallel grooves* called **striations** (Fig. 1.20). The grooves are present on only one of the two sets of cleavages and are best seen with a hand lens. They may not be visible on all parts of a cleavage surface. Before you decide there are no striations, look at all parts of all visible cleavage surfaces, moving the sample around as you look so that light is reflected from these surfaces at different angles.

Until you have seen striations for the first time, you may confuse them with the small, somewhat irregular, differently colored intergrowths or veinlets seen on cleavage faces of some specimens of potassium feldspar (Fig. 1.21). However, these have variable widths, are not strictly parallel, and are not grooves, so they are easily distinguished from striations.

APPLICATIONS

GEOLOGISTS, LIKE OTHER SCIENTISTS, MUST LEARN TO BE GOOD OBSERVERS. OBSERVATIONS PROVIDE THE BASIC DATA ON WHICH MOST SCIENCE IS BASED. GOOD OBSERVATION IS A SKILL ACQUIRED THROUGH TRAINING, AND EXPERIENCE TEACHES YOU WHAT KINDS OF OBSERVATIONS ARE MOST IMPORTANT. OBSERVATION MAY MEAN LOOKING WITHOUT DOING ANYTHING ELSE, IT MAY REQUIRE MAKING SOME MEASUREMENTS, OR IT MAY MEAN OBSERVING THE RESULTS OF A TEST OR EXPERIMENT. THE PROBLEMS IN THIS FIRST EXERCISE ILLUSTRATE THE KINDS OF OBSERVATIONS NEEDED TO IDENTIFY MINERALS.

OBJECTIVES

If you complete all the problems, you should be able to:

1. Determine the color, luster, streak, and hardness of a mineral.

2. Determine whether a mineral has a cleavage or a fracture.

3. Determine the number of cleavages and the angular relations between the cleavages, if a mineral has a cleavage.

4. Distinguish between calcite and dolomite using the "acid test."

5. Determine whether a mineral is magnetic.

6. Distinguish between minerals on the basis of specific gravity.

7. Determine whether a sample of feldspar has striations.

PROBLEMS

1. Determine whether the luster of each sample provided for you is metallic or nonmetallic. If nonmetallic, indicate whether it is vitreous, pearly, greasy, waxy, resinous, or dull.

2. Determine the streak of each of the samples designated by your instructor.

3. Arrange the designated samples in order of increasing hardness. Determine the Mohs hardness of each.

4. Determine whether each of the samples provided has a cleavage. If it does, determine whether it has 1, 2, 3, 4, or 6 cleavages. If it has 2 or 3 cleavages, estimate whether or not adjoining cleavages intersect at right angles (90°). If only a fracture is present, indicate whether it is a conchoidal fracture.

5. Compare the samples provided by your instructor and list them in order of increasing specific gravity. This is done most easily by holding equal-size samples in each hand and estimating, on the basis of feel, which is the heavier; if the samples are the same size, the heaviest one has the highest specific gravity. Alternative methods for determining specific gravity are described in Problem 11.

6. Determine which of the samples provided is attracted to a magnet.

7. Describe the feel of the samples provided.

8. Describe the odor of the mineral supplied by the instructor after rubbing it on a streak plate.

9. Put a small drop of dilute hydrochloric acid (HCl) on the designated samples. Describe what happens and explain why.

 10. Go to *http://mineral.galleries.com* or link to author's *homepage* (see Preface) and link to *Fluorescent Minerals* through *By Groupings*.

    a. What is fluorescence?

    b. Can minerals that fluoresce always be identified with certainty? Explain.

    c. Look up the mineral willemite and describe its fluorescence.

    d. What is phosphorescence and how does it differ from fluorescence?

    e. What is tribolescence?

IN GREATER DEPTH

11. Using a well-developed crystal of the mineral provided, measure and record the angles at which adjoining crystal faces intersect. Your instructor will specify the interfacial angles to be measured, but for example, you might measure the angles at which the side faces of a quartz crystal intersect. Use a simple contact goniometer to make the measurements, as shown in Figure 1.22. Make several measurements of each interfacial angle on the crystal and record them in Table 1.2. To ensure that everyone understands how the angles are measured, and to avoid mistakes, each member of the group should make a few measurements.

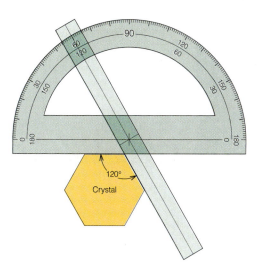

Calculate the mean (average) value for each angle and, if applicable, the mean value for all similar interfacial angles on the crystal. The mean is found by adding the values in each column and dividing by the number of measurements. For example, if the values in column 1 are 44, 48, 43, and 45 degrees, the mean is $(44 + 48 + 43 + 45)/4 = 45°$.

Based on the data from one crystal, formulate a *hypothesis* regarding the constancy or inconstancy of similar interfacial angles on the same mineral. Then suggest a way in which your hypothesis might be tested.

FIGURE **1.22**
A simple contact goniometer consists of a protractor and a rotatable arm. The angle between adjacent crystal faces is measured as shown.

MEASUREMENTS OF INTERFACIAL ANGLES OF A CRYSTAL

ANGLE	1	2	3	4	5	6
Measurement 1						
Measurement 2						
Measurement 3						
Measurement 4						
SUM						
AVERAGE						

Compare the results from one crystal with the results from other crystals of the same mineral. Formulate a *theory* regarding the constancy or inconstancy of similar interfacial angles on the same mineral.

The theory you developed from a few measurements was first proposed by Nicolaus Steno in 1669. It has been tested again and again with many minerals and is now referred to as a *law* (Steno's law). Since minerals are formed of atoms, what does Steno's law suggest about the regularity or irregularity of the arrangement of these atoms?

12. Your instructor may want you to determine the specific gravity of a mineral from the results of a few simple measurements.

Method 1 1. Weigh a dry sample of the mineral to be tested. 2. Weigh the dry dish to be used to catch the overflow in step 3. 3. Put the mineral into a beaker that is brim-full of water, causing the water to overflow into the previously weighed dish. 4. Weigh the dish with the overflow water in it. 5. Subtract the weight of the dry dish from the weight of the overflow water; the result is the weight of water that has a volume equal to that of the mineral. 6. Calculate the specific gravity of the mineral by dividing the weight of the dry mineral by the weight of the equal volume of water.

Method 2 A more accurate method utilizes Archimedes' principle that "a body floating or submerged in a fluid is buoyed by a force equal to the weight of the fluid displaced." It was this discovery that caused Archimedes to run naked through the streets shouting "Eureka," because it enabled him to prove that the King's crown did not have a high enough specific gravity to be pure gold. A simple instrument that utilizes this principle is called a Jolly balance. Using a Jolly balance, first weigh a sample in air, then submerge it in water, and weigh it again. The specific gravity is calculated as:

$$\text{specific gravity} = \frac{\text{weight in air}}{\text{weight in air} - \text{weight in water}}$$

2
Mineral Identification

Overview

Most of the Earth is made of rocks, and rocks are made of minerals. Just as you had to learn the alphabet before you could read, you must learn about minerals before you can understand rocks. An important step is learning to identify Earth's minerals. Chapter 1 shows that minerals have a number of rather easily determined physical properties and that different minerals have different properties. This chapter explains how to use these properties to identify minerals. Another purpose of this chapter is to familiarize you with some of the common minerals, so that when you read about them or hear them discussed, you can formulate a mental picture of them.

Materials Needed

Set of mineral samples • Equipment for determination of mineral properties: streak plate, copper penny, glass plate or steel knife, magnet, hand lens, dilute hydrochloric acid

INTRODUCTION

More than 3000 minerals are known to exist, yet only a comparative few are common. The **rock-forming minerals** are the most abundant minerals found in rocks at the Earth's surface. **Minor minerals** may be important constituents of certain rocks, but are not common in most rocks. **Accessory minerals** usually are present in rocks, but generally in small amounts. **Ore minerals** have economic value when concentrated and are mined for the elements they contain. Table 2.1 lists some of the more important minerals within each of these groups; descriptions follow.

DESCRIPTIONS OF SOME IMPORTANT MINERALS

Rock-Forming Minerals

Quartz (SiO₂)

Quartz is one of the most abundant minerals in the crust of the Earth and is common in soils and rocks, and even in atmospheric dust. It is hard and tough, so it survives well on the Earth's surface, and it occurs in a wide variety of colors and forms.

Distinctive properties are hardness (H = 7), conchoidal fracture, and vitreous to somewhat greasy luster, but color varies. Commonly encountered varieties of quartz include well-formed crystals (colorless *rock crystal quartz,* violet *amethyst,* and gray to black *smoky quartz* are common in rock shops and collections but not in your backyard; see Fig. 1.1, Chapter 1), coarsely crystalline types (pinkish *rose quartz* and the abundant white *milky quartz;* Fig. 2.1), and extremely fine-grained (microcrystalline) varieties of many colors (varicolored banded *agate* and *onyx,* brown to gray translucent *chalcedony,* red *jasper,* white to gray *chert,* and black *flint;* Fig. 2.2).

Table 2.1

SOME IMPORTANT MINERALS

ROCK-FORMING MINERALS	ACCESSORY AND MINOR MINERALS	ORE MINERALS
Potassium feldspar	Corundum	Native Copper
Plagioclase feldspar	Halite	Graphite
Quartz	Fluorite	Sulfur
Hornblende (an amphibole)	Apatite	Galena
Augite (a pyroxene)	Garnet	Sphalerite
Biotite	Sillimanite	Chalcopyrite
Muscovite	Kyanite	Pyrite
Chlorite	Topaz	Hematite
Talc	Staurolite	Magnetite
Clays (e.g., kaolinite)	Epidote	Chromite
Olivine	Beryl	Goethite
Calcite	Tourmaline	Malachite
Dolomite	Serpentine	Azurite
Gypsum	Opal	Barite
Halite		

FIGURE 2.1
Common milky quartz.

A.

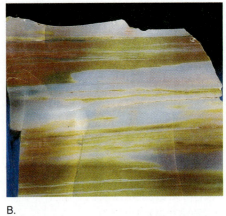

B.

C.

D.

E.

F.

FIGURE 2.2
Microcrystalline varieties of SiO_2: A. agate; B. petrified wood; C. chalcedony; D. jasper; E. chert; F. flint.

FIGURE 2.3

Various colors of plagioclase. Light color varieties are usually richer in sodium. Darker varieties are richer in calcium.

Quartz has many uses. It is a major constituent in glass, concrete, mortar, and in some kinds of building stones. Quartz watches keep good time, because a quartz crystal, when hooked to the watch battery, vibrates at a constant rate. Some varieties of quartz are used in jewelry, and quartz crystals have ceremonial significance in some religious practices.

Feldspars (K, Na, Ca, Al Silicates)

As a group, the feldspars are by far the most abundant minerals in the Earth's crust. They are the principal constituents of many kinds of igneous and metamorphic rocks and are abundant in some kinds of sedimentary rocks. Feldspar is fairly easy to identify, because it has two good cleavages that intersect at essentially 90°, a Mohs hardness of 6, and a vitreous luster. There are two principal varieties of feldspar—plagioclase and potassium feldspar.

Plagioclase has a variable chemical composition, ranging from sodium aluminum silicate ($NaAlSi_3O_8$) to calcium aluminum silicate ($CaAl_2Si_2O_8$). The color varies as well: sodium-rich varieties tend to be light colored—commonly white, cream to buff, or light gray—whereas calcium-rich varieties tend to be gray to dark gray (Fig. 2.3). However, **striations,** not color, are the distinctive feature of plagioclase (see p. 11 and Fig. 1.20, Chapter 1).

Potassium feldspar is a general name that includes several varieties of $KAlSi_3O_8$ (e.g., *orthoclase, microcline, sanidine*). It comes in white, cream, light gray, and even green, but salmon pink is a common and often distinctive color (Fig. 2.4). Some varieties contain semi-parallel veinlets, which, as discussed on p. 11 in Chapter 1,

are not to be confused with the striations in plagioclase.

Some varieties of plagioclase show a play of colors as the angle of reflected light changes, and have been used for ornamental purposes, as has the green variety of microcline (*amazonstone*). The principal use for feldspar is in the production of porcelain and other ceramics.

Amphiboles (Complex Hydrous Na, Ca, Mg, Fe, Al, Silicates)

Hornblende is the most common amphibole. As you can see in Figures 1.7 and 1.10B, Chapter 1, hornblende is a black mineral with two directions of good cleavage that are not at 90°; the cleavage gives it a splintery look, which helps to identify it. Hornblende occurs most commonly as small crystals in combination with other minerals in igneous and metamorphic rocks, but even there, it is distinctive when viewed with a hand lens.

The amphibole group also includes several varieties that occur as asbestos fibers. While airborne asbestos fibers can become imbedded in various parts of the lung or abdomen and cause asbestosis, only one variety (the amphibole crocidolite) is known to be especially carcinogenic. Asbestos is a poor conductor of heat and electricity and can be made into a variety of fabrics. These properties make it useful in many industrial applications, including cements, flooring, friction material, roofing, coatings, packings, and gaskets.

Pyroxenes (Ca, Mg, Fe Silicates)

Like the amphiboles, most pyroxenes occur as small, crystalline components of

FIGURE 2.4

Microcline, a variety of potassium feldspar.

FIGURE 2.5

Augite, a variety of pyroxene.

rocks, especially certain igneous rocks. The most common pyroxene is *augite,* a greenish-black to black mineral that is easily confused with hornblende. Figure 2.5 shows that augite has two cleavages that intersect at about 90°, and it is not splintery-looking like some hornblende. The cleavages may be difficult to see and are not nearly as good as those on hornblende. Another pyroxene, *diopside,* is commonly light green, and the samples you are likely to see in the lab do show good cleavage. A few varieties of pyroxene, such as *jadeite,* are prized for their ornamental value.

Micas (Hydrous K, Mg, Fe, Al Silicates)

The two most frequently encountered micas, *biotite* and *muscovite,* are both common constituents of igneous and metamorphic rocks. Biotite is black, and muscovite commonly is colorless to pale gray, as you can see in Figure 1.2, Chapter 1. Both have one excellent cleavage and a vitreous luster, so they are usually easy to identify. Muscovite, as an excellent insulator for both heat and electricity, is used widely in the electrical industry.

Clays (Hydrous Al Silicates)

Clays are a significant component of minerals in soils, sediment, and sedimentary rocks. But they are not the sort of thing that you would want to put in a display case; in fact, when you dust the display case, you commonly are getting rid of clay. The word clay is used in two different ways: it refers to a specific group of minerals, and to particles that are extremely small (<0.004 mm [<0.0002 in] diameter). Clay minerals usually comprise clay-size particles. Clay is used for all kinds of things, from manufacture of china and bricks, to stopping leaks in dams, to a filler in paper, to medicine (the clay mineral kaolinite is used in Kaopectate™), to a filler in chocolate (doesn't that make your mouth water?).

Clays are recognized by their extremely small crystal size, dull luster, softness, and earthy odor. Their colors vary, but the variety most commonly encountered in physical geology labs is the white clay *kaolinite* (Fig. 2.6).

Olivine (Mg, Fe Silicate)

Figure 2.7 illustrates a typical sample of olivine. It is yellowish green to green and is granular, being made of a multitude of small grains (crystals). You have to look at individual crystals to see that it has a conchoidal fracture but no obvious cleavage. And if you check the hardness, make sure that little

FIGURE 2.6
Kaolinite, a type of clay.

pieces aren't breaking off, because it is hard (H = 6.5 to 7). Most olivine occurs in dark-colored igneous rocks. There is a gem variety of olivine (*peridot*), and some is used for refractory bricks.

Calcite (CaCO$_3$) and Dolomite (CaMg(CO$_3$)$_2$)

These two carbonates are the major minerals in the common sedimentary rocks limestone and dolostone, and their metamorphosed equivalent, marble. Calcite is also one of the most common minerals to form in veins and open cavities in rocks. When formed in such places, large cleavable pieces and beautiful crystals may develop, such as those shown in Figure 1.18, Chapter 1.

If you have a piece of calcite that shows cleavage (chances are you will), the three directions of cleavage not at 90° to each other are a giveaway. Add to that the Mohs hardness of 3, and calcite is easy to identify. The easiest way to identify these carbonate minerals is to put a drop of dilute hydrochloric acid (HCl) on them to see if they bubble; calcite bubbles strongly, dolomite not so much.

Accessory and Minor Minerals

Halite (NaCl)

This is an easy mineral to identify. Just taste it—it's salt. If you don't like the idea of licking where others have licked before, however, check the other properties. It is generally colorless or white, although

FIGURE 2.7
Olivine.

impurities may add color. It has perfect cubic cleavage (see Fig. 1.6, Chapter 1) and is soft (H = 2.5). Halite occurs in sedimentary rocks that formed as a result of evaporation of seawater and commonly is associated with other *evaporite* minerals, such as gypsum. Its principal use is not for a flavoring or for melting ice, but as a source of sodium and chlorine for the chemical industry.

Fluorite (CaF$_2$)

Fluorite is a pretty mineral (see Fig. 1.12, Chapter 1), but with a Mohs hardness of only 4, it is not suitable as a gemstone. Large cubic crystals are common, and it has a perfect octahedral cleavage. Fluorite is commonly light to dark purple, colorless, pale yellow, or light green. It occurs in vein deposits. It is used as a flux in making steel and for the manufacture of hydrofluoric acid.

Gypsum (CaSO$_4$ • 2H$_2$O)

Gypsum ranges from a massive, white to gray, carvable variety, called *alabaster,* to clear crystals or cleavage fragments (Fig. 2.8). Its hardness—softer than your fingernail—gives it away. The clusters of crystals can be quite beautiful, but they are so fragile that you have to display them in a safe place. Gypsum is used extensively as the basis for plaster and plaster board. If you have a clay-rich soil in your garden, you may have used it as a soil conditioner. Gypsum is a common evaporite mineral found as thick layers in sedimentary rocks.

A.

B.

FIGURE 2.8
(A.) Crystals and (B.) cleavage fragments of gypsum.

FIGURE 2.9
Inuit carving of walrus mother and baby in soapstone, a rock composed mostly of talc.

FIGURE 2.10
Serpentine asbestos. Note the fibers.

Talc (Hydrous Mg, Al Silicate)

As number one on the Mohs hardness scale, talc is suitable for use even on a baby's bottom, in its familiar form, talcum powder. It forms by metamorphism or similar alteration of preexisting minerals such as olivine, and is the major ingredient in the rock soapstone. You may be familiar with the soapstone used in carvings, especially by the Inuits of the Canadian Arctic (Fig. 2.9). Talc crystals, if large enough, show one perfect cleavage, but in more massive varieties, this is difficult to see. The pearly luster (Fig. 1.2F), greasy feel, and hardness are distinctive.

Serpentine (Hydrous Mg Silicate)

The most familiar type of serpentine is serpentine asbestos (Fig. 2.10), but massive, green serpentine is most common. The greasy to waxy luster of massive serpentine is distinctive; its Mohs hardness (H = 3–5) serves to distinguish it from massive talc. Massive serpentine is commonly used as a decorative building stone. More than 95 percent of the asbestos produced in the world is serpentine. Studies have shown that health risks from serpentine asbestos are considerably less than from certain types of amphibole asbestos, although the distinction is rarely recognized by the public.

Chlorite (Hydrous Mg, Al Silicate)

This dark green mineral exhibits one perfect cleavage, but unlike the micas, thin pieces don't spring back to their original position after being bent; pieces are flexible, but not elastic. Most chlorite occurs as small grains in metamorphic rocks, and you may have to use a hand lens to see that it looks like a dark green mica.

Garnet (Ca, Mg, Fe, Al Silicate)

Garnet is a metamorphic mineral that almost always occurs as a perfect crystal (Fig. 2.11). However, some crystals are so large that the sample you may see in the lab is just a piece of a crystal. Mohs hardness (H = 6.5–7.5), vitreous to resinous luster, and crystal form help to identify garnet, but be careful of color. A common color is brownish red, but garnet can be black, brown, yellow, green, or other colors. Garnet is used for abrasives (such as in garnet sandpaper) and gemstones.

Staurolite (Hydrous Fe, Al Silicate)

Like garnet, staurolite almost always occurs as "well-formed" crystals in metamorphic rocks. Dark brownish crystals (Fig. 2.12) are characteristic of this mineral. Intergrown crystals called *twins* are common and are sometimes sold as "fairy crosses."

Kyanite (Al₂SiO₅)

This metamorphic mineral is light blue and forms blade-shaped crystals. It is used in the manufacture of spark plugs, among other things.

Ore Minerals

Hematite (Fe₂O₃)

The red to red-brown streak is the key to identifying hematite, because, as shown in Figure 2.13, it occurs as several strikingly different forms. One form is a reddish brown, soft, earthy material. Another is dull black and hard. A third type, known as *specular hematite,* has a metallic luster and occurs as a mass of small, mica-like scales. However, scratch any hematite, or rub it across a streak plate, and you will see the same red powder. Most hematite occurs in sedimentary rocks (or their metamorphic equivalents) called *iron formations.* It is the most important ore of iron, which is the principal ingredient in steel.

Magnetite (Fe₃O₄)

The only magnetic mineral you are likely to see in this lab is magnetite (Fig. 2.14). But just in case you don't have a magnet, other properties (such as black color and streak, metallic luster on fresh breaks, and moderately high specific gravity) should identify it. Magnetite is a common accessory mineral in igneous and metamorphic rocks. It is an important ore of iron in iron formation and where concentrated by igneous processes.

FIGURE **2.11**
Twelve-sided crystals (dodecahedrons) of different varieties of garnet.

FIGURE **2.12**
Crystals of staurolite. Intergrown crystals on right are *twinned.*

FIGURE **2.13**
Three varieties of hematite: specular hematite, pisolitic hematite, and botryoidal hematite.

FIGURE **2.14**
The lodestone variety of magnetite is a natural magnet strong enough to hold the paper clip.

Limonite

Limonite is a general name for a mixture of poorly crystalline, hydrated, and non-hydrated iron oxides. Although really a rock, it was once considered to be a distinct mineral. The name applies to yellow, yellow-brown, red or black, dull to metallic, soft to hard material. An example is shown in Figure 2.15. It usually forms by oxidation of iron-bearing minerals, so in essence, it is a rust. It has a yellow-brown streak, which serves to distinguish it from hematite, and it is a minor ore of iron.

Galena (PbS)

Figures 1.2A and 1.11A, Chapter 1, show that galena has an obvious metallic luster and three perfect cleavages that intersect at 90°. It also has a very high specific gravity (H = 7.5). Galena is almost the only ore of lead, which is used in batteries and metal products, and it occurs most commonly in vein-like deposits.

Sphalerite (ZnS)

This may be the only mineral you will ever see with six directions of cleavage, although you probably will have a difficult time seeing them all (see Fig. 1.13, Chapter 1). If the cleavage is hard to see, look for the resinous luster; the yellow, brown, or black color; and the yellow-brown streak with its faint, rotten-egg-like odor of hydrogen sulfide. Sphalerite is the most important ore of zinc and commonly is found with galena. Zinc is used in galvanizing iron or as a metal by itself; it is also used in batteries, paints and dyes, and in alloys.

Pyrite (FeS₂)

Pyrite is pale brass-yellow, has a metallic luster, and is called *Fool's gold* by those in the know (who, presumably, are not themselves fools). Crystals in the form of cubes and pyritohedrons (see Fig. 1.17, Chapter 1) are common. Given these properties, a Mohs hardness of 6 to 6.5 will enable you to distinguish it from other yellow sulfide minerals such as chalcopyrite or, should the chance arise, gold. It is a common accessory mineral in igneous, sedimentary, and metamorphic rocks and is concentrated in veins or irregular masses in many places. Pyrite is mined along with the more valuable sulfide minerals with which it occurs, and some of it is used in the production of sulfuric acid.

Chalcopyrite (CuFeS₂)

This brass-yellow mineral is distinguished from massive pyrite by its slightly more yellow color and its Mohs hardness of 3.5 to 4. It is one of the most important ores of copper. Copper is used for electrical conductors (such as wire), brass (an alloy with zinc), bronze (an alloy with tin), fittings, sheathings, and other manufactured items. Chalcopyrite occurs in veins and various types of more massive deposits.

Graphite (C)

Graphite is dark gray or black, greasy, soft, and soils your fingers or whatever it touches. Mixed with clay, it forms the "lead" in pencils, and mixed with oil, it is a good lubricant. It has other uses as well, including generator brushes, batteries, and electrodes. It is usually found in metamorphic rocks.

Bauxite (Hydrous Al Oxides)

Bauxite is really a rock made of several minerals, but it once was regarded as a mineral. It is characterized by pea-shaped structures (Fig. 2.16); it may be white, gray, or red, and usually is soft, dull, and earthy. It forms by extensive weathering in tropical to subtropical climates. It is important because it is the only ore of aluminum, which is used wherever a light, strong metal is needed.

HOW TO IDENTIFY MINERALS

Minerals are identified on the basis of their physical properties. Therefore, the first thing to do is to determine those properties described in Chapter 1, in order to collect the basic information or data for that mineral. Given these data, one or more hypotheses regarding the identity of the mineral can be formulated. In most cases, determination of these properties will provide enough data to identify the mineral, but in some instances, you will have to make an additional test or two before a single hypothesis (mineral identification) can be selected.

The mineral identification tables that follow will help you "key out" an unknown mineral, given the proper data. The tables are arranged according to physical properties, and you can quickly narrow the number of possibilities for a given sample to a few choices. Individual mineral descriptions enable you to identify the mineral correctly.

FIGURE **2.15**
Limonite.

FIGURE **2.16**
Bauxite with pea-shaped structures.

Use of the tables is illustrated with the following example. Suppose you have identified the following properties in a mineral:

Luster	Hardness	Streak
non-metallic, vitreous	3.5 to 5	white

Cleavage	Color	Other
4 perfect	clear	purplish tinge

The first separation is on the basis of luster. Table 2.2A contains minerals with metallic luster; Tables 2.2B through 2.2G contain minerals with nonmetallic luster. Because the example has nonmetallic luster, the mineral is not in Table 2.2A, and that table can be eliminated.

The second separation is on the basis of hardness. Tables 2.2B through 2.2E contain minerals with hardnesses less than 5.5 (that is, softer than glass or most knives), so our unknown, with hardness between 3.5 and 5, must be in one of them.

The next thing to look at is streak. The white streak eliminates Table 2.2B, which contains minerals with colored streaks. Tables 2.2C through 2.2E are all that remain.

The last major distinction is on the basis of cleavage. Table 2.2C is for minerals with no cleavage, 2.2D is for minerals with one or two cleavages, and 2.2E contains minerals with three or four cleav-ages. Because the example has four good cleavages, it must be in Table 2.2E; there is only one choice, fluorite.

NOTE: The most useful properties for identifying a given mineral are underlined in the tables. "H" in the descriptions refers to *hardness*, and "S.G." to *specific gravity*. Some minerals appear in more than one table, because they may or may not show a particular property in the sample you examine. For example, hematite appears in Table 2.2A, because one variety has a metallic luster, and in Table 2.2B, because another variety does not. Hornblende, augite, and bauxite are listed in two places, because their hardness may be greater or less than 5.5.

Table 2.2A

METALLIC LUSTER

Hardness	Streak	Cleavage	Other	Mineral
Greater than 5.5	Greenish black	None	Brass-yellow; H = 6 to 6.5; S.G. = 5.0; massive or as crystals (cube, pyritohedron).	Pyrite FeS_2
	Black	None	Dark gray to black; H = 6; S.G. = 5.2; may have dull luster if surface not fresh; attracted to magnet.	Magnetite Fe_3O_4
	Red-brown	None	Steel gray to dull red; H = 6; S.G. = 5.0; may be micaceous (tiny flakes) or massive.	Hematite Fe_2O_3
Less than 5.5	Black, greenish black, brownish black, or gray	1 (but not always visible)	Gray to black; H = 1; S.G. = 2.5; may have dull luster; greasy feel, soils paper and fingers.	Graphite C
		3 (excellent cubic cleavage)	Silvery gray; H = 2.5; S.G. = 7.5; cubic crystals common.	Galena PbS
		None	Black to brownish black; H = 5.5; S.G. = 4.6; luster may be submetallic; brownish-black streak; weakly magnetic.	Chromite $FeCr_2O_4$
		None	Golden yellow or greenish yellow; H = 3.5 to 4; S.G = 4.2; massive.	Chalcopyrite $CuFeS_2$
	Yellow-brown	None	Dark brown to black; H = 5 to 5.5; S.G. = 4; massive or rounded forms with radiating, fiber-like layers.	Goethite FeO(OH)
	Copper-red	None	Copper to brown, may have green coating; H = 2.5 to 3; S.G. = 8.9; malleable, sectile.	Native Copper Cu

Underlined properties are most useful in mineral identification. H = hardness (Mohs). S.G. = specific gravity.

Table 2.2B

NON-METALLIC LUSTER
HARDNESS LESS THAN 5.5
STREAK—COLORED

Streak	Cleavage	Other	Mineral
Red-brown	None	Red to reddish brown; H = 1.5 to 5.5; S.G. = 5.0; dull luster; earthy or oolitic (containing spherical structures having diameters of 0.25 to 2 mm) masses.	Hematite Fe_2O_3
Yellow-brown	None	Yellow-brown, orange-brown to dark brown; H = 1.5 to 5.5; S.G. = 3.6 to 4.0; dull luster, earthy, may be a rust-like coating. Not a true mineral. Limonite is a general name for several rust-like, hydrous iron oxides.	Limonite Various hydrous iron oxides
Yellow	None	Yellow; H = 1.5 to 2.5; S.G. = 2.1; resinous luster.	Sulfur S
	6-may be difficult to count	Light yellow, yellowish brown, black; H = 3.5 to 4; S.G. = 4.0; resinous luster.	Sphalerite ZnS
Blue	None	Azure blue; H = 3.5 to 4; S.G. = 3.8; occurs as coatings, masses, or tiny crystals, commonly with malachite.	Azurite $Cu_3(CO_3)_2(OH)_2$
Green	None	Bright green; H = 3.5 to 4; S.G. = 4.0; occurs as coatings, masses, or tiny crystals, commonly with azurite.	Malachite $Cu_2CO_3(OH)_2$

Underlined properties are most useful in mineral identification. H = hardness (Mohs). S.G. = specific gravity.

Table 2.2C

NON-METALLIC LUSTER
HARDNESS LESS THAN 5.5
STREAK—WHITE, WHITISH, OR FAINTLY COLORED

Cleavage	Other	Mineral
None or Indistinct	White; H = 1 to 2.5; S.G. = 2.6; dull luster; greasy feel, earthy odor, powdery.	Kaolinite $Al_2Si_2O_5(OH)_4$
	White, gray, or apple green; pearly luster; H = 1 (or 2 in some impure varieties); S.G. = 2.7 to 2.8; greasy feel; may have light gray streak.	Talc $Mg_3Si_4O_{10}(OH)_2$
	Reddish brown to brown color; H = 1 to 6; S.G. = 2.0 to 2.5; dull to earthy luster; massive or with pea-shaped structures; may scratch glass with difficulty. Not a true mineral, but a rock consisting of several similar minerals. May have faint, red-brown streak.	Bauxite Mixture of hydrous aluminum oxides
	Multicolored green, gray, black; H = 3 to 5; S.G. = 2.5 to 2.6; dull to greasy luster; slight greasy feel; massive to fibrous (asbestos).	Serpentine $Mg_3Si_2O_5(OH)_4$
	Light green to Coke-bottle green, brown, yellow; vitreous luster; H = 5; S.G. = 3.2; six-sided crystals common; may show 1 poor cleavage.	Apatite $Ca_5(PO_4)_3(F,Cl,OH)$
	Buff, gray, white, pinkish; H = 3.5 to 4; S.G. = 2.8 to 2.9; small, rhombohedral crystals or massive; three cleavages, not at 90°, may be indistinct in massive varieties; reacts slowly with dilute hydrochloric acid unless powdered.	Dolomite $CaMg(CO_3)_2$

Underlined properties are most useful in mineral identification. H = hardness (Mohs). S.G. = specific gravity.

Table 2.2D

NON-METALLIC LUSTER
HARDNESS LESS THAN 5.5
STREAK—WHITE, WHITISH, OR FAINTLY COLORED

Cleavage	Other	Mineral
One	Light apple green, gray, white; pearly luster; H = 1; S.G. = 2.7 to 2.8; cleavage not evident in finely crystalline, massive varieties; may have light gray streak; greasy feel.	Talc $Mg_3Si_4O_{10}(OH)_2$
	Green to blackish green; dull to pearly luster; H = 2 to 2.5; S.G. = 2.6 to 3.3; may have faint green-yellow streak; crystal flakes are flexible but not elastic; finely crystalline aggregates common.	Chlorite Hydrous Mg-Fe-Al silicate
	Black to brownish black; vitreous luster; H = 2.5 to 3.0; S.G. = 2.8 to 3.2; may have faint brown-gray streak; perfect cleavage; transparent, flexible and elastic in thin sheets.	Biotite Hydrous K-Mg-Fe-Al silicate
	Colorless, silvery white, brownish silvery white; vitreous luster; H = 2.0 to 2.5; S.G = 2.8 to 2.9; perfect cleavage; transparent, flexible, and elastic in thin sheets.	Muscovite Hydrous K-Al silicate
	Clear, white, light gray; H = 2; S.G. = 2.3; vitreous to pearly luster; brittle sheets; one perfect cleavage, two poor cleavages; *alabaster* is massive, *satin spar* is fibrous.	Gypsum $CaSO_4 \cdot 2H_2O$
Two	Black; H = 5 to 6; S.G. = 3.0 to 3.4; vitreous luster; may have faint green-gray streak; two perfect cleavages meet at 124° and 56°; cleavage faces stepped rather than smooth; splintery appearance; may have greenish-black to black streak. An amphibole.	Hornblende Hydrous Na-Ca-Mg-Fe-Al silicate
	Black to dark green; H = 5 to 6; S.G. = 3.2 to 3.4; vitreous to dull luster; two imperfect cleavages meet at nearly 90°. A pyroxene. Another pyroxene, *diopside*, is similar but is light gray to light grayish green.	Augite Ca-Mg-Fe silicate

Underlined properties are most useful in mineral identification. H = hardness (Mohs). S.G. = specific gravity.

Table 2.2E

NON-METALLIC LUSTER
HARDNESS LESS THAN 5.5
STREAK—WHITE OR WHITISH

Cleavage	Other	Mineral
Three	Clear to gray; H = 2.5; S.G. = 2.2; three perfect cleavages meet at 90°; salty taste.	Halite NaCl
	Colorless or white; H = 3 to 3.5; S.G = 4.5—heavy for a non-metallic; tabular crystals, rose-like array of crystals, or massive.	Barite $BaSO_4$
	Clear, white; vitreous luster; H = 3; S.G. = 2.7; three perfect cleavages form rhombic cleavage fragments; double image seen through clear pieces; reacts strongly with dilute hydrochloric acid.	Calcite $CaCO_3$
	Buff, gray, white, pinkish; H = 3.5 to 4; S.G. = 2.8 to 2.9; small, rhombohedral crystals or massive; three cleavages, not at 90°, may be indistinct; reacts slowly with dilute hydrochloric acid unless powdered.	Dolomite $CaMg(CO_3)_2$
Four	Purple, green, yellow, clear; H = 4; S.G. = 3.2; vitreous luster; perfect cleavage up to four directions may yield octahedral cleavage fragments; may occur as cubic crystals.	Fluorite CaF_2

Underlined properties are most useful in mineral identification. H = hardness (Mohs). S.G. = specific gravity.

NON-METALLIC LUSTER
HARDNESS GREATER THAN 5.5
STREAK—WHITE OR WHITISH IF SOFTER THAN STREAK PLATE

CLEAVAGE	OTHER	MINERAL
None	Brown, pink, blue, gray; <u>H = 9</u>; S.G. = 4.0; <u>six-sided prismatic crystals</u>; *ruby* (red) and *sapphire* (commonly blue) are gem varieties.	Corundum Al_2O_3
	<u>Black</u>, pink, blue, green, brown; vitreous luster; H = 7 to 7.5; S.G. = 3.0 to 3.3; <u>slender crystals with triangular cross sections and striated sides.</u>	Tourmaline Complex hydrous silicate
	Reddish brown, yellowish tan; vitreous to resinous luster; <u>H = 6.5 to 7.5</u>; S.G. = 3.6 to 4.3; <u>twelve-sided crystals</u> common.	Garnet Ca-Mg-Fe-Al silicate
	<u>Red-brown to brownish-black</u>; vitreous, resinous, or dull luster; H = 7 to 7.5; S.G. = 3.7; <u>prismatic and X- or cross-shaped crystals.</u>	Staurolite Hydrous Fe-Al silicate
	Coarsely crystalline varieties: clear, milky, white, purple, smokey, pink; transparent to translucent; vitreous luster; <u>H = 7</u>; S.G. = 2.7; <u>conchoidal fracture</u>; usually massive; sometimes occurs as six-sided crystals; milky = *milky quartz*, purple = *amethyst*, smokey = *smokey quartz*, pink = *rose quartz*. Microcrystalline varieties: *chert* (gray, dull luster), *flint* (black, dull luster), *chalcedony* (brown to gray, translucent, waxy luster), *agate* and *onyx* (varicolored bands, vitreous luster).	Quartz SiO_2
	<u>Olive green to yellow green</u>; vitreous to dull luster; H = 6½ to 7 but difficult to test because <u>granular</u>; S.G. = 3.3 to 4.4.	Olivine $(Mg,Fe)_2SiO_4$
	Colorless, white, or pale shades of yellow, green, red, or blue; may show <u>play of colors</u>; vitreous to resinous luster; H = 5 to 6; S.G. = 2.0 to 2.3; <u>rounded forms common</u>, but also massive; conchoidal fracture. A mineraloid.	Opal $SiO_2 \cdot nH_2O$
	Reddish brown to brown color; H = 1 to 6; S.G. = 2.0 to 2.5; dull to earthy luster; massive or with <u>pea-shaped structures</u>; may scratch glass with difficulty. Not a true mineral, but a rock consisting of several similar minerals.	Bauxite Mixture of hydrous aluminum oxides

<u>Underlined</u> properties are most useful in mineral identification. H = hardness (Mohs). S.G. = specific gravity.

Table 2.2G

NON-METALLIC LUSTER
HARDNESS GREATER THAN 5.5
STREAK—WHITE OR WHITISH IF SOFTER THAN STREAK PLATE

CLEAVAGE	OTHER	MINERAL
One	Colorless, yellow, brown, pink, bluish; <u>H = 8</u>; S.G. = 3.4 to 3.6; vitreous luster; <u>elongate crystal prisms</u> with pointed ends and striated side faces.	Topaz $Al_2SiO_4(OH,F)_2$
	<u>Bluish-green</u>, yellow, white, pink; <u>H = 7.5 to 8</u>; S.G. = 2.7 to 2.8; elongate, six-sided crystal prisms with flat ends common.	Beryl $Be_3Al_2(Si_6O_{18})$
	<u>Light blue to greenish-blue</u>; vitreous luster; H = 5 parallel to long direction of crystal, 7 across crystal; <u>blade-shaped crystals</u>; 1 cleavage.	Kyanite Al_2SiO_5
	White, pale green, brown; H = 6 to 7; S.G. = 3.2; <u>long, slender crystals</u>, <u>commonly as groups of parallel crystals</u>.	Sillimanite Al_2SiO_5
Two	<u>Salmon-pink, white</u>, gray, green; vitreous luster; H = 6; S.G. = 2.5 to 2.6; <u>two cleavage directions meet at nearly right angles</u>; <u>no striations</u>.	*orthoclase* Potassium feldspar $KAlSi_3O_8$
	<u>White to dark gray</u>, sometimes buff; vitreous luster; H = 6; S.G. = 2.6 to 2.8; <u>two cleavage directions meet at nearly right angles</u>; some cleavage faces have perfectly <u>straight, parallel striations</u>, which show up in reflected light.	Plagioclase feldspar $NaAlSi_3O_8$ to $CaAl_2Si_2O_8$
	<u>Pistachio green</u>, yellowish green; H = 6 to 7; S.G. = 3.3 to 3.5; elongate crystals or finely crystalline masses.	Epidote Hydrous Ca-Fe-Al silicate
	<u>Black</u>; H = 5 to 6; S.G. = 3.0 to 3.4; vitreous luster; may have faint green-gray streak; <u>two perfect cleavages meet at 124° and 56°</u>; cleavage faces stepped rather than smooth, giving splintery appearance; may have faint greenish-gray streak. An amphibole.	Hornblende Hydrous Na-Ca-Mg-Fe-Al silicate
	<u>Black to dark green</u>; H = 5 to 6; S.G. = 3.2 to 3.4; vitreous to dull luster; <u>two imperfect cleavages meet at nearly 90°</u>. A pyroxene. Another pyroxene, *diopside*, is similar but is light gray to <u>light grayish green</u>.	Augite Ca-Mg-Fe silicate

<u>Underlined</u> properties are most useful in mineral identification.
H = hardness (Mohs).
S.G. = specific gravity.

APPLICATIONS

JUST AS BACTERIOLOGISTS NEED TO BE ABLE TO RECOGNIZE OR IDENTIFY DIFFERENT KINDS OF BACTERIA, GEOLOGISTS NEED TO KNOW WHAT THE MOST COMMON MINERALS LOOK LIKE AND HOW TO IDENTIFY ONES THEY DON'T ALREADY KNOW. IN CHAPTER 1, YOU LEARNED HOW TO DETERMINE SOME OF THE PHYSICAL PROPERTIES OF MINERALS. THE PRESENT CHAPTER ILLUSTRATES HOW THE BASIC PHYSICAL PROPERTIES, ARRANGED IN THE FORM OF AN IDENTIFICATION KEY OR DETERMINATIVE TABLES, PROVIDE A RATIONAL, STRAIGHTFORWARD METHOD FOR MINERAL IDENTIFICATION. IDENTIFICATION KEYS OR TABLES, LIKE THE ONES USED FOR MINERALS, ARE A MEANS OF ORGANIZING LARGE AMOUNTS OF INFORMATION. THEY CAN BE USED TO IDENTIFY TREES, FLOWERS, BIRDS, ANIMALS, OR ANY GROUP OF THINGS THAT CAN BE ORGANIZED ON THE BASIS OF SIMILARITIES AND DISSIMILARITIES. ORGANIZATION OF KNOWLEDGE IS ONE OF THE WAYS IN WHICH SCIENCE PROGRESSES; IT WOULD BE TERRIBLY INEFFICIENT IF NO GENERAL PROCEDURES WERE AVAILABLE TO IDENTIFY THINGS LIKE MINERALS.

OBJECTIVES

When you have completed this laboratory, you should be able to identify those minerals specified by your instructor.

PROBLEMS

1. Determine the physical properties of the mineral samples provided by your instructor and identify each sample using the mineral identification tables.

2. Distinguish among similar minerals:

    a. Both pyrite and chalcopyrite have a metallic luster, a greenish-black streak, and no cleavage; they are heavy and can be very difficult to distinguish on the basis of color. If no crystals are visible, how would you tell them apart?

    b. Magnetite and hematite can both be dark gray to black and look similar. How could you tell them apart if you did not have a magnet?

    c. Hornblende and augite commonly are black, have the same hardness, and exhibit two directions of cleavage. They are difficult to distinguish when they occur as small crystals in a rock. What would you look for to tell them apart?

d. Calcite, halite, and fluorite all have perfect cleavages, and they can all be the same color. How would you distinguish among them?

e. Talc and serpentine can be the same color, have similar greasy to pearly lusters, and both can have a greasy feel. How do they differ?

f. If chlorite and biotite occur as small crystals in rocks, it may be difficult to tell them apart. What properties might be helpful?

g. What single property is most useful for distinguishing between potassium feldspar and plagioclase?

3. Go to *http://mineral.galleries.com* or link to author's *homepage* (see Preface) and link to *Birthstones* through *By Groupings*. Look up your birthstone.

a. What is your birthstone?

b. What is its chemical formula?

c. What is its color? Luster? Hardness? Cleavage? Streak?

d. With what other minerals does it commonly occur?

e. Where are its notable occurrences?

f. Is it a variety of another type of mineral or a member of a mineral group?

chapter

3

Igneous Rocks

Overview

*Most **rocks** are aggregates of grains or crystals of one or more kinds of minerals. The individual mineral particles in rocks generally are small, with average dimensions of less than 1 centimeter (cm). Even so, the minerals are identifiable because they have the same physical properties as the larger mineral specimens that you have already studied. Deciphering Earth's history begins with the study of rocks, for they bear testimony to their origin and subsequent history. Geologists recognize the informative characteristics of rocks, and when you have completed the next three chapters, you too will be able to reconstruct at least some of the history of common rocks.*

Three classes of rocks are distinguished on the basis of origin: igneous, sedimentary, and metamorphic. In this chapter, you will see examples of igneous rocks and learn how to interpret and identify them. In addition, you will learn how minerals form in sequence from magma, how textures are used to determine this sequence, and how the magma changes as new minerals form.

Materials Needed

Hand lens • Samples of igneous rocks

INTRODUCTION

Igneous rocks form when **magma** (molten rock material)) freezes and solidifies. Volcanic eruptions take place when magma reaches the surface before it solidifies. The magma may flow onto the surface as **lava** or erupt explosively as rapidly expanding gas propels bits of lava and rock outward. The rocks resulting from volcanic eruptions are called **extrusive** igneous rocks, because magma was extruded from the Earth. **Intrusive** igneous rocks form from magma that solidifies below the Earth's surface following intrusion into other rocks.

Because igneous rocks form under diverse conditions, they differ widely in texture, mineral composition, and appearance.

TEXTURE

Texture refers to the size, shape, and arrangement of the crystals or grains composing the rock. The texture of a rock is a consequence of the physical and chemical conditions under which it formed and, perhaps, some of the processes that have acted on the rock since that time.

Most igneous rocks have a **crystalline** texture (Fig. 3.1), in which the various mineral crystals are interlocked with one another. This texture develops when crystals grow together as a magma solidifies to form an igneous rock.

The rate at which a magma cools has the greatest effect on the sizes of the crystals in an igneous rock. In general, the more slowly a magma cools, the larger the mineral crystals will be, because slow cooling

Photomicrograph
×10

Hand specimen

Crystalline texture. Individual crystals are intergrown. Potassium feldspar, quartz, biotite, and plagioclase are visible. Colors in this and most subsequent photomicrographs are not actual colors of minerals, but are due to the interference of polarized light rays that have passed through the mineral.

Photomicrograph
×10

Hand specimen

A fine-grained, or aphanitic, texture is typical of basalt.

Obsidian is easy to identify because of its glassy texture. The excellent conchoidal fracture produces edges that are sharper than any made from metals. This characteristic made obsidian a favorite among Native Americans for arrows and cutting tools and, more recently, it has even been used for very delicate eye surgery. It is most commonly black but may be reddish. Snowflake obsidian has white spots that formed where the glass devitrified (crystallized slightly), usually to a variety of silica (SiO_2).

provides more time for the chemical constituents to migrate to the growing mineral. The chemical composition of the magma also affects crystal size and other elements of texture. For example, abundant water in magma allows chemical constituents to migrate more rapidly to growing mineral crystals. And rapid expansion of gases during the eruption of a volcano has a profound effect on the texture of some volcanic rocks.

Textures Based on Crystal Size

Glassy. A rock with a **glassy** texture is made of glass and has few or no crystals. An example, shown in Figure 3.2, is obsidian. Some igneous rocks are partly crystals and partly glass.

Fine grained (aphanitic). An igneous rock (or part of a rock) with a **fine-grained** or **aphanitic** texture has crystals, but they are too small to be recognizable with the unaided eye, as in Figure 3.3. In practice, this means that the average dimension of the crystals is less than 1 mm (0.04 in).

Coarse grained (phaneritic). An igneous rock consisting mostly of crystals large enough to see without a

hand lens (>1 mm [0.04 in]) has a **coarse-grained** or **phaneritic** texture (see Fig. 3.1).

Pegmatitic. Igneous rocks with exceptionally large crystals (>3 cm [1.2 in]) have a **pegmatitic** texture and are called *pegmatites.* These large crystals result from slow cooling of water-rich magmas.

Porphyritic. Crystals of two different sizes are present in rocks with a **porphyritic** texture (Fig. 3.4). The larger crystals, or **phenocrysts,** are surrounded by many smaller crystals, collectively called the **groundmass.** The groundmass can be either fine grained or coarse grained, but the phenocrysts are typically coarse grained. Both intrusive and extrusive rocks can be porphyritic, but it is more typical of extrusive rocks.

Other Textures of Igneous Rocks

Vesicular. As magma approaches and flows onto the surface, abundant gas bubbles may form. If the lava solidifies before the gas escapes, the resulting rock will contain numerous cavities, or **vesicles,** at the sites of former gas

FIGURE **3.4**
Porphyritic texture in hand specimen of basalt. Larger crystals (phenocrysts) are plagioclase.

FIGURE **3.5**
Pyroclastic texture in hand specimen of rhyolite tuff.

RECOGNIZING MINERALS IN IGNEOUS ROCKS

MINERAL	PROPERTIES
Potassium feldspar	Usually white or pink Two cleavages at 90° Equidimensional crystals
Plagioclase feldspar	Usually white (Na-plag) or gray (Ca-plagioclase) Two cleavages at 90° Elongate crystals Striations
Quartz	Colorless to gray Glassy with conchoidal fracture Irregular crystals in intrusive rocks Equidimensional phenocrysts in extrusive rocks
Biotite	Shiny and black One perfect cleavage Thin crystals
Muscovite	Shiny and silvery white One perfect cleavage Thin crystals
Hornblende (amphibole)	Black with shiny, splintery appearance Two cleavages at 56° and 124° Elongate crystals
Augite (pyroxene)	Black, greenish black, or brownish black Vitreous, but rather dull, luster Two cleavages at 90° Blocky crystals
Olivine	Light green to yellow-green Glassy luster Small (few millimeters), equidimensional crystals

bubbles, giving it a **vesicular texture.** Vesicles vary from a few millimeters to several centimeters in diameter. A lava containing a great deal of gas may froth out onto the surface, solidify very rapidly, and form a glassy rock with lots of very small (a fraction of a centimeter) vesicles; *pumice* and *scoria,* described later, are examples.

Amygdaloidal. Vesicles in some extrusive rocks are filled with minerals such as calcite or agate. The minerals were deposited from water solutions that trickled through the rock after it solidified. Filled vesicles are **amygdules,** and the texture **amygdaloidal.**

Pyroclastic. Volcanoes that erupt explosively expel lots of gas, bits of lava, and fragments of rocks and minerals. The volcanically produced clasts (fragments), or *pyroclasts,* either are blown into the air before falling to the ground or are carried away from the volcano as a gas-charged surface flow. Because the rocks formed from these types of eruptions are made of *pyroclasts,* their texture is termed **pyroclastic.** Extrusive rocks with pyroclastic textures usually contain small bits of broken glass (shards formed from disruption of rapidly cooled lava), mineral crystals, and various sizes of rock fragments. The rock may be soft and lightweight for its size and have a rather "open" texture, or it may be quite hard and dense if it has been compacted and fused or *welded* together. An extrusive igneous rock with a pyroclastic texture, like that shown in Figure 3.5, is a **tuff.** Dense, compacted tuffs are *welded tuffs* or *ignimbrites.*

MINERAL COMPOSITION

Relatively few minerals make up most igneous rocks. To correctly identify the rock, you must identify the major minerals. This is fairly easy for coarse-grained rocks, more difficult (but not necessarily impossible) for fine-grained rocks, and impossible for glassy rocks (unless phenocrysts are present). Table 3.1 and the following may be helpful:

1. Most dark-colored igneous rocks are rich in calcium plagioclase and ferromagnesian (iron-magnesium) minerals such as pyroxene or olivine.

CLASSIFICATION OF IGNEOUS ROCKS

MINERAL COMPOSITION	Felsic	Intermediate	Mafic	Ultramafic
TEXTURE	Quartz, Potassium feldspar, Na-plagioclase, ± biotite, muscovite, hornblende	Na/Ca-plagioclase, hornblende, ± biotite, augite	Ca-plagioclase, augite, ± olivine	Olivine, ± pyroxene
Pegmatitic	Granite pegmatite	Diorite pegmatite	Gabbro pegmatite	
Coarse grained (phaneritic)	Granite	Diorite	Gabbro	Peridotite (Dunite, if mostly olivine)
Coarse grained and porphyritic	Porphyritic granite	Porphyritic diorite	Porphyritic gabbro	
Fine grained (aphanitic)	Rhyolite	Andesite	Basalt	
Fine grained and porphyritic	Porphyritic rhyolite	Porphyritic andesite	Porphyritic basalt	
Pyroclastic	Rhyolite tuff or breccia	Andesite tuff or breccia	Basalt tuff or breccia	
Vesicular			Vesicular basalt	
Amygdaloidal			Amygdaloidal basalt	
Many small vescicles	Pumice		Scoria	
Glassy	Obsidian		Tachylyte or palagonite	

The word **mafic** refers to such rocks. **Ultramafic** igneous rocks are composed entirely of ferromagnesian minerals.

2. Light-colored or **felsic** igneous rocks commonly contain potassium feldspar, sodium plagioclase, and quartz, and only minor amounts of other minerals.

3. **Intermediate** igneous rocks are neither dark nor light and generally contain light-colored minerals (feldspars, some quartz) and dark minerals such as hornblende or biotite.

4. Quartz has a "glassy" appearance and conchoidal fracture.

5. A pink feldspar is usually potassium feldspar; white or gray feldspars may be either potassium feldspar or plagioclase—if striations are present, it is plagioclase.

6. Cleavage and general appearance help to identify amphiboles and pyroxenes.

Amphibole cleavages do not intersect at 90°, but pyroxene cleavages do; amphiboles typically are elongate and have a splintery appearance, whereas pyroxenes look blocky.

CLASSIFICATION AND IDENTIFICATION OF IGNEOUS ROCKS

Igneous rocks are classified by their texture and mineral content. The classification scheme in Table 3.2 lists textures in the left-hand column and mineral composition in the top row.

To identify a rock, determine the texture and mineral content, and find the name in the "pigeonhole" that satisfies both. For example, a rock that is coarse grained, porphyritic, and contains quartz and potassium feldspar is a porphyritic granite.

It makes no difference whether texture or mineral content is determined first. Texture is more easily recognized, though, so start with that, as follows:

A. If crystals are not visible or are barely visible with a hand lens, and:

1. The rock looks like glass, the texture is glassy.

2. It is not glassy, but has a texture similar to fine or medium sandpaper, the texture probably is fine grained.

3. Vesicles are present, the texture is vesicular.

4. The rock consists of fragments, the texture is pyroclastic.

B. If crystals are visible without a hand lens, and:

1. Most crystals are larger than 3 cm, the texture is pegmatitic.

2. Most crystals are smaller than 3 cm, the texture is coarse grained.

Photomicrograph
×10

Hand specimen

FIGURE 3.6

Granites are coarse-grained igneous rocks that make up the bulk of the large bodies of intrusive rock called batholiths. Most are light gray or pinkish, but gray and red granite are common too. Gray, glassy-looking quartz is an essential mineral, as is potassium feldspar. Potassium feldspar is commonly pink, but may be white. If feldspars of two colors are present, the pink one is probably potassium feldspar, and the white or light gray one is probably sodium plagioclase. Most granites contain some black biotite or hornblende, and some contain muscovite. The fine-grained equivalent of granite is rhyolite. Some granites are very pretty when cut and polished, so are quarried for use in monuments and buildings.

Photomicrograph
×10

Hand specimen

FIGURE 3.8

Diorite is a coarse-grained, intrusive rock usually made of white to light gray sodium-calcium plagioclase and black, splintery-looking hornblende. Finely crystalline diorites look like a mixture of salt and pepper. Some diorites contain biotite as well as hornblende, and possibly a little quartz. Diorite is common in both large (batholith) and small intrusive bodies. Andesite is the fine-grained equivalent of diorite.

3. Crystals are of two sizes, the texture is porphyritic.
   a. If the groundmass is fine grained, the texture is fine grained and porphyritic.
   b. If the groundmass is coarse grained, the texture is coarse grained and porphyritic.
4. There are features that look like phenocrysts but have rounded shapes (like the shapes of large

vesicles) and the groundmass is fine grained, the texture is amygdaloidal.

Next identify the major minerals. The important mineralogic criteria are:

1. Presence or absence of quartz (only the granites and rhyolites in column 2 contain quartz);

2. The type of feldspar (whether potassium feldspar or plagioclase,

FIGURE 3.7

Pegmatites are very coarse crystalline intrusive rocks that typically occur as small intrusions. Granite pegmatite, like that shown here, is light colored and commonly has large, irregular-shaped crystals of gray quartz and large crystals of potassium feldspar. Some varieties contain white, sodium-rich plagioclase. Big "books" of both biotite and muscovite mica are common. Some pegmatites contain rare minerals such as the silicate minerals tourmaline and beryl, both of which have several gem varieties (for example, emerald is a variety of beryl).

and if plagioclase, whether sodium-rich or calcium-rich); and

3. The proportions and kinds of dark-colored minerals (biotite, hornblende, pyroxene, and olivine).

Minerals in the groundmass of fine-grained rocks commonly cannot be identified; however, phenocrysts in such rocks may be helpful. According to Table 3.2, if quartz phenocrysts are present, the rock is rhyolite; if only hornblende phenocrysts are present, it is andesite; if only pyroxene or olivine phenocrysts are present, it is basalt; and if feldspar phenocrysts are present, the feldspar must be identified on the basis of the presence or absence of striations.

Some common igneous rocks are shown in Figures 3.6 through 3.15. Remember, the photographs are representative examples only, and your samples may differ considerably in color and general appearance. The texture and mineral content should be similar, so concentrate on those, not on color.

Photomicrograph
×10

Hand specimen

FIGURE 3.9
Gabbro is a coarse-grained intrusive rock composed of light to dark gray calcium plagioclase and black pyroxene. Olivine may be present as well. In many gabbros, the plagioclase is easy to identify, because the striations are readily visible on the large crystals. Gabbro is the coarse-grained equivalent of basalt, and commonly forms thin (dikes and sills) or irregular-shaped, small bodies of intrusive rock. The plagioclase in some gabbros causes a peculiar play of colors that makes the rock valuable for monument or building purposes.

Photomicrograph
×10

Hand specimen

FIGURE 3.10
Peridotite is generally a dark-colored rock, because its principal constituents are pyroxene and olivine. Peridotite consisting almost entirely of olivine is called dunite, which usually has the green color of olivine. In fact, the mineral sample of olivine you saw in the lab was probably a piece of dunite. Peridotite rich in pyroxene may have fairly large crystals; in many peridotites, however, the crystals are small, though large enough to identify. Although peridotite is not abundant in the crust, it is the most abundant rock beneath the crust, in the upper part of the Earth's mantle.

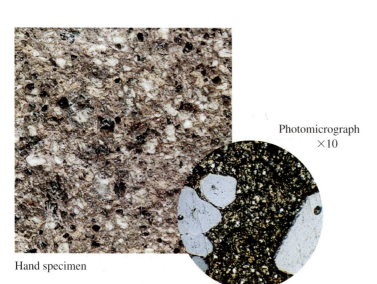

Photomicrograph
×10

Hand specimen

FIGURE 3.11
Rhyolite is a fine-grained extrusive rock that comes in a variety of colors. It is commonly light gray or pink, but red or even black rhyolite is not rare. Many rhyolites, like the one shown here, are porphyritic, so are identifiable by the glassy, quartz phenocrysts and white to salmon potassium feldspar phenocrysts in an aphanitic or even glassy groundmass. In the absence of phenocrysts, a light color suggests, but does not prove, rhyolite. Granite is its coarse-grained equivalent.

Photomicrograph
×10

Hand specimen

FIGURE **3.12**
Andesite is the common extrusive rock of volcanoes formed at convergent plate boundaries. It is typically gray and porphyritic, with phenocrysts of white to light gray plagioclase and black hornblende or biotite. It is the fine-grained equivalent of diorite.

FIGURE **3.13**
Basalt is the most abundant extrusive igneous rock, forming the base of the seafloor and large plateaus on land. It is dark gray and fine grained, as in Figure 3.3, vesicular, as shown here, or amygdaloidal. Porphyritic basalt may have phenocrysts of calcium plagioclase, olivine, or both. Olivine phenocrysts are usually small (1 to 5 mm [0.04 to 0.2 in]) and are recognizable by their green color and glassy luster. Basalt is the fine-grained equivalent of gabbro.

Photomicrograph
×10

Hand specimen

FIGURE **3.14**
Tuff is a textural name indicating a pyroclastic texture and particle size less than 64 mm (2.5 in). A similar rock with particles bigger than 64 mm (2.5 in) is a volcanic breccia. Compositional varieties of tuff or breccia are indicated by using rhyolite, andesite, or basalt as prefixes when possible; for example, rhyolite tuff, basalt breccia. Tuffs vary widely in appearance. Some are made entirely of compacted volcanic glass shards, others have abundant mineral crystals, and others consist of many angular fragments of extrusive rock, especially pumice; most commonly, all three components are present. Just as components and composition are variable, so is color, although many tuffs are light colored, like the rhyolite tuff shown here.

A.

B.

FIGURE 3.15

A. Pumice is the rock that floats because it has so many tiny holes in it. It is a froth of volcanic glass. Look at it with a hand lens to appreciate its character. Pumice is commonly, but not always, white or light gray. It is a good abrasive and is used in Lava® soap, in blocks used to scrape calluses off your feet, or for cleaning griddles. B. Scoria is a basaltic rock with many small holes. You can think of it as being extremely vesicular with small vesicles. It is dark in color, ranging from black to reddish brown. Although it looks as if it might float in water, it is usually too dense. Scoria is sold under the general name "lava rock" for outdoor decorative purposes.

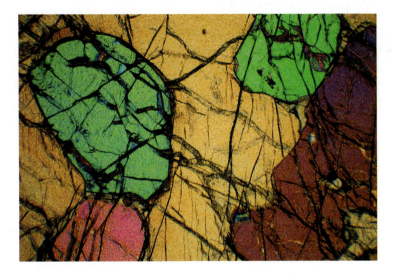

FIGURE 3.16

Photomicrograph of rounded olivine (green and red) in pyroxene (yellow). The early-formed olivine reacted with the magma to form the pyroxene that now surrounds it (×10).

CRYSTALLIZATION OF MAGMA, AND BOWEN'S REACTION SERIES

At high temperatures, a magma is completely liquid, but as the temperature drops, crystals begin to form. They don't all form at once, however. At first, crystals of only one or two minerals begin to grow. As the temperature continues to drop, these early-formed crystals may grow to become nicely shaped, they may react with the magma and be partially or wholly dissolved, or new minerals may begin to grow around them, preventing further growth. Continued crystallization increases the proportion of crystals and decreases the amount of liquid. Crystals grow up against one another and form an interlocking, crystalline network. Other minerals may crystallize with further temperature decreases; these will grow in the liquid remaining between earlier-formed crystals.

Because minerals form in a sequence, it is possible to figure out that sequence by carefully studying the textural relations in the rock. This kind of process was done by N. L. Bowen during the early part of the 20th century, but he carried it a giant step further by duplicating the crystallization process in the laboratory. He paid particular attention to reactions between minerals as indicated by textural relations. For example, he commonly found rounded crystals of olivine surrounded by pyroxene (Fig. 3.16) and concluded that olivine formed first, then at a lower temperature, reacted with the liquid (magma) to form pyroxene. He called this a *discontinuous reaction,* because it results in formation of a completely different mineral (pyroxene). As the pyroxene grows around the olivine, it sometimes prevents further reaction from taking place, and a partially reacted-upon, rounded olivine is left surrounded by pyroxene.

Bowen also observed that plagioclase feldspars gradually changed their chemical composition as magma temperature decreased and plagioclase crystallized. He attributed this to a *continuous reaction* between the magma and the growing crystals of plagioclase: as the temperature of the magma drops and plagioclase crystals grow, they continuously change their chemical composition and become richer in sodium.

Bowen summarized his studies with a diagram, now called *Bowen's Reaction Series,* which includes all the major rock-forming minerals of common igneous rocks (Fig. 3.17). The *discontinuous reaction series* contains the common iron-magnesium (ferromagnesian) silicate minerals. Plagioclase makes up the *continuous reaction series.* The three minerals at the bottom do not form by reactions. They simply crystallize last, sometimes in the sequence in which they are listed and sometimes simultaneously.

The reaction series shows that, when a natural magma cools, certain minerals crystallize earlier and at higher temperatures than others. The common rock-forming

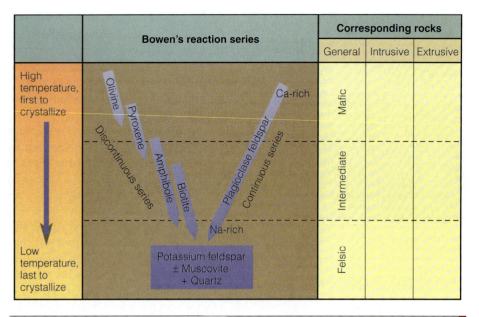

	Bowen's reaction series			Corresponding rocks		
				General	Intrusive	Extrusive
High temperature, first to crystallize	Olivine / Pyroxene / Amphibole / Biotite (Discontinuous series)	Plagioclase feldspar (Continuous series) Ca-rich / Na-rich		Mafic		
				Intermediate		
Low temperature, last to crystallize	Potassium feldspar ± Muscovite + Quartz			Felsic		

FIGURE 3.17

Bowen's Reaction Series. Minerals at top crystallize first from a magma at high temperatures (1200–1300°C); those at bottom crystallize last at lower temperatures (600–900°C). Ferromagnesian minerals are related by discontinuous reactions with the magma as temperature decreases, whereas plagioclase reacts continuously to form crystals progressively richer in sodium. For names of *Corresponding Rocks,* see Problem 5a.

minerals form in a regular sequence, with olivine and calcium-rich plagioclase, at the top of the list, crystallizing at temperatures some 500–600° C higher than quartz, which is at the bottom of the list. The crystallization temperatures of the plagioclase feldspars fall in the same range as the ferromagnesian minerals. Note, however, that the more calcium-rich plagioclases crystallize at higher temperatures than the less calcium- and more sodium-rich plagioclases. The most sodium-rich one crystallizes at the lowest temperature.

Cautionary note: Although the reaction series is quite instructive, it is not universally applicable in detail; it was designed to summarize the most common reactions and the most common order of crystallization for a specific type of magma (tholeiitic basalt). The overlapping arrows in the discontinuous series suggest the possibility of variation in the sequence.

WHERE AND HOW MAGMAS FORM

Most volcanoes are associated with tectonic plate boundaries (Fig. 3.18). What happens at plate boundaries that allows magmas to form? Are there simply cracks that open into a huge reservoir of magma? Not likely; although part of the Earth's core is molten, the silicate magmas from which igneous rocks form are very different from the molten material in the core.

The most likely way for magmas to form is by **partial melting** of existing rocks. What would happen if a rock made of quartz, potassium feldspar, sodium-plagioclase, biotite, and hornblende began to melt? We can use Bowen's Reaction Series to predict the results. The minerals lowest in the series, such as quartz and potassium feldspar, would melt at lower temperatures than minerals higher in the series. Imagine that the temperature got just high enough to melt the quartz and feldspar, but not high enough to melt the rest of the minerals. Because quartz and potassium feldspar are the main minerals in granite, the partial melt would have a felsic composition similar to that of granite. The liquid, because it is less dense than the surrounding rock, would migrate upward. Drops of liquid would eventually coalesce to form a body of magma, which would continue to move upward until it either erupted on the surface or crystallized below the surface.

At *divergent plate boundaries* (where plates move away from one another), hot, buoyant peridotite from the mantle rises and, because of its ultramafic composition, partially melts to form a mafic magma that is chemically similar to basalt or gabbro (Fig. 3.18). The magma rises and intrudes into the surrounding lithosphere or is extruded onto the seafloor.

The process at *convergent plate boundaries* (where plates move toward one another) is more complex. When the down-going slab of lithosphere gets hot enough, water-rich fluids are released that trigger partial melting in both the slab and the overlying wedge of mantle. As the magma rises, high-temperature minerals such as olivine may crystallize and be left behind, and the magma may partially dissolve some of the rocks through which it passes. These and other complex processes yield abundant intermediate magma (chemically like andesite or diorite), which erupts at the surface, and felsic magma (chemically like granite or rhyolite), which is trapped below the surface to form large bodies of intrusive igneous rock (Fig. 3.18).

CHANGES IN MAGMAS

Like the magma formed at convergent plate boundaries, most magmas undergo some kind of change between the time they form and the time they solidify into rock. The changes may result from partially dissolving or melting the surrounding rock, from mixing with other magmas, or from processes within the magma itself.

As an example of a process within the magma, consider what might happen when a gabbro magma crystallizes. According to Bowen's Reaction Series, olivine is the first mineral of the discontinuous series to crystallize. Olivine is denser than the magma, so if there are no currents in the magma, the olivine would sink and form a mush of olivine crystals and magma. This would leave a smaller amount of magma at the top of the chamber, and what is left would be depleted in the components of olivine (magnesium and iron). Should the magma rise, the olivine crystals would be left behind, and the rising magma would be different from the original one. Thus, two different igneous rocks would be formed: one from the olivine-rich mush left behind, and the other from the remaining magma.

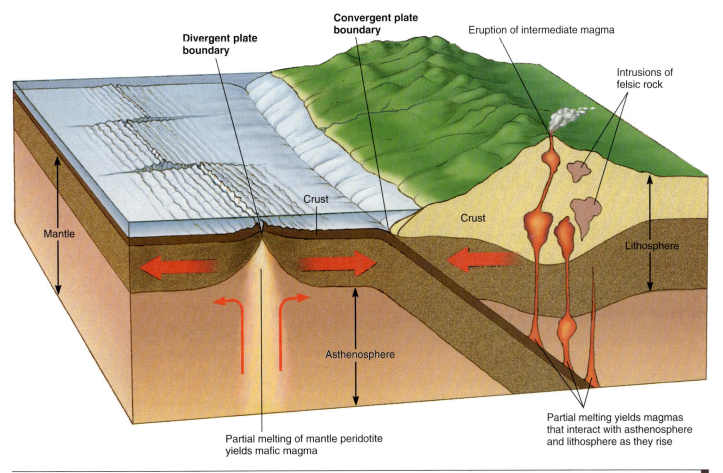

Divergent plate boundary

Convergent plate boundary

Eruption of intermediate magma

Intrusions of felsic rock

Crust

Mantle

Crust

Lithosphere

Asthenosphere

Partial melting of mantle peridotite yields mafic magma

Partial melting yields magmas that interact with asthenosphere and lithosphere as they rise

FIGURE 3.18

Most igneous activity takes place at divergent and convergent plate boundaries. Partial melting of peridotite mantle at divergent boundaries yields mafic magmas. Convergent boundaries are more complex; magmas formed by partial melting may interact with a variety of rocks to produce abundant intermediate and felsic magmas.

APPLICATIONS

WHEN A GEOLOGIST PICKS UP A ROCK, SHE OR HE WANTS TO KNOW WHAT KIND OF ROCK IT IS AND HOW IT FORMED. THIS INFORMATION IS THE BASIS FOR MOST GEOLOGICAL ENDEAVORS, WHETHER IT BE A SEARCH FOR MINERAL RESOURCES OR AN ENVIRONMENTAL IMPACT STUDY. TO "READ" THE ROCK, IT IS NECESSARY TO IDENTIFY ITS CONSTITUENTS AND INTERPRET THE BASIC TEXTURAL RELATIONS. THIS LAB WILL SHOW YOU HOW TO IDENTIFY IGNEOUS ROCKS AND HOW TO USE THEIR TEXTURES TO INTERPRET THEIR MODE OF FORMATION. YOU ALSO WILL LEARN HOW IT IS POSSIBLE TO MAKE LOGICAL INFERENCES OR PREDICTIONS FROM EXISTING INFORMATION. THIS IS ONE OF THE MOST IMPORTANT THINGS THAT GEOLOGISTS, OR ANY SCIENTISTS, DO.

If you complete all the problems, you should be able to:

1. Recognize the following textures in igneous rocks: glassy, aphanitic, phaneritic, pegmatitic, porphyritic, vesicular, amygdaloidal, and pyroclastic.

2. Explain how each texture formed.

3. Distinguish between intrusive and extrusive igneous rocks on the basis of texture.

4. Recognize quartz, potassium feldspar, plagioclase, biotite, hornblende, augite, and olivine in an igneous rock.

5. Explain how igneous rocks are classified.

6. Identify an igneous rock.

7. Use textural relations to determine the sequence in which minerals formed in an igneous rock.

8. Determine the relative temperature (high, intermediate, low) at which a common igneous rock formed, using Bowen's Reaction Series.

9. Relate the common igneous rocks to their plate-tectonic setting.

PROBLEMS

1. Which of the rock samples provided for you have the following textures? (All of these may not be available to you.)

   Glassy:

   Fine grained (aphanitic):

   Coarse grained (phaneritic):

   Pegmatitic:

   Porphyritic:

   Vesicular:

   Amygdaloidal:

   Pyroclastic:

2. Of the rocks with glassy, fine-grained, coarse-grained, pegmatitic, and porphyritic textures, which:

   a. Formed from the most rapidly cooled magma?

   b. Probably resulted from two stages of cooling—slow, then rapid?

   c. Crystallized from a water-rich magma?

3. Which of the rocks in the previous question are likely to be extrusive, and which are likely to be intrusive igneous rocks? Why?

4. Identify the samples of igneous rocks provided by your instructor.

5. Use Bowen's Reaction Series (see Fig. 3.17) to answer the following.

   a. Based on the mineral content and texture of the following rocks, place them in the appropriate blank spaces under *Corresponding Rocks* in Figure 3.17: granite, gabbro, andesite, diorite, rhyolite, basalt. Which forms at the highest temperature, granite or gabbro? Andesite or basalt?

b.   Based on your knowledge of the colors of rock-forming minerals, are rocks at the top of Bowen's Reaction Series likely to be darker- or lighter-colored than those at the bottom? Which of the rocks in 5a are felsic? Intermediate? Mafic?

c.   If you identified quartz in a sample of igneous rock, what other minerals might you expect to find in the rock?

d.   If you identified olivine in a sample of igneous rock, what other minerals might you expect to find in the rock?

e.   Would you expect to find olivine and quartz in the same igneous rock? Explain your answer.

6.   Most igneous rocks form at divergent and convergent plate boundaries. Of granite, gabbro, andesite, diorite, rhyolite, and basalt, which are most likely to form at a divergent boundary?

At a convergent boundary?

7.   Go to *http://volcano.und.nodak.edu/vw.html,* or link from author's *homepage* (see Preface). Link to *Volcanoes of the World* then to *Exploring Earth's Volcanoes* and explore various links to find information on Mount Rainier, a volcano near Seattle, Washington, and answer the following questions. Some links have **highlighted** terms in their text that lead to a glossary link. The glossary may also be useful for text that does not provide glossary links. Its address (at present) is *http://volcano.und.nodak.edu/vwdoc/glossary.*

   a.   In what mountain range is Mount Rainier?

   b.   Is Mount Rainier a shield volcano, composite volcano (also called stratovolcano), or cinder cone?

   c.   Does it have a caldera at the summit?

   d.   What types of rock materials or deposits make up Mount Rainier, and what rock composition (such as basalt or granite) is most common?

   e.   When did Mount Rainier first become active, and when was its most recent eruption?

f. Volcanoes grow when the rate of addition of new volcanic materials exceeds the rate of erosion. Is Mount Rainier currently getting larger or smaller?

g. Mount Rainier is considered a dangerous volcano. Why?

h. What other volcanoes in the same mountain range are considered active and worthy of being monitored?

i. Explain, in terms of plate tectonic theory, why the mountain range containing Mount Rainier and other volcanoes is there.

IN GREATER DEPTH

8. The *density* of a substance is its mass divided by its volume ($\rho = m/V$, or density $=$ mass/volume) and is commonly expressed in grams per cubic centimeter (g/cm^3). *Specific gravity* is the weight of a substance divided by the weight of an equal volume of water, and it has no units. The numerical values of density and specific gravity are effectively the same, because the density of water is 1 g/cm^3. For most scientific purposes, it is more convenient to use density than specific gravity.

    a. Calculate the density of an igneous rock with the following volume percentages of minerals: 18% quartz, 57% plagioclase, 11% hornblende, 14% biotite. Assume the following densities: quartz $= 2.65$ g/cm^3, plagioclase $= 2.69$ g/cm^3, hornblende $= 3.20$ g/cm^3, and biotite $= 3.00$ g/cm^3. Show your calculations in outline form.

    b. What would be the total mass of this rock, in kilograms, if it were used to face the lower 7 m of a 75-m by 45-m building? Assume the facing stone is 2 cm thick. Show calculations.

    c. If 1 kg $= 2.2046$ pounds, what is the weight of all the facing stone in 8b in pounds? Show calculations.

9. Indicate which chemical elements increase in abundance toward the bottom of Bowen's Reaction Series (refer to the chemical compositions given in Chapter 2 or in your textbook).

10. Chemical analyses of representative extrusive rocks from the Cascade Range in Oregon are given in Table 3.3. Notice that SiO_2 is by far the most abundant oxide and that it makes up from about 50 to 75% of these rocks. These values are typical of most igneous rocks. The amount of SiO_2 can be used to identify and name extrusive rocks, many of which are too fine grained to identify on the basis of mineral content. For one category of extrusive rocks, the following names are assigned on the basis of percent SiO_2:

Percentage SiO$_2$	Name of Extrusive Rock
<52	Basalt
52–57	Basaltic andesite
57–63	Andesite
63–70	Dacite
>70	Rhyolite

a. Prepare a graph on Figure 3.19 with percent SiO$_2$ as the abscissa (*x* or horizontal axis), with values ranging from 45 to 80%, and with percent oxide as the ordinate (*y* or vertical axis), with values ranging from 0 to 12%. Draw vertical lines at 52, 57, 63, and 70% SiO$_2$ to mark the fields of the rock types listed above.

b. Plot the percent FeO against percent SiO$_2$ for each of the analyses in Table 3.3. Connect the points to show how FeO changes with SiO$_2$, and label the curve FeO. Do the same with MgO, CaO, Na$_2$O, and K$_2$O.

c. According to Bowen's Reaction Series, in what order should these rocks have formed?

d. What elements increase with increasing abundance of SiO$_2$, and which decrease? Why?

Table 3.3

CHEMICAL ANALYSES OF REPRESENTATIVE EXTRUSIVE ROCKS FROM THE CASCADE RANGE, OREGON

DATA FROM D. J. GEIST. ET AL. 1985, GPP (GEOCHEMICAL PROGRAM PACKAGE), CENTER FOR VOLCANOLOGY, UNIVERSITY OF OREGON

	1	2	3	4	5	6	7	8
SiO$_2$	48.60	50.68	63.81	59.85	53.50	69.36	74.12	55.92
FeO	10.31	9.27	4.05	5.39	8.31	2.32	1.26	7.32
MgO	6.73	6.12	2.28	3.58	5.42	1.14	0.16	4.04
CaO	10.65	9.67	4.87	5.95	7.63	3.07	1.23	6.77
Na$_2$O	2.47	2.70	3.72	3.65	3.18	3.92	4.08	3.49
K$_2$O	0.25	0.48	1.96	1.28	0.73	3.02	4.47	0.97
OTHER	20.99	21.08	19.31	20.30	21.23	17.17	14.68	21.46

Numbers are weight percentages of constituents in the rock.

Percent Oxides

Percent Si O_2

FIGURE **3.19**
Graph for Problem 8.

chapter

4

Sedimentary Rocks

Overview

Sedimentary rocks form from sediment that accumulates on the Earth's surface. The components of the sediment, derived from weathering and erosion of preexisting rocks, are transported as solid particles or ions to their site of deposition. Sedimentary rocks contain many kinds of clues about their history and the history of the Earth's surface. In this lab, you will learn how to recognize and interpret some of these clues and to identify common sedimentary rocks.

Materials Needed

Hand lens • Binocular microscope (optional) • Metric ruler, dilute hydrochloric acid, and glass or knife to test hardness • Samples of sedimentary rocks to be identified • Samples of sediment • Rock samples illustrating cross-stratification, ripple marks, graded beds (or jars of sediment in water), and mud cracks (optional) • Samples of weathered igneous rocks

INTRODUCTION

Rocks and soils at the Earth's surface disintegrate and decompose as they interact with the surface environment. The solid particles and *ions* (atoms or molecules with an electrical charge) freed by this action are carried by water, wind, or glaciers from their source to the place where they are deposited, the **basin of deposition.** There, under the influence of gravity, they accumulate as layers, or **strata,** on the Earth's surface (though commonly underwater). This collection of loosely packed, unconsolidated minerals or rock fragments is **sediment.** In time, sediment is buried, hardened, and consolidated to form **sedimentary rock.**

The mineral composition and texture of many sedimentary rocks provide clues to the:

1. original source of the sediment;

2. type and extent of the weathering processes by which the source rock was broken down;

3. type of agent (water, wind, ice) that transported the sediment and, perhaps, the duration of transport;

4. physical, chemical, and biological environment in which the sediment was deposited; and

5. changes that may have occurred after deposition.

By carefully examining the clues in the rock and knowing how to interpret them, it may be possible to learn a good deal about the history of the rock and the environment in which it formed.

WEATHERING

Weathering is the process by which rocks at the surface disintegrate or decompose; it results in most of the components of sedimentary rocks. **Mechanical weathering**—for example, by frost action—causes rocks to disintegrate into pieces, or into minerals. During **chemical weathering,** chemical reactions cause the rock to decompose. As a result, ions of some elements, such as

COMMON WEATHERING PRODUCTS

MECHANICAL WEATHERING	CHEMICAL WEATHERING
Quartz	Secondary minerals
K-feldspar	Clay
Na-plagioclase	Iron oxide
Rock fragments	Fine quartz
Other silicate minerals are less common	In solution
	K^+, Na^+, Ca^{2+}, Mg^{2+}
	SO_4^{2-}, CO_3^{2-}, HCO_3^-
	Silicic acid

calcium, sodium, potassium, and magnesium, are freed from the rock and carried off in water. The reactions also produce new minerals, principally clays and iron oxides, and release nonreactive ones, especially quartz, from the rock. Table 4.1 lists the common mineral products of weathering.

EROSION AND TRANSPORTATION

As a rock is broken down by weathering, the freed materials move downhill under the influence of gravity and are eroded and transported away by water, glaciers, or air. The effectiveness of erosion is determined in part by the relief—or differences in elevation—of the area. In areas of high relief, large pieces of rock can roll down steep slopes into fast, turbulent streams that move them still farther down the valley. If slopes are gentle, gravity is less effective, streams are more sluggish, and only small pieces of rock or mineral can be carried any appreciable distance. Dissolved chemicals (ions) are carried away regardless of the slope, as long as water is present to carry them.

Nature of the Transporting Agent

Water is the most important transporting agent, and fast-moving, turbulent streams do the best job of it. It's like cleaning off a sidewalk or a car; you can do a better job with a hose and nozzle than with buckets of water. The greater the velocity and the more turbulent the water, the larger the particle that can be transported. Turbulent streams moving down steep slopes are capable of moving big boulders, albeit slowly and for very short distances at a time. But even slowly moving streams can carry small particles in suspension; that's what gives some streams their muddy look.

Moving glaciers carry any size particle that gets trapped in the ice. Most of the surface sediment in Canada and the northern part of the United States was carried there by massive glaciers.

Wind transports small particles only. The smallest particles (clay- and silt-size; defined below) will stay in the air as long as the wind keeps blowing, and they can be carried long distances. Sand grains can be moved by the wind, but they move close to the surface and don't go very far in a single trip.

Effects of Transportation on Particles

Particle or Grain Size

Larger pieces of rocks break into smaller pieces as they roll or bounce downhill or are banged about in a turbulent stream. Thus, the size of a particle may be an indication of how far it has traveled; large pieces do not get very far (unless they are moved by ice).

An easy and useful way to describe a sediment or sedimentary rock is by its particle size, or **grain size.** Familiar words for the size of pieces of rock, such as *sand* or *pebble,* are defined in terms of their diameters in millimeters, as illustrated in Figure 4.1A. Larger particles include cobble (diameter of 64 to 256 mm [2.5 to 10.25 in]) and boulder (diameter greater than 256 mm [10.25 in]).

Grain Shape

Like grain size, the **roundness** of a grain tells something of its history. The corners and sharp edges of grains moved by wind or water are smoothed and rounded as they bump against each other or the ground. Relative terms such as *well-rounded, subrounded,* and *angular* refer to the degree of roundness (Fig. 4.1B). The amount of rounding is an indication of the amount of abrasion the grain has undergone. Angular particles have not been abraded and, unless they were transported by ice, are likely to be found near their source. Grains are rounded most effectively in *high-energy environments,* like a beach or fast-moving stream.

Sorting

Whereas grain size and grain roundness refer to individual grains, **sorting** refers to the similarity in size of all grains composing the sedimentary rock. If all grains are essentially the same size, the rock is *well sorted;* if grains differ greatly in size, it is *poorly sorted* (Fig. 4.1C). A poorly sorted sedimentary rock may have a range of grain sizes, or, as shown in Figure 4.1D, may consist of larger **grains** (particles) imbedded in a **matrix** (the mass of smaller particles filling the space between the larger grains). The degree of sorting is a good indication of how particles were transported. Glaciers carry particles of all sizes; when the glacier melts, the particles are deposited together, producing a distinctive, very poorly sorted sediment. Wind is capable of transporting only sand and finer particles; sand moves close to the ground, whereas silt and clay separate from the sand and can travel great distances high in the air. Wind deposits tend to be very well sorted. The size of particles transported by water depends principally on its velocity and turbulence; most water-laid deposits are moderately well to well sorted.

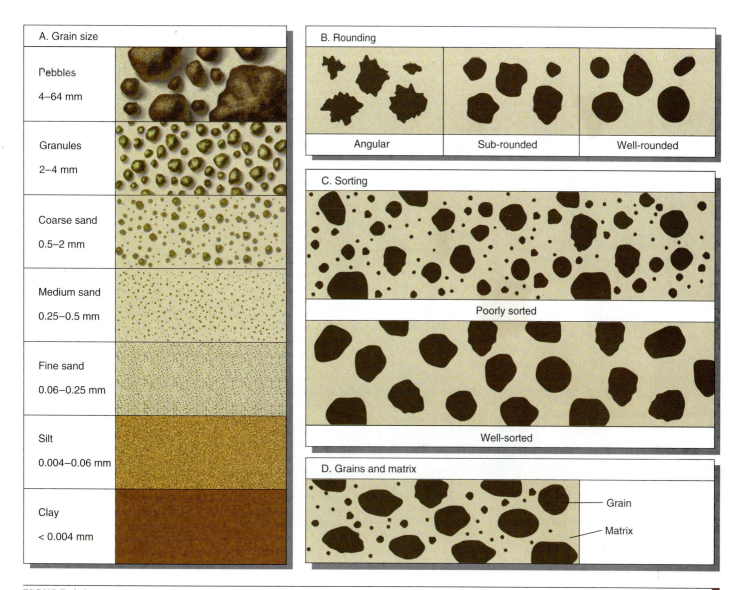

FIGURE 4.1

Characteristics of sedimentary grains. A. Scale of grain-size names for sediment and sedimentary rocks. B. Angular, sub-rounded, and well-rounded particles. Note that rounding is not synonymous with, nor does it imply, sphericity. C. Poorly sorted and well-sorted sediment. D. Larger particles, or *grains,* are surrounded by a *matrix* of much smaller particles.

DEPOSITION AND LITHIFICATION

Types of Sediment

The transporting agent, whether it be water, wind, or ice, eventually slows (or melts, in the case of ice), and responding to the force of gravity, the particles in transport settle and accumulate on the Earth's surface as a *detrital sediment* (after *detritus,* a word referring to the loose material resulting from rock disintegration). A **detrital sediment** consists of rock or mineral fragments that have been transported into the basin of deposition. Such a sediment is also known as a *clastic sediment,* because it consists of *clasts*—broken fragments of preexisting rocks. In the case of particles transported by water or air, the rate of settling depends mainly on particle size and density; large, heavy particles settle faster than small, light ones. Silt and larger particles begin to settle when the velocity and turbulence of the water or wind decrease enough for them to do so. Clay-size particles may remain suspended for long periods before settling, or may clump together to form larger particles that settle sooner. Particles transported by ice are deposited when the ice melts, although they may be moved farther by meltwater streams.

Chemical sediment is deposited when the ions that are dissolved in the water join together to form a solid. The solid, usually a mineral crystal, is a *chemical precipitate,* and the process of forming a solid from the ions in solution is *chemical precipitation.*

Inorganic chemical precipitation takes place within the basin of deposition where the chemistry of the water is appropri? For example, evaporation of seawater ? es the concentration of ions in the r ing liquid to become so great tha?

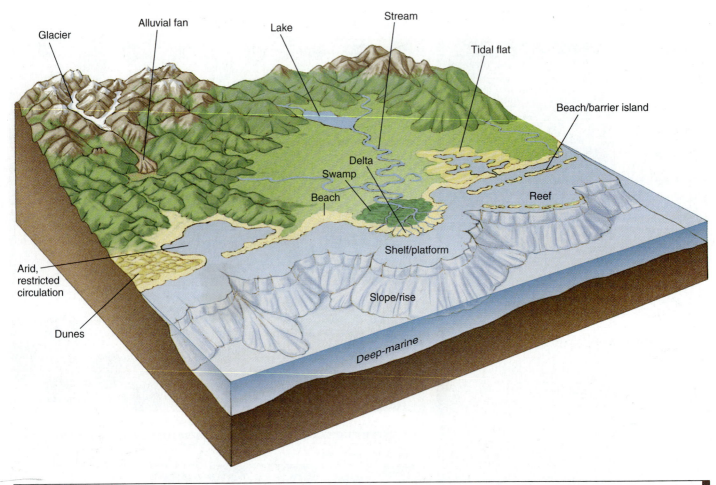

FIGURE 4.2
Typical sedimentary depositional environments.

are precipitated; this is how rock salt and rock gypsum are formed. Inorganic precipitation from *normal* seawater seems to be restricted to certain environments, and the only common rock formed in this manner is *oolitic limestone.*

Biochemical precipitation, involving organisms living in the basin of deposition, is common. Animals and plants extract calcium carbonate and silica from seawater to build skeletal structures such as shells or, in the case of some algae, tiny carbonate crystals that hold up the soft organic parts. When the organisms die, the skeletal parts sink and become part of the bottom sediment.

Biochemical sediment formed directly from the decay of plant or, less commonly, animal materials is rich in carbon. The most common example is peat, which is formed of partially decayed vegetation that accumulated in a swamp.

Environments of Deposition

One of the most important goals of rock study is to learn exactly how, and under what conditions, a particular rock formed. If we know these things, we can learn or predict many others. For example, we can reconstruct the geologic past, we can predict where we might find ore deposits or fossil fuels, we can see how climatic conditions have varied in the past, and we can even formulate hypotheses about the future.

We can learn how and under what conditions ancient sediments formed by studying the environments in which modern sediments accumulate. The most important depositional environments are illustrated in Figure 4.2; Table 4.2 is a summary of the sedimentary rocks and their characteristics for each of these environments.

Sedimentary Structures and Their Significance

Sedimentary structures are features in sedimentary rocks that formed during or after deposition of the sediment, but before lithification (see below), and are large enough to be visible in the field. They are important because they provide evidence of the transporting agent and the environment of deposition. Figures 4.3 through 4.7 illustrate some common sedimentary structures.

Lithification

Lithification is the conversion of sediment to sedimentary rock. It is accomplished by *compaction* caused by the weight of the overlying sediment, *cementation* by new minerals that fill the open spaces between grains (Fig. 4.8), and various kinds of *crystallization* processes that form interlocking crystals.

Table 4.2

DEPOSITIONAL ENVIRONMENTS OF COMMON SEDIMENTARY ROCKS

DEPOSITIONAL ENVIRONMENT			CHARACTERISTIC SEDIMENTARY ROCKS
CONTINENTAL	Stream	Channel	Conglomerate and sandstone with current ripple marks, cross-beds, and discontinuous beds
		Floodplain	Shale, some mud cracks
	Alluvial fan		Conglomerate, arkosic sandstone, poor sorting, cross-bedded
	Dune		Sandstone, well-sorted, large cross-beds
	Glacier	Till	Tillite—poorly sorted, unstratified conglomerate; particles angular to rounded and may have scratches
		Stratified drift	Sandstone and conglomerate, similar to stream channel deposits
	Swamp		Coal
	Lake		Shale, freshwater limestone away from shore; sandstone, conglomerate near shore
TRANSITIONAL	Delta		Complex association of marine and nonmarine sandstone, siltstone, and mudstone, possibly with cross-beds and ripple marks; coal common
	Beach/barrier island		Fine- to medium-grained, well-sorted sandstone; cross-beds common
	Tidal flat		Mudstone, siltstone, and fine-grained sandstone; ripple marks, cross-beds, mud cracks; evaporites possible
MARINE	Shelf/platform		Sandstone: cross-beds, ripple marks common; shale; various limestones (or dolostone)
	Reef		Fossiliferous limestone, massive; coral and algae fossils common
	Slope/rise		Mudstone, graywacke (graded bedding common)
	Deep marine		Chert, chalk, micritic limestone, mudstone
	Arid, hot, restricted circulation		Evaporites (gypsum, anhydrite, halite)

SEDIMENTARY ROCKS

Classification

The two types of sediment provide the basis for two categories of sedimentary rocks (Fig. 4.9): (1) **detrital**—those made of solid particles derived from outside the basin of deposition, and (2) **chemical and biochemical**—those formed by precipitation of ions or accumulation of organic materials within the basin of deposition. Classification within these groups is based on texture and mineral composition.

Texture

Most sedimentary rocks have a **clastic texture** (Fig. 4.10), characterized by discrete clasts or grains of rocks, minerals, or fossils. Grains are not intergrown with each other, but are generally bound together or cemented by a chemical precipitate of silica, calcite, dolomite, or iron oxide. This **cement** generally has a crystalline texture (see Fig. 4.8). The particles commonly were derived from outside the basin of deposition, but in the instance of fossils or reworked chemical sediment, may have

formed in the basin itself. If the clastic texture is due to abundant fossils or fossil fragments, the rock is called **bioclastic,** and as you have already learned, **pyroclastic** refers to volcanic rocks with a clastic texture.

Clay-size particles of platy or flat minerals, such as clays or micas, generally are deposited with their flat sides parallel to the depositional surface. The rock formed from this type of sediment easily splits into sheets parallel to the flat faces of the tiny grains and is said to be **fissile** or to possess **fissility.**

FIGURE 4.3

Stratification (bedding or **layering) is** the most distinctive feature of sedimentary rocks. Strata accumulate by deposition on horizontal or gently inclined surfaces. The strata shown here, exposed near Arkadelphia, Arkansas, have tilted from their original horizontal position. If the characteristics of the sediment being deposited remain constant for a long enough period, a thick stratum, or **bed** (1 cm or thicker), is formed; if characteristics fluctuate, thin strata, or **laminations** (less than 1 cm thick), accumulate. The top or bottom surface of a bed or lamination is a **bedding plane.**

FIGURE 4.4

Cross-stratification (*cross-bedding*) in a sedimentary rock indicates that sediment was transported by a current (water or wind) before being deposited. Cross-strata not only indicate the presence of a current, but can be used to tell the direction of the current. Most of the cross-strata shown here were deposited as wind transported sand from right to left. This is the Navajo Sandstone.

A.

B.

FIGURE 4.5

Ripple marks are wave-like features found on bedding planes. A. **Current ripple marks,** like these in Illinois, form when wind or water currents shape the loose sediment into asymmetrical wave forms whose gentle slope is on the side from which the current came. In this example, the current moved from right to left. B. **Oscillation ripple marks** commonly are formed where the back-and-forth motion of water waves shapes the bottom sediment into symmetrical wave forms. This can occur only in relatively shallow water—less than 10 m for normal waves and up to 200 m for large storm waves. (Loraine Formation, near Desbarets, Ontario, Canada.)

FIGURE 4.6

In **graded beds,** larger grains on the bottom usually grade to finer ones on the top, as shown here. Graded beds may form when a sediment-laden (turbidity) current slows after moving down an underwater slope; larger grains settle first, smaller ones last. (Lake Vermillion Formation, near Tower, Minnesota.)

FIGURE 4.7

Mud cracks form when mud (fine-grained sediment) dries and shrinks. The cracks are wide at the top and taper downward. When sediment is deposited on top of cracked mud, some of the sediment falls into the cracks, and they are preserved in the sedimentary record. The presence of mud cracks in sedimentary rocks indicates a depositional environment in which periodic wetting and drying occurred. (Moenkopi Formation, near Moab, Utah.)

FIGURE 4.8

Lithification by cementation occurs when larger grains are "glued" together with crystalline **cement,** as shown in this "microscopic" view.

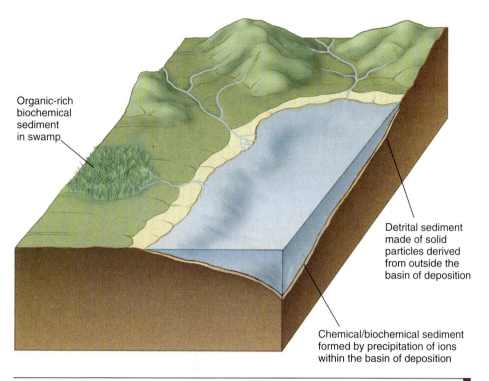

Organic-rich biochemical sediment in swamp

Detrital sediment made of solid particles derived from outside the basin of deposition

Chemical/biochemical sediment formed by precipitation of ions within the basin of deposition

FIGURE 4.9
The two means by which sediment accumulates serve as a basis for classifying sedimentary rocks into detrital and chemical/biochemical.

FIGURE 4.10
Clastic texture in conglomerate. Conglomerates contain rounded granules, pebbles, cobbles, even boulders, in a matrix of finer-grained particles. The largest fragments can be any kind of rock that is durable enough to be transported by water or ice. Some conglomerates contain a variety of large fragments that are made up principally of one rock type. Granule- and pebble-size pieces in some conglomerates are made entirely of quartz because it is so resistant to abrasion. Common matrix minerals are sand-size quartz and feldspar. Conglomerates form from gravels, and common depositional environments are alluvial fans, stream channels, and beaches along rocky coasts.

Chemical and biochemical sedimentary rocks that do not have a clastic texture commonly exhibit recognizable **crystalline textures.** A primary crystalline texture, like that in rock gypsum or rock salt, forms during or shortly after deposition. Secondary crystalline textures result from recrystallization of existing minerals or replacement of existing minerals by new ones after lithification; an example is dolostone.

Some sedimentary rocks, such as chert, are so fine grained or dense that it is impossible to see much detail within them. These rocks are typically **microcrystalline;** they are made of crystals so small that they can be seen only with a microscope.

Mineral Composition

The major mineral components of sedimentary rocks, and their salient properties, are listed in Table 4.3.

Identification of Sedimentary Rocks

Common sedimentary rocks are listed in Table 4.4 (Detrital Sedimentary Rocks)

and Table 4.5 (Chemical and Biochemical Sedimentary Rocks). Choose which table to use as follows:

A. If grains or crystals are visible without a hand lens, and:

1. The rock is harder than 5.5 (scratches glass), use Table 4.4 (the common detrital minerals in rocks containing silt-size and larger grains will scratch glass).
2. The rock is softer than 5.5, use Table 4.5.

B. If grains or crystals are not visible, or barely visible with a hand lens:

1. The rock is a shale or mudstone (Table 4.4), or
2. It is a chemical or biochemical rock (Table 4.5).

If Table 4.4 is chosen, determine the grain size (Fig. 4.1 may be helpful) and mineral composition (see Table 4.3) to identify the rock. The terms *claystone, siltstone, shale, mudstone, sandstone,* and *conglomerate* refer to grain size; *graywacke, quartz sandstone,* and *arkose* are sand-

stones with specific mineral compositions. When naming a detrital sedimentary rock, it is most informative if both mineral composition and grain size are given (e.g. *quartz sandstone, granite-pebble conglomerate,* and *coarse-grained graywacke*). In addition, the textural names for sedimentary rocks made of grains larger than about 2 mm are based on grain roundness: if the rock or mineral fragments are angular, the rock is a *breccia;* if they are rounded, it is a *conglomerate.*

Table 4.5 lists chemical and biochemical sedimentary rocks. They are named primarily on the basis of mineral composition, although the names of *limestones* (sedimentary rocks composed mostly of calcite) also depend on texture or fossil content. If the rock is listed in Table 4.5, first determine the mineral composition (Table 4.3 will help), then, in the case of limestones, the texture.

Some common sedimentary rocks are shown in Figures 4.10 through 4.21. The photographs are representative examples only, and your samples may differ considerably in color and general appearance from those illustrated here.

Table 4.3

RECOGNIZING MINERALS IN SEDIMENTARY ROCKS

MINERAL	PROPERTIES
Quartz	White to light gray Exterior may be frosted; broken grains glassy Hardness = 7
Chert	White, gray, black, red, green Microcrystalline (extremely fine grained) Smooth, conchoidal fracture Hardness = 7
Feldspar	K-feldspar most common White or pink 2 cleavages Commonly angular
Biotite	Shiny and black (gold if altered) 1 perfect cleavage
Muscovite	Shiny and silvery 1 perfect cleavage
Clay (e.g., kaolinite)	White, gray, green, red, black Clay-size particles Soft (H = 1–2.5)
Iron oxide	Yellow, orange, red, brown—strong coloring agent Usually very small particles Mainly a cement
Calcite	White, gray, black Hardness = 3 Bubbles readily in acid
Dolomite	Gray, buff Hardness = 3.5 Bubbles slowly in acid
Gypsum	White to gray Soft (H = 1)
Halite	White, light gray, pinkish white Hardness = 2.5 Tastes like salt

CLASSIFICATION OF DETRITAL SEDIMENTARY ROCKS

GRAIN SIZE	COMMENTS		ROCK NAME	
Gravel (>2 mm)	Rounded grains. Rock name may be modified by type and size of dominant grains (e.g., quartz-pebble conglomerate) or composition of grains and matrix (e.g., arkosic conglomerate).		Conglomerate	
	Angular grains. Rock name may be modified by type of dominant grains (e.g., chert breccia).		Breccia	
Sand (0.06–2 mm)	Mostly quartz grains. Grains commonly well rounded and well sorted.		Quartz sandstone	Sandstone
	Mostly quartz and feldspar. Grains commonly angular and poorly sorted. Commonly reddish.		Arkose	Sandstone
	Composed of quartz, feldspar, rock fragments, and clay. Grains angular and poorly sorted. Dark color common. "Graywacke" is an imprecise general term for such sandstones.		Graywacke	Sandstone
Silt (0.06–0.004 mm)	Commonly massive. Feels gritty on teeth.	Grains too small to identify composition. Mudstone and shale are made of silt and clay; mudstone breaks with conchoidal fracture, shale is fissile.	Siltstone	Shale (fissile) or mudstone
Clay (<0.004 mm)	Commonly laminated. Feels smooth on teeth.		Claystone	Shale (fissile) or mudstone

CLASSIFICATION OF CHEMICAL AND BIOCHEMICAL SEDIMENTARY ROCKS

COMPOSITION	COMMENTS		ROCK NAME
Calcite (CaCO$_3$)	All effervesce in dilute HCl	Texture variable; clastic to crystalline.	Limestone
		Microcrystalline. Breaks with conchoidal fracture.	Micritic limestone
		Sand-size spheres with concentric layers.	Oolitic limestone
		Abundant fossils in crystalline matrix; well cemented.	Fossiliferous limestone
		Poorly cemented shell fragments.	Coquina
		Fine-grained, massive, earthy, poorly cemented.	Chalk
		Banded, finely crystalline to microcrystalline.	Travertine
Dolomite (CaMg(CO$_3$)$_2$)	Effervesces weakly with dilute HCl; powder effervesces more strongly. Commonly crystalline. May contain small, crystal-lined cavities.		Dolostone
Quartz (SiO$_2$)	Dense, microcrystalline texture, hardness = 7.		Chert
Gypsum (CaSO$_4$ · 2H$_2$O)	Massive, crystalline, soft.		Rock gypsum
Halite (NaCl)	Massive, crystalline, tastes like salt.		Rock salt
Plant debris (mostly C)	Plant fossils may be visible. Combustible.		Coal

FIGURE 4.11
Breccia is similar to conglomerate, but has angular, rather than rounded, particles. Because the angular pieces have not moved far from their source, they commonly are all of the same rock type, rather than a mixture of rock types. The matrix of a breccia may be finer particles or cement with a crystalline texture. The angular particles imply a nearby source, so common depositional environments include the bases of cliffs or steep slopes, landslide areas, or places where caves have collapsed.

FIGURE 4.12
Quartz sandstone is made almost entirely of sand grains of quartz. The quartz grains are commonly rounded and cemented together by calcite, iron oxide, or secondary quartz. Some quartz sandstone is white, but more commonly, it is buff or rusty brown, because of a small amount of iron oxide cement. Relatively pure quartz sand can accumulate only when other minerals have weathered away, or where quartz is the only material available in the source area. Depositional environments are beach and near-shore areas, extensive sand-dune fields, and some stream channels.

FIGURE 4.13
Arkose. While most sandstones contain a small amount of feldspar, arkose is unusual, because feldspar is a prominent component along with quartz. The grains are commonly less rounded and more poorly sorted than in quartz sandstone. The combination of pink potassium feldspar and rusty iron oxide gives arkoses their typical reddish color. Many arkoses were derived from erosion of granite in steep terrains and were deposited in nearby alluvial fans or beach-nearshore settings.

FIGURE 4.14
Graywacke is a general, or field, term for gray to dark gray sandstones with fine-grained matrices. Common components include quartz, feldspar (especially plagioclase), rock fragments (especially volcanic rocks), and micas. The angular grains tend to be fine sand, rather than coarse, and the combination of small grain size and even finer-grained matrix give graywacke a rather nondescript appearance. A hand lens may be needed to recognize the clastic texture. Graywackes are common sedimentary rocks in the vicinity of volcanic island arcs or off the margins of continental shelves, where they were deposited from dense, sediment-laden (turbidity) currents that periodically flow downslope into deeper water.

A.

B.

FIGURE 4.15
(A.) Shale and (B.) mudstone. The finest particles end up as components of these mud rocks. Shale is fissile; mudstone is not. Both tend to break up easily in outcrops, but mudstone is more likely to break into larger pieces with a conchoidal fracture. If rich in organic matter, shales and mudstones are black; if not, they are gray, red, or green. Some geologists distinguish two types of mudstone—claystone and siltstone—based on particle size. Siltstone feels gritty when chewed; claystone does not. (So, if you like, take a chew.) Mudstone and shale are, by far, the most abundant kinds of sedimentary rocks. They accumulate in the quiet water of floodplains, lagoons, and lakes, and offshore in shallow marine environments. The shale shown here contains fossils of ferns.

A. B. C.

D. E.

FIGURE **4.16**

Limestone is made of calcite, so it has a hardness of 3 and bubbles readily in acid. There are many varieties, only some of which are mentioned here. *Crystalline limestone* (A.) is made of small, but recognizable, interlocking crystals of calcite. *Micritic* or *microcrystalline limestone* (B.) has such extremely small crystals that its surface is smooth. It formerly was used extensively for making lithographs because of its dense texture and the fact that it dissolves in acid. *Oolitic limestone* (the *oo* is pronounced like the two o's in too or moo) (C.) consists of sand-size spheres (called *ooids*) made of concentric layers of calcite. The spheres are held together with a calcite cement. Ooids form by inorganic precipitation around small particles in shallow, agitated water. *Fossiliferous limestone* (D.) and *coquina* (E.) both contain abundant calcareous (made of calcium carbonate) fossils or fragments of fossils. In fossiliferous limestone, the fossils are cemented together with crystalline calcite to form solid rock; in coquina, the fossil fragments are only loosely cemented, and the rock is porous. *Chalk* is also composed of calcareous fossils, but they are microscopic. Most limestones form in warm, shallow seas where the water is relatively clear and free from land-derived detritus. In the past, shallow seas covered extensive parts of the continents, but today, such environments are confined to the continental shelves.

FIGURE **4.17**

Dolostone is made of the mineral dolomite. It is distinguishable from limestone by its lesser reaction with acid. It is sometimes necessary to use a hand lens to see that it actually bubbles, or to scratch the rock (to make a powder) before testing with acid. Dolostone has a crystalline texture and commonly is light gray or buff. A common, but not invariable, feature is the presence of crystal-lined cavities, or *vugs*. The sample shown here contains a crystal of clear quartz (a "Herkimer diamond"). Most dolostone is formed by replacement of limestone.

FIGURE **4.18**

Chert is made of microcrystalline quartz. It breaks with a conchoidal fracture and has a smooth surface. It most resembles micritic limestone, but its hardness (7) gives it away. It can be white, black, gray, brown, green, red, or other colors. Some chert forms by replacement of limestone and may retain some of the textural elements, such as fossils or ooids. Most chert forms by biochemical extraction of silica from ocean water. Such chert consists of microscopic plant (diatoms) or animal (radiolaria and sponge spicules) fossils.

FIGURE **4.19**

Rock gypsum is usually white, light gray, or pale shades of red or orange. It has a crystalline texture and can be scratched with your fingernail. Dark impurities may be present as layers or outlines around irregular areas. Gypsum is an evaporite mineral; when seawater evaporates, gypsum is one of the minerals that precipitates. Thick deposits accumulate in bays in arid regions that have restricted access to the open ocean; deposition also occurs in salty inland seas or lakes.

FIGURE 4.20
Rock salt is made of halite. It has a crystalline texture and is white, light gray, or tinged with red or orange. It forms in the same manner and depositional environment as does gypsum. The reddish mineral is sylite (KCl).

FIGURE 4.21
Coal is brown to black, depending on the amount of carbon it contains. Brown to brownish black *lignite* contains less carbon than black *bituminous coal,* which in turn, contains less carbon than metamorphic coal, *anthracite.* Coal is noncrystalline and has a luster that ranges from dull to vitreous as carbon content increases. Coal forms by burial and compaction of vegetation that accumulated in swamps.

APPLICATIONS

THIS LAB CONTINUES THE PROCESS OF LEARNING TO IDENTIFY AND INTERPRET ROCKS BY FOCUSING ON SEDIMENTARY ROCKS. YOU ALSO WILL LEARN HOW OBSERVATION, MEASUREMENT, AND COMPARISON HELP TO ESTABLISH GENERAL RELATIONS BETWEEN CHARACTERISTICS OF ROCKS AND THEIR HISTORY OF TRANSPORT AND DEPOSITIONAL ENVIRONMENT.

OBJECTIVES

If you complete all the problems, you should be able to:

1. Determine whether a sedimentary rock has a clastic, bioclastic, or crystalline texture.

2. Distinguish grains from matrix in a clastic sedimentary rock.

3. Identify a sedimentary rock and the major minerals contained in it.

4. Estimate grain size, degree of rounding of grains, and degree of grain sorting of a clastic sediment or sedimentary rock.

5. Recognize the following sedimentary structures in a rock sample or a picture: stratification, cross-bedding, ripple marks, graded bedding, and mud cracks.

6. Study selected sedimentary rock samples and suggest possible depositional environments from which they might have come.

PROBLEMS

1. To help you learn to recognize the principal textures of sedimentary rocks, sort the samples provided for you into the following textural categories:

    Clastic

    Bioclastic

    Crystalline

    Microcrystalline or very fine grained

2. Identify the samples of sedimentary rocks provided by your instructor.

3. Examine the sample of sediment provided for you using a hand lens or a binocular microscope.

    a. Identify the major minerals and estimate their proportions.

    b. By comparison with the roundness scale (Fig. 4.1B), determine the roundness of the grains. Are some minerals rounder or more angular than others? If so, which are most angular and which are most rounded? Suggest an explanation for the similarities or differences.

    c. Use a metric ruler to measure the diameters of several grains in millimeters. What is the range in size, and what sizes are the most prevalent? What grain-size names (for example, sand or granules) best describe the sediment (see Fig. 4.1A)? (For an alternative and more accurate way to estimate grain-size distribution, see Problem 8.)

    d. Describe the degree of sorting. For the purposes of this lab, consider the sediment to be well sorted if most grains have the same grain-size name, and poorly sorted if they do not.

    e. Make the unlikely assumption that all of the grains were freed from the source area by mechanical weathering (that is, no chemical weathering took place). Based on the mineral content of the sediment, what igneous rock(s) could have been the source of this sediment?

    f. What was the probable transporting agent for the materials in the sediment? Explain.

    g. Is the sediment detrital chemical/biochemical? Why?

    h. What lithification process would probably be the most effective for this sediment?

i.   What kind of sedimentary rock would be formed by lithification of this sediment?

4.   Examine samples of the following sedimentary structures, and answer the questions.

   a.   Graded bedding.

      (1)   Before looking at the rock sample, shake a jar of water containing a mixture of clay, silt, sand, and granules; let it sit on the lab bench while you do something else for 5 or 10 minutes.

      (2)   Look at the sediment that has settled on the bottom of the jar. How are different-size grains distributed? Are similar-size grains segregated in layers, or are they distributed randomly throughout the sediment?

         If in layers, are the layers parallel or perpendicular to the bottom of the jar?

         Does the grain size change in any regular manner from top to bottom or from one side of the jar to the other?

         Is there still sediment in suspension that has not settled? If so, what grain-size term would describe it most aptly?

      (3)   Next, look at the rock sample, paying particular attention to the distribution of its grains. Make a rough sketch of the sample showing stratification and emphasizing grain-size differences within or across strata. By analogy with the sediment in the jar, determine the original top of the sample and label it in your sketch.

   b.   Ripple marks. Sketch the sample and identify the type of ripple mark. For current ripple marks, indicate with an arrow the current direction at the time the ripples formed. For oscillation ripple marks, label the top and bottom if you can. Explain your answer.

   c.   Cross-stratification. Sketch the sample and indicate with an arrow the current direction at the time of deposition. Explain your answer.

   d.   Mud cracks. Sketch the original top of the sample. How did you know which side was the original top?

5. Go to www.kaibab.org:80/gc/geology/gc_layer.htm or link to author's *homepage* (see Preface) which contains descriptions of the rocks exposed in the Grand Canyon.

    a. What kinds of Paleozoic sedimentary rocks occur in the Grand Canyon (types of rocks, not names of formations)?

    b. Use the maximum thicknesses given for each of the formations and calculate the relative proportions of each rock type.

    c. Based on their descriptions and Table 4.2 in this lab manual, what are the depositional environments of the following:

        (1) Kaibab Limestone

        (2) Coconino Sandstone

        (3) Hermit Shale

        (4) Supai Formation

        (5) Temple Butte Limestone

IN GREATER DEPTH

6. Examine the samples of partially weathered igneous rocks provided in the lab.

    a. Assuming the samples were all subjected to the same degree of chemical weathering, arrange and list the rocks in order of susceptibility to weathering.

    b. Next, indicate which *minerals* seem to be weathered the least, and which the most, in each sample; list the minerals in order of increasing susceptibility to weathering.

    c. How does the order of your list compare with the sequence of crystallization according to Bowen's Reaction Series (Chapter 3)? Suggest a relation between crystallization temperature of igneous minerals and susceptibility to chemical weathering.

7. Following are descriptions of some common sedimentary rocks, possibly similar to ones provided in lab. Using the descriptions, suggest possible depositional environments for each. (See Fig. 4.2 and Table 4.2)

    a. Reddish rock made principally of angular to sub-rounded, coarse sand- to granule-size grains of quartz and feldspar. In the field, structures like those in Fig. 4.4 are evident, and the sandstone is seen to be interbedded with pebble and cobble conglomerate.

    b. Gray rock with numerous fossil shells cemented firmly together (similar to Fig. 4.16D). Rock readily fizzes with dilute HCl.

c. White rock with crystalline texture. Very soft, does not taste funny or fizz with acid.

d. A conglomerate with grain sizes ranging from clay to boulder. No apparent bedding, unsorted; particles range from angular to rounded, and some have flat, scratched faces.

e. Light tan rock with well-rounded, 0.5 mm (0.02 in), frosted quartz grains. Well sorted, with large cross-beds.

f. Hard, microcrystalline rock with thin beds and laminations.

g. Reddish-gray, very fine-grained, detrital rock, with features like those in Fig. 4.7.

h. Dark gray rock that looks somewhat like basalt, but with a hand lens, the texture appears to be clastic, not crystalline. Some white feldspar grains are present, as are dark grains of something unidentifiable; the matrix is too fine to recognize anything. One sample has thin beds that show a progressive change in grain size from top to bottom.

i. Four sedimentary rock samples were collected while walking up a high hill. 1. A quartz-pebble conglomerate with well-rounded quartz sand and pebbles was found at the bottom. 2. The number of pebbles decreases a short distance up the hill until the rock becomes a quartz sandstone with cross-beds and ripple marks. 3. This changes gradually upward to a gray shale, but no mud cracks were found. 4. At the top of the hill is a gray rock with a crystalline texture and small cavities lined with crystals; the rock bubbles slightly in acid. How did the environments of deposition change at this locality, and how did that most likely take place?

8. Visual estimates of grain size and sorting of sediments are *subjective* and not appropriate for most scientific purposes. For this reason, *objective* methods of measuring these properties must be used, and the results must be reproducible to be valid. A common way to measure size distribution is to pass the sediment through a series of screens or sieves with progressively smaller, pre-measured openings. Grains pass through holes larger than themselves but are caught in sieves whose openings are smaller. The proportion of grains falling within a certain size range is determined by weighing the amount of sediment trapped by each sieve and comparing that to the weight of the entire sediment sample. The activities required for this problem can be done as a group.

a. Weigh the sample of dry sediment provided by your instructor.

b. Pass the sediment through a series of sieves and weigh the sediment trapped in each sieve. Along with the weight of sediment, record the opening sizes of the sieve that trapped the sediment and the sieve with the next largest opening. The grain size of the sediment is between these two values.

c. Determine the percent of grains in each size range by dividing the weight of each by the weight of the entire sample and multiplying by 100. Make a histogram or bar graph of these data by plotting grain-size increments on the horizontal (x) axis and weight percent on the vertical (y) axis.

d. Sorting can be evaluated numerically from these data, but that aspect will not be pursued here. Instead, use your histogram to determine whether a few bars are markedly higher than others. If so, the sample is well sorted; if not, it is poorly sorted.

e. What factors could affect the accuracy of your results?

5

Metamorphic Rocks

Overview

A rock that is subjected to heat and pressure may change in appearance, because its mineralogy and texture change. The characteristics of the original rock, and the type and intensity of metamorphism, determine how drastic the change is. This chapter will introduce some of the common varieties of metamorphic rocks, illustrate their typical textural and mineralogic features, show how they relate to the original or parent rock, and introduce the idea of metamorphic zones.

Materials Needed

Hand lens, dilute hydrochloric acid, glass or knife to test hardness • Samples of metamorphic rocks to be identified.

INTRODUCTION

Metamorphic rocks form from other rocks. **Metamorphism** takes place at varying depths within the Earth's crust, where both temperature and pressure are higher than at the surface. The results of metamorphism include the creation of new minerals, the development of bands or layers of like minerals (for example, layers of quartz or mica), and the parallel alignment of new and old mineral crystals.

CONDITIONS OF METAMORPHISM

Metamorphism takes place at temperatures and pressures that fall between those in which sediments are lithified and those in which rocks begin to melt to form mag-

mas. Figure 5.1 shows approximate lower and upper boundaries of metamorphism. A rock will be metamorphosed if it is heated and buried deeply enough to fall within the areas labeled "metamorphism" on Figure 5.1. Some rocks, such as granite or quartz sandstone, are not affected much by metamorphism until temperatures and pressures are well within the field of metamorphism on Figure 5.1; others, such as porous tuffs consisting of shards of volcanic glass, are highly reactive and undergo changes at temperatures and pressures near those of the field labeled "lithification of sediment." Similarly, melting begins at different temperature-pressure conditions for different rocks.

Water and other fluids, such as carbon dioxide, are generally present during metamorphism, because they are either con-

tained in the rocks undergoing metamorphism or are released by metamorphic reactions. These fluids are important in the metamorphic process. Among other things, they allow ions to move about more readily, thereby speeding up metamorphic reactions and enabling the growth of mineral crystals.

TYPES OF METAMORPHISM

There are two common types of metamorphism: regional and contact.

Regional metamorphism occurs over areas of hundreds or thousands of square kilometers. Regionally metamorphosed rocks were buried beneath thick accumulations of other rocks at some time during their history (Fig. 5.2). During this

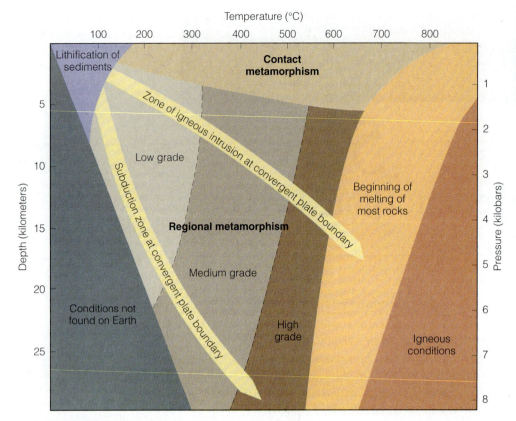

FIGURE 5.1
Temperature and depth (or pressure) conditions for lithification, contact metamorphism, grades of regional metamorphism, and melting. Arrows show how temperature changes with depth at convergent plate boundaries.

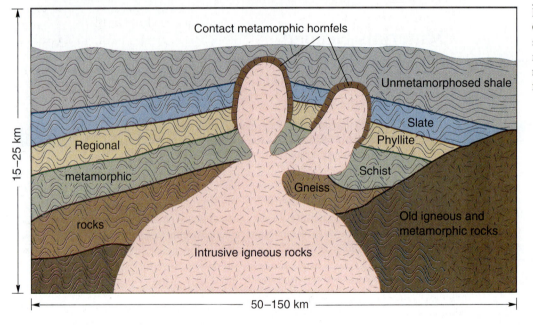

FIGURE 5.2
Cross section illustrating possible setting for regional and contact metamorphism. The metamorphic rocks shown are those that would form from a shale parent.

period of burial, they were subjected to the higher temperatures and pressures found beneath the surface (Fig. 5.1). The weight of the overlying rocks provided most of the pressure, but forces within the Earth, such as those at convergent tectonic-plate boundaries, also may have contributed.

Heat from an intruding body of magma causes **contact metamorphism** of the surrounding rocks (Fig. 5.2). Fluids released from the magma or the surrounding rocks may accentuate the changes. As suggested in Figures 5.1 and 5.2, contact metamorphism is most evident around igneous intrusions that formed within a few kilometers of the surface; at greater depths, it becomes difficult to differentiate contact and regional metamorphism.

CHANGES DURING METAMORPHISM

The important changes that may take place during metamorphism are:

1. Recrystallization of existing minerals, especially into larger crystals (Fig. 5.3);

A.   B.

FIGURE 5.3
These photomicrographs (×10) compare (A.) quartz sandstone and (B.) quartzite. The original sand grains are evident in the sandstone, but recrystallization has removed traces of them in the quartzite.

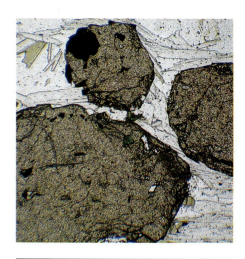

FIGURE 5.4
A new mineral, garnet (brown in photograph), formed during metamorphism, as seen in this photomicrograph of garnet in schist (×10).

FIGURE 5.5
This photomicrograph of schist in polarized light shows the distinctive orientation of the mica crystals caused by directed pressure during metamorphism (×10).

2. Development of new minerals and disappearance of some old ones (Fig. 5.4); and

3. Deformation and reorientation of existing mineral crystals and growth of new ones with a distinctive orientation (Fig. 5.5).

The net result is a rock with a different texture and, commonly, a different mineral content.

Recrystallization and development of new minerals (and disappearance of old ones) take place during both contact and regional metamorphism. In general, the crystal size and kinds of minerals in metamorphic rocks indicate the intensity, or **grade,** of metamorphism (see Fig. 5.1). *High-grade* metamorphic rocks are the most intensely metamorphosed. They are characteristically coarsely crystalline and contain minerals that are stable under higher temperatures and pressures. *Low-grade* metamorphic rocks are the least intensely metamorphosed; they are generally finely crystalline (although crystal size also depends on the size of the grains or crystals in the rock prior to metamorphism) and contain minerals that are stable under lower temperatures and pressures.

Deformation and reorientation of mineral crystals is primarily a result of **directed pressure,** pressure that is greater in one direction than in others. Directed pressure is prevalent during the type of regional metamorphism associated with mountain building and tectonic-plate convergence.

On the other hand, the pressure during contact metamorphism is usually **lithostatic—** equally great in all directions—and results in little deformation or reorientation.

METAMORPHIC TEXTURES

Metamorphic rocks are subdivided into two main textural groups: **nonfoliated** and **foliated. Foliation** is the arrangement of mineral crystals in parallel or nearly parallel planes. An example is shown in Figure 5.5, where the flat crystals of mica are aligned in the same direction.

Nonfoliated metamorphic rocks lack flat mineral crystals with parallel alignment. Instead, crystals typically are all about the same size and are interlocked in a crystalline texture (see Fig. 5.3).

Foliated metamorphic rocks contain minerals that are aligned in parallel planes (Figs. 5.6 through 5.9). The **foliated texture** commonly results from the parallelism of micas, but crystals of other metamorphic minerals, such as chlorite or hornblende, also can form parallel planes. Foliated rocks commonly appear to be layered or banded. In addition to the parallel arrangement of minerals, the crystals also are grown together in an interlocking fashion, as in other rocks with crystalline textures. Foliation results from directed pressures usually present during regional metamorphism.

Four distinct types of foliation are recognizable, primarily on the basis of crystal size.

Slaty (Fig. 5.6): a foliation in metamorphic rocks made of platy (flat) minerals too small to be seen without a microscope. Such rocks readily split or cleave along almost perfectly parallel planes and are said to have a **slaty cleavage.**

Phyllitic (Fig. 5.7): metamorphic rocks made of platy minerals just visible to the unaided eye have a shiny or glossy luster. The rock cleavage is along nearly parallel surfaces, but these surfaces may be wrinkled to varying degrees. This type of foliation represents a state of textural development intermediate between slaty and schistose.

Schistose (Fig. 5.8): a foliation in rocks composed of mineral crystals large enough to be seen with the unaided eye. Rock cleavage is parallel to the foliation,

FIGURE 5.6
Slaty cleavage is the most distinctive feature of slate. Slate is so finely crystalline—mineral crystals are indistinguishable—and breaks along such flat surfaces that the really good slate was used for blackboards before about 1945. It makes a durable roof too, and is still used for that purpose. Slate comes in a variety of colors, but is commonly black, bluish black ("slate blue"), gray-green, purple, or dull red. Slate forms from metamorphism of shale and is usually distinguishable from it by its slightly shinier luster.

FIGURE 5.7
Metamorphic intensity slightly higher than that required to form slate turns shale into phyllite, a rock characterized by its phyllitic foliation. On the low-metamorphic-grade side, phyllite is distinguished from slate by a shinier, more satin-like luster and by crystals that are large enough to give just a hint of their presence. On the high-grade side, phyllite is distinguished from schist by crystals that are almost, but not quite, large enough to identify. Phyllite has a good rock cleavage, but the cleavage surfaces are commonly wrinkled and the cleavage is not as good as slate's.

but cleavage surfaces are usually irregular. Platy minerals commonly predominate, but other minerals, such as quartz, are also present.

Gneissic (Fig. 5.9): a coarse foliation in which like minerals are more or less segregated into roughly parallel bands or layers. A common example is a rock in which nonfoliated layers made of quartz and feldspar alternate with foliated layers made of biotite. Mineral crystals are coarse and can be identified with the unaided eye.

METAMORPHIC MINERALS

Metamorphic rocks contain many of the same minerals found in igneous and sedimentary rocks. Certain minerals, however, occur almost exclusively in metamorphic rocks. In addition, some minerals of metamorphic rocks are present only in rocks of a specific metamorphic grade. Table 5.1 summarizes the properties of common metamorphic minerals, and Figure 5.10 shows the metamorphic grades at which they occur.

CLASSIFICATION AND IDENTIFICATION OF METAMORPHIC ROCKS

Metamorphic rocks are classified on the basis of texture and the kinds and proportions of minerals present. However, microscopic examination may be necessary to determine the mineralogic content. Therefore, the primary basis of the classification to be used here is texture, as shown in Table 5.2.

To identify a metamorphic rock, first identify the texture and:

A. If the rock is foliated, use a name consistent with the type of foliation: *slate* (Fig. 5.6), *phyllite* (Fig. 5.7), *schist* (Fig. 5.8), or *gneiss* (Fig. 5.9). This name is then modified by mineral names when possible, as in quartz-biotite schist.

B. If the rock is nonfoliated, its name is determined by mineral content when

A.

B.

FIGURE 5.8
Schists have crystals that are large enough for individual minerals to be identified. Their distinctive schistose foliation is caused by the presence of platy minerals such as talc, chlorite, biotite, and muscovite. The larger crystal size and the presence of other minerals, such as quartz and feldspar, cause schists to split along rough, irregular surfaces that more or less parallel the foliation. Schists derived from the metamorphism of shale may contain typically metamorphic minerals such as garnet, staurolite, kyanite, or even sillimanite, depending on the metamorphic grade. Schists derived from other parent rocks contain other minerals, such as blue, green, or black amphibole. Because the mineral content of schists is variable, the characterizing minerals are used in the rock name: (A.) quartz-biotite schist, and (B.) garnet-muscovite-staurolite schist are examples. More intensely metamorphosed schists may be somewhat banded due to segregation of dark and light minerals. Commercially valuable minerals such as talc and kyanite are obtained from schists in which these minerals are sufficiently concentrated.

FIGURE **5.9**

Gneissic foliation, or banding, characterizes gneiss. For example, *granitic gneiss* has light-colored layers of quartz and feldspar that alternate with dark layers of biotite and/or hornblende. Other minerals such as garnet or sillimanite may be present as well. Layers are a few millimeters to a few centimeters thick. During very high-grade metamorphism, local partial melting of quartz and feldspar may result in the formation of a *migmatite,* a mixed igneous-metamorphic rock with swirled light and dark bands. Some gneisses and migmatites are beautiful when cut and polished and are used extensively for monuments and decorative building stones.

possible, as in *quartzite* (Fig. 5.11), *marble* (Fig. 5.12), *soapstone* (Fig. 2.9), or *serpentinite.* Other nonfoliated rocks include *greenstone* and *amphibolite,* which are dark gray, greenish black, or black rocks formed by metamorphism of basalt, gabbro, or andesite. The parentage of the lower-grade greenstone (or *greenschist*) is usually recognizable, because it still looks like the parent rock; amphibolites are more difficult to interpret, because metamorphism has erased the original traits and they may be slightly foliated. *Anthracite* coal, or hard coal, is recognizable by its color, vitreous luster, and conchoidal fracture. *Metaconglomerate, metagraywacke,* and *metachert* are the metamorphosed equivalents of the sedimentary rocks conglomerate, graywacke, and chert, and are usually recognizable as such. **Hornfels** is a general name used for nonfoliated, contact-metamorphic

Table 5.1

RECOGNIZING MINERALS IN METAMORPHIC ROCKS

MINERAL	PROPERTIES
K-feldspar	Usually white or pink 2 cleavages at 90° Equidimensional to ovoid crystals
Plagioclase	Usually white (Na-plagioclase) or gray (Ca-plagioclase) 2 cleavages at 90° Elongate to equidimensional crystals Striations may or may not be present
Quartz	Colorless to gray Glassy with conchoidal fracture Irregular crystals or lens-shaped masses
Biotite	Shiny and black 1 perfect cleavage Thin crystals parallel to foliation
Muscovite	Shiny and silvery white 1 perfect cleavage Thin crystals parallel to foliation
Hornblende (amphibole)	Black with shiny, splintery appearance 2 cleavages at 56° and 124° Elongate, commonly parallel, crystals
Garnet	Pink, red, reddish brown Vitreous to resinous luster Equidimensional, 12-sided crystals with diamond-shaped faces are common
Staurolite	Brown, red-brown to brownish-black Vitreous to dull luster Prismatic and X- or cross-shaped crystals
Kyanite	Light blue to greenish blue Vitreous luster Blade-shaped crystals
Talc	White, gray, apple green Pearly luster Soft (H = 1), with greasy feel
Chlorite	Green to blackish green 1 good cleavage Crystals commonly small, flaky, and parallel to foliation

rocks in which individual minerals cannot be readily recognized.

PARENT ROCK

Metamorphic rocks are formed by the effects of temperature and pressure on a preexisting rock, the *parent.* It is always helpful in working out the geologic history of a metamorphic area to determine what the parent rock was, although this is not always possible. Common parent rocks and their metamorphic derivatives are listed in Table 5.2.

ZONES OF METAMORPHISM

During regional metamorphism, rocks subjected to higher temperatures and pressures are squeezed and cooked more than those exposed to less intense conditions. The result is that high-grade metamorphic rocks differ from intermediate- and low-grade rocks in texture and mineral content. For example, slate, phyllite, schist, and gneiss are the metamorphic products of progressively more intense temperature and pressure conditions.

Table 5.2

CLASSIFICATION OF METAMORPHIC ROCKS

FOLIATED METAMORPHIC ROCKS

Parent rock	Comments	Rock name
Shale or mudstone	Low metamorphic grade. Slaty cleavage, cleavage surfaces dull to slightly shiny. Very fine grained. Generally dark (e.g., black, green, red).	Slate
	Low metamorphic grade. Phyllitic foliation. Fine grained. Silky, shiny luster.	Phyllite
	Intermediate metamorphic grade. Schistose foliation. Medium to coarse grained. Name modified by mineral name (e.g., quartz-biotite schist).	Schist
Shale, mudstone, granite	High metamorphic grade. Gneissic foliation. Medium to coarse grained. Light and dark layers common. Name modified by mineral names.	Gneiss

NONFOLIATED OR WEAKLY FOLIATED METAMORPHIC ROCKS

Parent rock	Comments	Rock name
Limestone or dolostone	Variable metamorphic grade. Chiefly calcite or dolomite; H = 3–4; effervesces in dilute HCl. Commonly coarsely crystalline.	Marble
Quartz sandstone	Variable metamorphic grade. Chiefly quartz; H = 7. Crystalline.	Quartzite
Basalt or gabbro	Low to medium metamorphic grade. Chlorite, green amphibole, epidote, and Na-plagioclase. Greenish gray or black.	Greenstone
Basalt or gabbro	Medium to high metamorphic grade. Hornblende, Na-Ca plagioclase, ± garnet. Dark gray to black.	Amphibolite
Peridotite	Low to medium metamorphic grade. Chiefly talc. Gray to dark greenish gray. Soft; can be carved. (see Fig. 2.9).	Soapstone
	Variable metamorphic grade. Mostly fine-grained serpentine; some may be fibrous. Greenish; commonly mottled or streaked.	Serpentinite
Coal	Low to medium metamorphic grade. Shiny, dark gray to black. Conchoidal fracture.	Anthracite coal
Various	A low-grade, contact-metamorphic rock. Commonly fine grained and dark color.	Hornfels

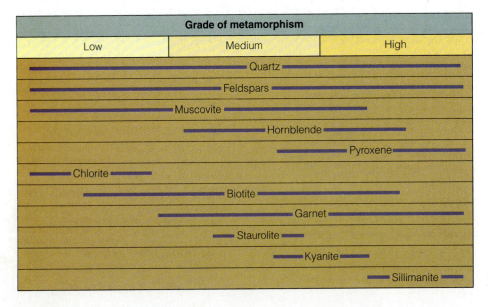

FIGURE 5.10
Most of the common minerals of metamorphic rocks form at different metamorphic grades. Thick lines indicate the range of metamorphic grade through which each mineral is possible.

FIGURE 5.11
Quartzite is a nonfoliated rock made mostly of quartz. It may be white to various shades of purple, green, gray, or pink. Its crystalline texture is evident in hand specimen, but a microscopic view (Fig. 5.3) better illustrates how the original grains of quartz sand have become intergrown. Impure quartzites contain minerals other than quartz, such as muscovite or kyanite. In many quartzites, original sedimentary features, such as bedding, cross-bedding or ripple marks, are still evident, having survived metamorphism.

FIGURE 5.12
Marble, a nonfoliated rock made mostly of calcite or dolomite, has a variety of colors and textural patterns. It has long been used for decorative purposes, but because of its susceptibility to acidic rain, it does much better in an interior setting. Its hardness (3 to 3.5), reaction with acid, and commonly coarse crystalline texture aid in identification. Some impure marbles, especially those formed by contact metamorphism, contain spectacular crystals of unusual minerals.

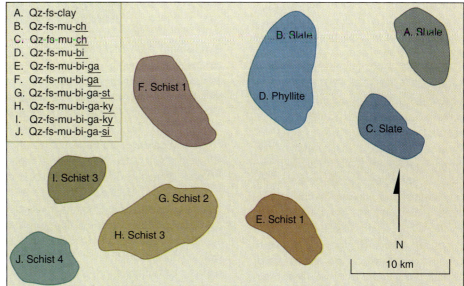

A.

A. Qz-fs-clay
B. Qz-fs-mu-ch
C. Qz-fs-mu-ch
D. Qz-fs-mu-bi
E. Qz-fs-mu-bi-ga
F. Qz-fs-mu-bi-ga
G. Qz-fs-mu-bi-ga-st
H. Qz-fs-mu-bi-ga-ky
I. Qz-fs-mu-bi-ga-ky
J. Qz-fs-mu-bi-ga-si

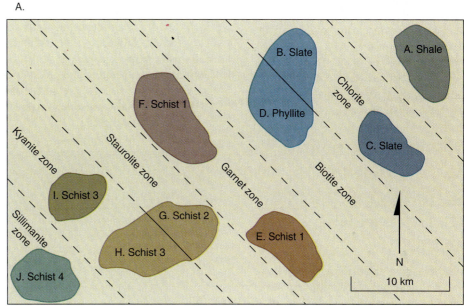

B.

FIGURE 5.13
A. Map of outcrops indicating rock types and mineral assemblages. *Qz* = quartz; *fs* = feldspar; *mu* = muscovite; *ch* = chlorite; *bi* = biotite; *ga* = garnet; *st* = staurolite; *ky* = kyanite; *si* = sillimanite. Index minerals are underlined. B. Same map showing isograds and mineral zones.

One way to learn how varying intensities of metamorphism affect rocks is to find a place where erosion has stripped away the overlying rocks to expose a sequence of metamorphic rocks ranging from low to high grade. It is then possible to study the entire sequence in the field and the laboratory to see exactly what textural and mineralogic changes have occurred, as illustrated by the following hypothetical example.

Figure 5.13A is a map of an imaginary area showing rock *outcrops,* places where

the bedrock is not covered by soil (circled and lettered areas). Most of the rocks in this area are metamorphosed to varying degrees, but the *parent rock* for all of them was shale. Furthermore, the whole area was tilted to the northeast before it was eroded to a nearly flat surface. Thus, the rocks that were most deeply buried during metamorphism are now exposed in the southwest part of the area. Unmetamorphosed shale, which was closest to the surface during metamorphism, is present only in the northeast corner. In between are rocks that were buried to intermediate depths. The minerals present in rocks from localities A through J also are listed.

The underlined minerals in the box in Figure 5.13A have special significance, because they formed by chemical reactions that take place only under certain temperature and pressure conditions. Thus, the presence or absence of these **index minerals** in a rock formed by the metamorphism of a shale indicates whether or not certain temperatures and pressures were attained during metamorphism (see Metamorphic Reactions box, on this page).

In Figure 5.13B, lines are drawn to outline zones based on the appearance of the index minerals. Because these lines approximate the same metamorphic grade, they are called **isograds.** This metamorphic map gives a clearer picture of the metamorphism of the shale. For example, it is easy to see at a glance that the metamorphic grade increases from northeast to southwest, and the index minerals indicate just how intense the metamorphism was.

DISTINGUISHING AMONG IGNEOUS, SEDIMENTARY, AND METAMORPHIC ROCKS

After finishing this lab, you will have looked at a number of specific examples of igneous, sedimentary, and metamorphic rocks. What if you were handed a rock and simply asked to tell whether it was igneous, sedimentary, or metamorphic? Could you do it? This seems like an easy thing, yet there are times when it is very difficult. Let's first review some of the major characteristics of each rock type, then consider some specific cases.

Most igneous rocks have nonfoliated, crystalline textures and are composed of several of the common minerals—feldspar, quartz, olivine, augite, hornblende, biotite, or muscovite—in varying proportions. Most common sedimentary rocks have clastic textures. They may contain some of the same minerals as igneous rocks (quartz and feldspar are the most common), but also may contain non-igneous minerals such as clay or calcite. Sedimentary rocks

An example of a metamorphic reaction is illustrated in Figure 5.14. That figure shows the temperature and depth (or pressure) conditions at which the reaction *muscovite + quartz = potassium feldspar + sillimanite + H_2O* can occur. Muscovite and quartz can occur together without reacting, as long as the combination of temperature and depth is to the left of the reaction curve; for example, at 500° and 10 km. If the temperature at that depth were higher, so that it was to the right of the reaction curve (for example, 700°), then muscovite and quartz could not occur together without reacting to form potassium feldspar, sillimanite, and water. When the rock cools again, the reverse reaction (forming muscovite + quartz) generally does not take place in nature, because the water necessary for the reaction has escaped. Because water and other fluids are driven off during metamorphism, the minerals present in metamorphic rocks generally are those formed at the highest temperature and pressure conditions that were attained.

It is sometimes possible to find rocks that contain both products and reactants. For the reaction shown in Figure 5.14, that rock would contain quartz, muscovite, potassium feldspar, and sillimanite. Such a rock would have been subjected to conditions right on the reaction curve (for example, 625° at a depth of 10 km). If a series of outcrops with the same products and reactants could be located, they would all represent the same metamorphic grade. A line on a map connecting them would be an isograd, a line of equal metamorphic grade, generally defined by a specific reaction or the appearance of a specific mineral.

By using numerous reactions like the one shown in Figure 5.14, it is possible to determine the temperature and pressure conditions that must have existed at the time of metamorphism. Because pressure, especially, can be related approximately to depth (see Fig. 5.1), it is also possible to figure out how deep the rocks must have been when they were metamorphosed.

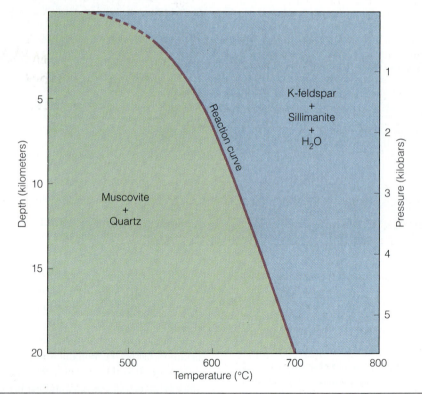

FIGURE **5.14**

The reaction *muscovite + quartz = K-feldspar + sillimanite + H_2O* occurs when the combination of temperature and depth (pressure) fall on the reaction curve. If temperature and depth are left of the curve, muscovite and quartz are present; if they are right of the curve, K-feldspar, sillimanite and H_2O are present.

with crystalline textures are composed of minerals, such as calcite, dolomite, gypsum, or halite, not usually found in igneous rocks. Foliated metamorphic rocks are distinguished from igneous and sedimentary rocks by foliation. Some contain typically metamorphic minerals like garnet, staurolite, or kyanite. Nonfoliated metamorphic rocks usually consist mostly of one mineral, such as quartz or calcite.

Some specific "problem rocks" are discussed below:

Shale vs. slate. Because a slate forms by low-grade metamorphism of a shale, these two rocks are very similar. In general, a slate is shinier and a bit harder and tougher than a fissile shale. Thin pieces of slate tend to "ring" when dropped on a table, whereas shale tends to "thunk."

Quartz sandstone vs. quartzite. Like shale and slate, a complete range, or gradation, can be seen from sandstone to quartzite, depending on the degree of metamorphism. One way to distinguish the two is based on the way the rock breaks. If the break cuts through individual quartz grains, the rock generally is quartzite; if it breaks around grains, it is quartz sandstone.

Limestone or dolostone vs. marble. Again, a complete range exists from limestone/dolostone to marble, depending on the degree of metamorphism. Marbles have readily distinguishable crystalline textures, with individual crystals large enough to see rather easily. Although some limestones have crystalline textures, the crystal size generally is quite small. Sometimes, however, it is nearly impossible to distinguish limestones from marbles in hand specimens. They usually can be distinguished in the field by looking at the surrounding rocks; if they are sedimentary, it is likely a limestone; if they are metamorphic, it is likely a marble. Marbles are *not* harder than limestones; they are equally hard. Why?

Graywacke vs. basalt. A dark gray, fine-grained graywacke may be difficult to distinguish from basalt. However, with a hand lens, you generally can see the clastic texture in graywacke and the crystalline texture in basalt. Graywackes may contain quartz, whereas basalts do not.

Mudstone vs. tuff. Some tuffs are so fine grained that they look like mudstones. Tuffs are commonly lighter in color and weight than mudstones, but examination with a microscope may be necessary to distinguish the two.

Conglomerate or sedimentary breccias vs. volcanic breccia. Volcanic breccias or tuffs with pyroclastic fragments may look like conglomerate or sedimentary breccia. The volcanic rocks usually contain only fragments of other volcanic rocks, whereas the particles in conglomerate and sedimentary breccia may be any rock type or a mixture of rock types. Tuffs commonly contain small crystals of common igneous minerals and almost always contain tiny bits of broken volcanic glass or shards (these may be difficult to see). Many tuffs and some volcanic breccias are lighter weight than conglomerates.

APPLICATIONS

THIS LAB CONTINUES THE APPLICATION OF THE SCIENTIFIC METHOD TO THE STUDY OF ROCKS. PROBLEMS FOCUS ON IDENTIFICATION AND INTERPRETATION BASED ON OBSERVATION OF METAMORPHIC ROCK SAMPLES. YOU WILL SEE HOW CORRECT IDENTIFICATION WILL ENABLE YOU TO INFER THE PARENT OF THE ROCK YOU HAVE IDENTIFIED AND THE APPROXIMATE TEMPERATURE-DEPTH CONDITIONS DURING METAMORPHISM.

OBJECTIVES

If you complete all the problems, you should be able to:

1. Identify the major minerals in metamorphic rocks.

2. Identify a metamorphic rock.

3. Distinguish between high- and low-grade metamorphic rocks.

4. Distinguish between foliated and nonfoliated metamorphic rocks.

5. Distinguish between slaty, schistose, and gneissic foliations.

6. Suggest a probable parent for a given metamorphic rock.

7. Determine whether a rock is igneous, sedimentary, or metamorphic.

8. Construct isograds on a map, given the appropriate data.

PROBLEMS

1. Metamorphic rocks are subdivided into two major textural groups: FOLIATED and NONFOLIATED. Divide the samples provided for you into these two categories.

   Foliated

   Nonfoliated

2. Which of the samples with foliated textures have the following specific types of foliation?

   Slaty

   Schistose

   Gneissic

3. Many foliated metamorphic rocks form by metamorphism of shale; the more intense (or *higher grade*) the metamorphism, the coarser the crystals contained in the rocks.

   a. Select from among your samples those that may have been derived from shale, and arrange them according to metamorphic grade (see Table 5.2).

   Lowest grade:

   Intermediate grade:

   Highest grade:

   b. Assuming that the rocks you selected formed in the vicinity of a magmatic zone at a convergent plate boundary, what were the approximate temperatures and depths of each grade at the time of metamorphism (see Fig. 5.1)?

4. What are the *parent rocks* of the samples specified by your instructor (see Table 5.2)?

5. Identify the samples of metamorphic rocks provided for you.

6. How can you distinguish between the following similar rocks?

   a. Marble and quartzite

   b. Slate and phyllite

   c. Schist and gneiss

IN GREATER DEPTH

7. When confronted with scattered outcrops of a variety of metamorphic rocks, a geologist must try to make some sense of their distribution in order to understand their origin. One of the first things to do is to make a map of the outcrops and indicate what rocks and minerals are present in each. Figure 5.15 shows outcrops of several kinds of rocks metamorphosed to varying degrees; colors indicate the type of parent rock. Between outcrops, the metamorphic rocks are covered with soil and other surficial sediment.

   a. Using Figure 5.15 and the following brief rock descriptions, construct isograds, and label the distinct metamorphic mineral zones.

Sample no.	Description	Sample no.	Description
1	Staurolite schist	17	Marble
2	Kyanite schist	18	Greenstone
3	Quartzite	19	Quartzite
4	Amphibolite	20	Greenstone
5	Greenstone	21	Chlorite phyllite
6	Staurolite schist	22	Marble
7	Quartzite	23	Chlorite slate
8	Garnet schist	24	Green shale
9	Marble	25	Chlorite slate
10	Garnet schist	26	Marble
11	Greenstone	27	Green shale
12	Garnet schist	28	Green shale
13	Quartzite	29	Quartz sandstone
14	Biotite schist	30	Basalt
15	Biotite schist	31	Red shale
16	Biotite phyllite		

   b. Where on the map was metamorphism most intense? Least intense?

   c. Which parent rocks best record differences in metamorphic grade?

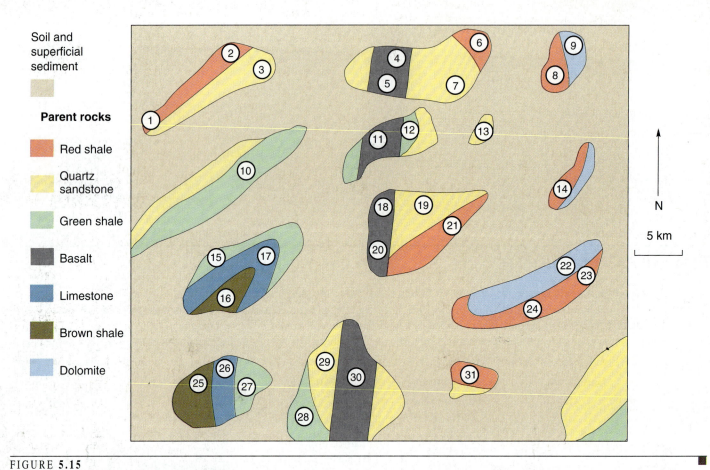

FIGURE 5.15

This map is to be used in conjunction with Problem 7. Rock outcrops are outlined, and different colors represent different parent rocks. Numbers refer to specific rocks described in Problem 7.

Part 3

MAPS AND IMAGES

6

Topographic Maps

Overview

Maps have many uses in everyday life as a means by which data of diverse kinds can be succinctly displayed. In geology, we are interested in the natural features on or below the Earth's surface. This chapter describes the basic features of maps, including the bases for describing locations and land areas; the standard map elements, such as scales and symbols; and the use and interpretation of contours to show the third dimension.

Materials Needed

Pencil and eraser • Calculator • Topographic quadrangle map
(provided by your instructor).

INTRODUCTION

A map is a representation of a surface and can be used to show a variety of features. **Planimetric** maps are a two-dimensional (length and width) representation of a part of the Earth's surface. They typically show roads, land boundaries, and some natural features, such as lakes and rivers. Road maps are examples. Planimetric maps commonly are used as the base for special-purpose maps that show the distribution of something of social, economic, or political interest. **Topographic maps** show the **topography** (three-dimensional configuration of the Earth's surface—hills, valleys, and so forth) with *contours,* as discussed later. Most topographic maps also show **cultural features** (those of human design), such as land boundaries, buildings, and roads. **Geologic maps** show the distribution of different rock units exposed at the Earth's

surface, as well as cultural features and, in many instances, topography. **Land-use maps** combine elements of topography and geology to indicate the locations of such features as stream floodplains, potential natural hazards, or good agricultural soil.

In short, maps are a means of displaying one or a number of features at a useful scale.

MAP COORDINATES AND LAND SUBDIVISION

Latitude-Longitude System

Because most maps show only a small portion of the Earth, they have to be located with respect to the rest of the Earth. This is done by means of some coordinate system or network of previously deter-

mined lines. A system used the world over is based on east-west lines called lines of latitude, or parallels, and north-south lines called lines of longitude, or meridians (Fig. 6.1A).

Lines of **latitude,** or **parallels,** form east-west circles around a globe; they are parallel, so never intersect. The equator is one of these and represents the 0° latitude line. Other parallels are at angular intervals north or south of the equator, with the angles being measured between lines extending from the center of the Earth to the surface (Fig. 6.1B). A latitude line 40° to the north of the equator is termed latitude 40° N. The **geographic poles** are at latitudes 90° N and 90° S.

Lines of **longitude,** or **meridians,** form circles on a globe that pass through, and intersect at, the geographic poles. They are separated from each other by angular intervals. The intervals are measured between

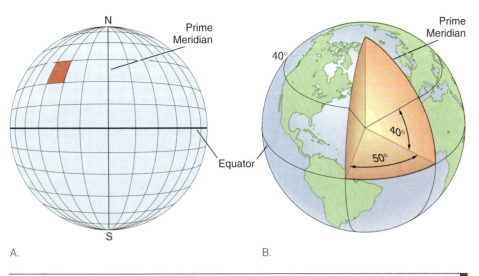

FIGURE 6.1

A. Lines of latitude parallel the Earth's equator, while lines of longitude intersect at north and south geographic poles. The shaded area, bounded by lines of latitude and longitude, is a quadrangle. B. Cutaway view of the Earth. The east-west line 40° north of the equator is latitude 40° N; the north-south line 50° west of the prime meridian is longitude 50° W.

lines extending from the center of the Earth to the surface in the plane of the equator (Fig. 6.1B). The north-south line passing through the Royal Observatory at Greenwich, England, is 0° longitude, and is known as the **Prime Meridian.** A longitude line 50° west of the Prime Meridian is termed longitude 50° W; one 20° east of the Prime Meridian is longitude 20° E. The meridian 180° away (on the opposite side of the Earth) corresponds, for the most part, to the International Date Line. In places, the date line deviates to avoid causing time-zone problems in islands or island groups cut by the 180° meridian.

Angular Measurements. The most commonly used angular measurement is a *degree* (symbol °). A circle is divided into 360°. One degree is divided into 60 *minutes* (60′), and each minute into 60 *seconds* (60″). Maps also may indicate angular measurements in *mils.* One mil = 1/6400 of 360°, or 0.05625°. When measured on the Earth's surface, one degree of latitude (as measured along a meridian) is approximately 111 km (69 miles), and one degree of longitude (as measured along a parallel) varies from about 111 km at the equator to 0 km at the poles, where the meridians intersect.

The maps used most frequently show small segments of the network of latitudes and longitudes. Because the Earth is nearly spherical, these segments (like the shaded one in Fig. 6.1A) are actually curved surfaces that are represented on maps as flat surfaces. To do this, the curved, three-dimensional surfaces must be projected onto a two-dimensional sheet of paper. Many ways have been developed to accomplish this—names such as "Mercator projection" or "polyconic projection" may be familiar to you—but all result in some sort of distortion, which is unavoidable.

The U.S. Geological Survey (USGS) has made most of the accurate maps of this country. These maps are bounded by latitudes and longitudes, both of which are usually separated by 1°, ½° (= 30′), ¼° (15′), or ⅛° (7½′) intervals. Such maps bound rectangular-shaped areas called **quadrangles,** which are generally named after the largest town or most prominent geographic feature in the area (for example, Baraboo, Wisconsin, 15′ Quadrangle). The numerical values of the latitudinal and longitudinal boundaries are given on each corner of the map (Fig. 6.2). In addition, intermediate values are indicated on the margins of a map (Fig. 6.2).

A point on a map can be located by referring to the latitude and longitude of the point (for example, latitude 43° 5½′N, longitude 132° 15½′W). By convention, latitude is given first, longitude second.

Three other coordinate or grid systems are in common use in the United States: the Universal Transverse Mercator (UTM), State Plane Coordinates, and U.S. Public Land Survey Systems. The first two of these are useful for locating points or describing land areas, and the third is useful for describing land areas.

Universal Transverse Mercator (UTM) and State Plane Coordinates Systems

The UTM system, which is used throughout the world, is based on a grid of 60 north-south zones, each 6° wide; Figure 6.3A shows these zones in the U.S. The zones are numbered from west to east, beginning at the International Date Line; the zone number is indicated in the small print in the lower left corner of a USGS quadrangle map (Fig. 6.2). The grid for each zone in the northern hemisphere has its origin at the intersection of the equator and its own central meridian, as shown for zone 15 (90° to 96°) in Figure 6.3B. A metric grid, with lines intersecting at right angles, is developed from this origin on a transverse-Mercator-type map projection. The grid is shown with black lines on newer USGS maps. Labels of east-west lines are based on the number of meters north of the equator. Labels of north-south lines use a *false origin* of 0 m, located 500,000 m west of the true origin; this makes it easier to locate features near zone boundaries.

Features are located by giving their UTM coordinates. UTM coordinates are shown in numbers of two sizes along the margins of USGS maps (Figs. 6.2, 6.4). The numbers are the distances in meters from the false origin. For example, $^{45}15^{000mN}$ describes an east-west line 4,515,000 m (4,515 km) north of the equator. A location description is always given in the same sequence, as shown in Figure 6.4: east-west coordinate, north-south coordinate, zone number, hemisphere (northern or southern). Because east-west coordinates increase in value from left to right on a map, and north-south coordinates increase in value from bottom to top of a map, the informal rule "read right up" is a useful mnemonic device. Locations that are not on intersections of labeled UTM coordinates are determined as shown in Figure 6.4.

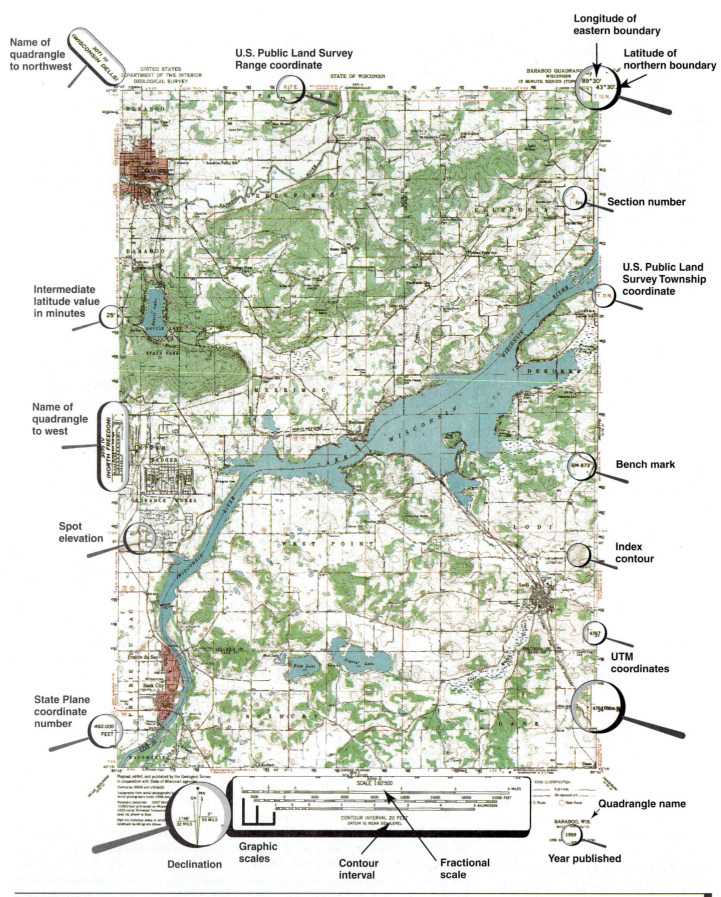

Name of quadrangle to northwest

U.S. Public Land Survey Range coordinate

Longitude of eastern boundary

Latitude of northern boundary

Section number

U.S. Public Land Survey Township coordinate

Intermediate latitude value in minutes

Name of quadrangle to west

Bench mark

Spot elevation

Index contour

UTM coordinates

State Plane coordinate number

Declination

Graphic scales

Contour interval

Fractional scale

Quadrangle name

Year published

FIGURE **6.2**

Reduced copy of the Baraboo, Wisconsin, 15-minute quadrangle, with principal features highlighted.

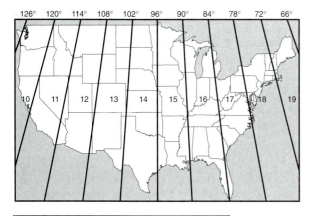

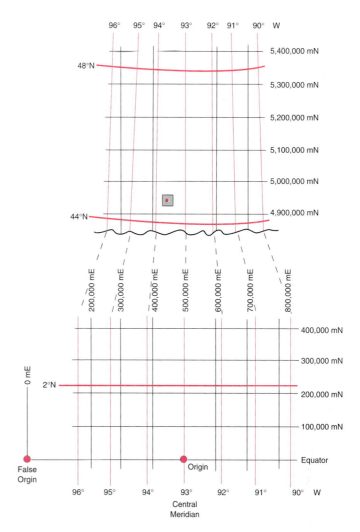

FIGURE 6.3
A. UTM-grid zones in the United States.
B. Example showing two parts of UTM grid (black lines) in zone 15 superimposed on lines of latitude and longitude (red lines). Note that lines on the two grids are parallel near the equator but diverge at higher latitudes. This is because most latitude and longitude lines are projected as curves, whereas UTM lines are drawn as straight lines on that projection. The point in the shaded area is located in Figure 6.4.

The State Plane Coordinates system is similar to the UTM system in that it utilizes coordinate systems with origins in each zone. It differs in that it is used only in the U.S., has many more zones (120), and uses a grid in units of feet, rather than meters. Points are located with the same "read right up" logic of the UTM, and grid values are given along the map border.

U.S. Public Land Survey System

In the western two-thirds of the United States, land is subdivided by the United States Public Land Survey System (also called the Township-Range System). The system was started in 1785, when the old Northwest Territory (Lake Superior region) was opened to homesteading, and has been used for ordinary and legal land descriptions since. This method subdivides land into 6- × 6-mile squares called *townships;* these are further subdivided into 1- × 1-mile squares called *sections.*

The starting point for subdivision is the intersection of selected latitude and longitude lines. The starting latitude is the **base line,** and the starting longitude is the **principal meridian.** Base lines and principal meridians are established for a number of areas in the United States, such as the one shown in Figure 6.5A. Lines drawn six miles apart and parallel to the baseline

form east-west rows called **tiers.** North-south lines parallel to the principal meridian and six miles apart form north-south columns called **ranges** (Fig. 6.5B). The squares formed by the intersection of tiers and ranges are called **townships.** Each township is approximately six miles square and has an area of about 36 square miles. Political townships, usually named after the largest town within the area at the time the land was surveyed (for example, Baraboo Township), may or may not coincide with Public Land Survey townships.

Tiers and ranges are numbered by reference to the baseline and principal meridian (Fig. 6.5B). A township in the first tier north of the baseline is Township 1 North (abbreviated T1N); one in the fifth tier to the north is T5N, and so forth. Ranges are numbered to the east of the principal meridian (for example, R5E) and to the west (R2W). A Public Land Survey township (like the shaded one in Fig. 6.5B) is located using tier-range coordinates—T3S, R4E. *NOTE:* Tier is always written first; Range second.

Because lines of longitude (meridians) converge toward the poles, it is impossible to maintain squares that are 6 miles on a side. Thus, a correction is made at every fourth tier line (labeled *correction line* on Fig. 6.5B), and new range lines 6 miles apart are established. The correction restores townships immediately north of the line to their proper size.

Each 6-mile-square township is subdivided into 36, 1- × 1-mile squares, called **sections,** which are numbered in a specific sequence (Fig. 6.5C). Each section consists of 640 **acres.** A section is subdivided into halves, quarters, eighths, sixteenths, and so on (Fig. 6.5D). A sixteenth of a section is 40 acres.

Points are located according to the smallest subdivision required. In Figure 6.5D, the star is located, to the nearest 40 acres, in the SE¼, NW¼, Sec.16, T3S, R4E. *Locations are always written from the smallest unit to the largest, and tier is written before range.*

77

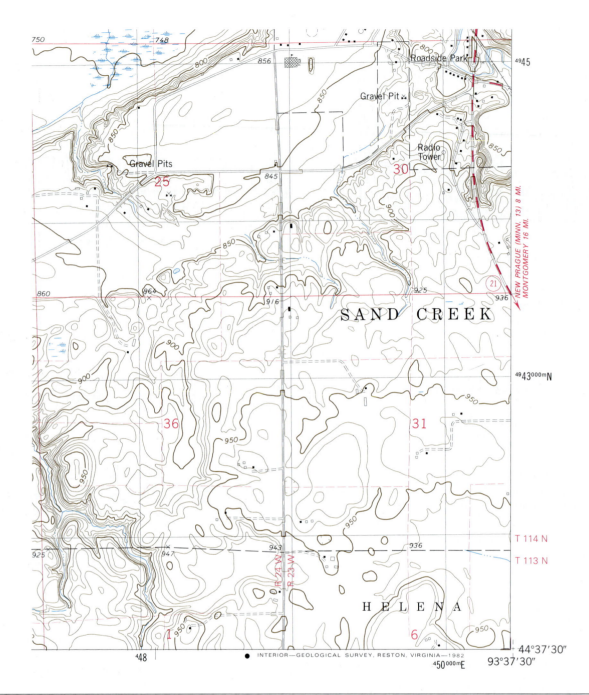

FIGURE 6.4

The house (shown as a small black square) falls within the 1-km square with UTM grid lines 450,000mE and 449,000mE on the east and west sides and 4,943,000mN and 4,942,000mN on the north and south sides. Using a ruler, you can find that the house is about 500 m east of the 449,000mE grid line at 449,500mE and about 930 m north of the 4,942,000mN grid line at 4,942,930mN. The location of the house, to within a 10-m square, is given as 449,500mE; 4,942,930mN; 15; N. ("15" means zone 15, and "N" means northern hemisphere.

Section numbers and tier and range values are written in red on USGS topographic maps (see Fig. 6.2).

ELEMENTS OF MAPS

Scale

For a map to be usable, it must have a **scale,** so the user can tell the size of the area represented or the distance between various points on a map. Three types of scales are in common use.

A **ratio,** or **fractional scale,** shown in Figure 6.2, is the ratio between a distance on a map and the actual distance on the ground. A common ratio scale is 1:24,000 (or 1/24,000), which means that one unit (for example, an inch) on the map equals 24,000 of the same units on the ground.

A **graphic scale** usually consists of a line or bar subdivided into divisions corresponding to a mile or kilometer, or fractions of a mile or kilometer (see Fig. 6.2). This scale is helpful because it is readily visualized and will stay in true proportions if the map is enlarged or reduced. The graphic scale provides a convenient way of measuring distances on a map. To find the distance between two points on a map,

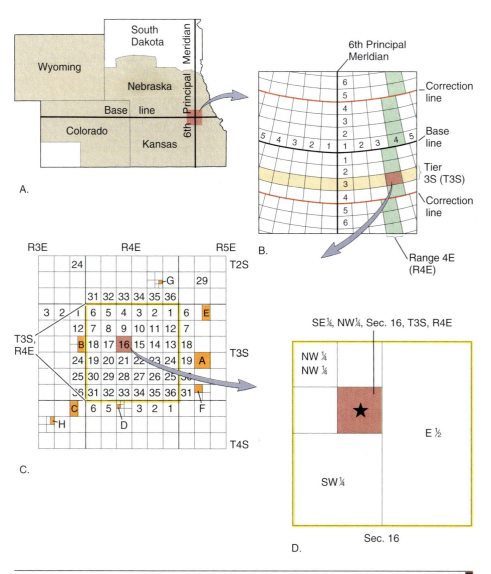

FIGURE 6.5

U.S. Public Land Survey subdivision, illustrated by successively smaller areas, A–D. A. Example of a baseline and principal meridian in the western U.S. B. From a starting point at the intersection of a principal meridian and a baseline, 6-mile-wide tier and range bands subdivide land into 36-square-mile townships. C. Townships are subdivided into 36 1-square-mile sections. D. Sections can be divided into halves, quarters, eighths, or other fractions.

To show larger or smaller areas of the Earth's surface on conveniently sized maps, different scales must be used. For example, it may be possible to show a small city on a map where 1 inch on the map represents 12,000 inches (1,000 ft) on the ground. This map would have a scale of 1:12,000. However, to show a mid-sized state, such as Indiana, on a map of similar size, the scale would have to be much smaller: 1 inch on the map to 500,000 inches (approximately 8 miles) on the ground. In general, the larger the area shown on a map, the smaller the scale of the map.

Converting among Scales

To convert from a verbal to a fractional scale:

1. Convert map and ground distances to the same units.

2. Write the verbal scale as the fraction:

$$\frac{\text{Distance on the map}}{\text{Distance on the ground}}$$

3. Divide both numerator and denominator by the value of the numerator:

$$\frac{\text{Distance on map/distance on map}}{\text{Distance on ground/distance on map}}$$

Example: Convert the following verbal scale to a fractional scale: 2.5 inches on the map represents 5,000 feet on the ground.

1. Convert both map and ground distances to the same units, inches: $5,000 \times 12'' = 60,000''$. The verbal scale is now 2.5 inches on the map represents 60,000 inches on the ground.

2. Write the verbal scale as a fraction:

$$\frac{2.5'' \ (\text{distance on map})}{60,000'' \ (\text{distance on ground})}$$

3. Divide the numerator and denominator by the value of the numerator:

$$\frac{2.5''/2.5''}{60,000''/2.5''} = \frac{1}{24,000} \text{ or } 1:24,000.$$

To convert from a fractional scale to a verbal scale:

1. Convert map and ground distances to the units you want for the map distance (for example, inches).

---

lay a strip of paper between the points and make pencil marks on the strip adjacent to each point. Then lay the paper along the desired graphic scale at the bottom of the map to determine the distance.

A **verbal scale** is used to discuss a map but is rarely written on it. An example is "one inch on the map represents, or is proportional to, one mile on the ground" or, simply, "one inch to one mile." Because one mile equals 63,360 inches, the common fractional scale 1:62,500 corresponds closely to the verbal scale "one inch to one mile." For the other common scales, 1:24,000, 1:31,680, 1:125,000, 1:250,000, one inch nearly represents ⅓ mile, ½ mile, 2 miles, and 4 miles, respectively. Many U.S. maps, and essentially all foreign maps, use metric scales, making common fractional scales easily convertible to verbal scales. For example, for scales of 1:50,000, 1:100,000, and 1:250,000, 1 centimeter represents 0.5, 1.0, and 2.5 kilometers, respectively.

USGS 4° quadrangle maps are drawn at a fractional scale of 1:1,000,000; 2° quadrangles at 1:500,000; 1° at 1:250,000; 15′ at 1:62,500 or 1:50,000; and 7½′ at 1:24,000 or 1:25,000. Both graphic and fractional scales are shown at the bottom center of the map (see Fig. 6.2).

2. Choose the units you want for the ground distance (for example, miles).

3. Convert the ground distance of the fractional scale to the desired units using the appropriate conversion factor (for example, convert inches to miles).

Example: Convert a fractional scale of 1:62,500 (or 1/62,500) to a verbal scale in which so many inches on the map represents so many miles on the ground.

1. One inch on the map represents 62,500 inches on the ground.

2. One mile = 5,280 feet, and 1 foot = 12 inches, so one mile = 5,280 × 12″ = 63,360″. To convert 62,500 inches to miles, divide by 63,360″:

$$\frac{62,500''}{63,360''/\,mi} = 0.986 \ mi$$

3. Express verbally as "1 inch to 0.986 mile" or "1 inch approximately represents one mile."

Magnetic Declination

By convention, maps are usually drawn with north at the top. North on a map refers to **true geographic north.** At most places on the Earth, however, a compass needle does not point toward the geographic north pole, but toward the **magnetic north pole.** The magnetic north pole is in the Canadian Arctic, but its exact position changes. For example, in 1955, it was located near Prince of Wales Island at latitude 73° N, longitude 100° W; its 1992 location was in the Queen Elizabeth Islands at about 78° N, 103° W.

The angular distance between true north and magnetic north is the **magnetic declination.** Because the location of the magnetic pole changes, the magnetic declination varies slightly with time at most locations. For very accurate work, these variations must be taken into account. Magnetic declination is shown at the bottom of most USGS maps by two arrows (see Fig. 6.2), one pointing to true north (commonly marked with a star, or T. N.) and one pointing toward magnetic north (commonly marked M. N.). The angular separation between them (the magnetic declination) also is given. When stating the magnetic declination of a map, it is always necessary to indicate whether the arrow pointing to the magnetic pole is east or

west of the geographic pole. If it is east, the declination is stated as so many degrees east, for example, 2½° E. Most maps also have an arrow pointing toward **G. N.,** the location of the **grid north** direction for the Universal Transverse Mercator (UTM) grid system (see Fig. 6.2).

Symbols

Standardized symbols and colors are used on USGS maps to designate various features. Cultural features (those made by people) are generally drawn in black; forests or woods are shown in green (they are not always represented); blue is used for bodies of water; brown shows elevation (contours), some mining operations, and beaches or sand areas; red is used for the better roads and some land subdivision lines. See Figure 6.6 for symbols and Figure 6.2 for some examples.

TOPOGRAPHIC MAPS

A topographic map shows the size, shape, and distribution of landscape features; that is, the **topography,** or the configuration of the land surface. It is a planimetric map to which a third dimension, elevation, has been added. Elevation differences are shown by means of contour lines, hachures (short, closely spaced lines), and shading or tinting. The emphasis here is on contour lines, because they provide the most precise means for depicting the third dimension. Figure 6.7 shows how the topography of the area sketched in the top diagram is depicted with contour lines in the bottom diagram.

Contour Lines

A contour line is a line on which all points have the same elevation. It is shown in brown on USGS topographic maps (see Fig. 6.2). The **elevation** or altitude of a point is the vertical distance between that point and a fixed **datum** (a point of reference). The datum from which most elevations are measured is mean or average sea level, which by definition, has an elevation of zero. Figure 6.7 shows an area along a sea coast, with the sea at its average, or mean, elevation of zero feet. Because the edge of the shore is everywhere at an elevation of zero feet, the shoreline coincides with the zero-foot contour. If sea level rose by 100 feet, the

shoreline would everywhere coincide with the 100-foot contour shown in Figure 6.7; if it rose 200 feet, it would coincide with the 200-foot contour.

Characteristics of Contour Lines

The following characteristics of contour lines govern the construction and reading of contour maps:

1. Every point on the same contour line has the same elevation.

2. A contour line always rejoins or closes upon itself to form a loop, although this may or may not occur within the map area. Thus, if you walked along a contour, you would eventually get back to your starting point.

3. Contour lines never split.

4. Contour lines never cross one another; however, if there is a steep cliff, they may appear to overlap because they are superimposed on one another.

5. Slopes rise or descend at right angles to any contour line.
   - Evenly spaced contours indicate a uniform slope
   - Closely spaced contours indicate a steep slope
   - Widely spaced contours indicate a gentle slope
   - Unevenly spaced contours indicate a variable or irregular slope.

6. Contours usually encircle a hilltop; if the hill falls within the map area, the high point will be inside the innermost contour (however, see discussion of depression contours).

7. Contour lines near the tops of hills or bottoms of valleys always occur in pairs having the same elevation on either side of the hill or valley.

8. Contours always bend upstream where they cross stream valleys.

9. If two adjacent contour lines have the same elevation, a change in slope occurs between them. For example, adjacent contours with the same elevation would be found on both sides of a valley bottom or ridge top.

10. Small depressions may be encircled by contours with hachures (short lines

Topographic Map Symbols

BOUNDARIES

National . . . .
State or territorial . . . .
County or equivalent . . . .
Civil township or equivalent . . . .
Incorporated—city or equivalent . . . .
Park, reservation, or monument . . . .
Small park . . . .

LAND SURVEY SYSTEMS

U.S. Public Land Survey System:

Township or range line . . . .
  Location doubtful . . . .
Section line . . . .
  Location doubtful . . . .
Found section corner; found closing corner . . .
Witness corner; meander corner . . . . WC | MC

Other land surveys:

Township or range line . . . .
Section line . . . .
Land grant or mining claim; monument . . . .
Fence line . . . .

ROADS AND RELATED FEATURES

Primary highway . . . .
Secondary highway . . . .
Light duty road . . . .
Unimproved road . . . .
Trail . . . .
Dual highway . . . .
Dual highway with median strip . . . .
Road under construction . . . .
Underpass; overpass . . . .
Bridge . . . .
Drawbridge . . . .
Tunnel . . . .

BUILDINGS AND RELATED FEATURES

Dwelling or place of employment: small; large . .
School; church . . . .
Barn, warehouse, etc.: small; large . . . .
House omission tint . . . .
Racetrack . . . .
Airport . . . .
Landing strip . . . .
Well (other than water); windmill . . . .
Water tank: small; large . . . .
Other tank: small; large . . . .
Covered reservoir . . . .
Gaging station . . . .
Landmark object . . . .
Campground; picnic area . . . .
Cemetery: small; large . . . . †|Cem

RAILROADS AND RELATED FEATURES

Standard gauge single track; station . . . .
Standard gauge multiple track . . . .
Abandoned . . . .
Under construction . . . .
Narrow gauge single track . . . .
Narrow gauge multiple track . . . .
Railroad in street . . . .
Juxtaposition . . . .
Roundhouse and turntable . . . .

TRANSMISSION LINES AND PIPELINES

Power transmission line: pole; tower . . . .
Telephone or telegraph line . . . .
Above-ground oil or gas pipeline . . . .
Underground oil or gas pipeline . . . .

CONTOURS

Topographic:

Intermediate . . . .
Index . . . .
Supplementary . . . .
Depression . . . .
Cut; fill . . . .

Bathymetric:

Intermediate . . . .
Index . . . .
Primary . . . .
Index Primary . . . .
Supplementary . . . .

MINES AND CAVES

Quarry or open pit mine . . . .
Gravel, sand, clay, or borrow pit . . . .
Mine tunnel or cave entrance . . . .
Prospect; mine shaft . . . .
Mine dump . . . .
Tailings . . . .

SURFACE FEATURES

Levee . . . .
Sand or mud area, dunes, or shifting sand . . . . . .
Intricate surface area . . . .
Gravel beach or glacial moraine . . . .
Tailings pond . . . .

VEGETATION

Woods . . . .
Scrub . . . .
Orchard . . . .
Vineyard . . . .
Mangrove . . . .

COASTAL FEATURES

Foreshore flat . . . .
Rock or coral reef . . . .
Rock bare or awash . . . .
Group of rocks bare or awash . . . .
Exposed wreck . . . .
Depth curve; sounding . . . .
Breakwater, pier, jetty, or wharf . . . .
Seawall . . . .

BATHYMETRIC FEATURES

Area exposed at mean low tide; sounding datum
Channel . . . .
Offshore oil or gas: well; platform . . . .
Sunken rock . . . .

RIVERS, LAKES, AND CANALS

Intermittent stream . . . .
Intermittent river . . . .
Disappearing stream . . . .
Perennial stream . . . .
Perennial river . . . .
Small falls; small rapids . . . .
Large falls; large rapids . . . .

Masonry dam . . . .

Dam with lock . . . .

Dam carrying road . . . .

Intermittent lake or pond . . . .
Dry lake . . . .
Narrow wash . . . .
Wide wash . . . .
Canal, flume, or aqueduct with lock . . . .
Elevated aqueduct, flume, or conduit . . . .
Aqueduct tunnel . . . .
Water well; spring or seep . . . .

GLACIERS AND PERMANENT SNOWFIELDS

Contours and limits . . . .
Form lines . . . .

SUBMERGED AREAS AND BOGS

Marsh or swamp . . . .
Submerged marsh or swamp . . . .
Wooded marsh or swamp . . . .
Submerged wooded marsh or swamp . . . .
Rice field . . . .
Land subject to inundation . . . .

FIGURE 6.6

Standard symbols on USGS maps. SOURCE: DATA FROM U.S. GEOLOGICAL SURVEY.

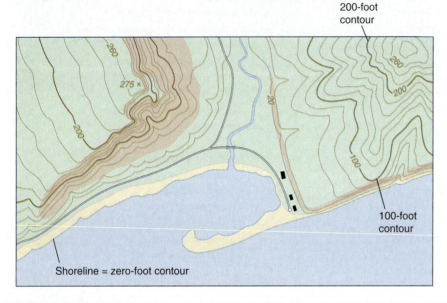

200-foot contour

100-foot contour

Shoreline = zero-foot contour

FIGURE 6.7

The area sketched in the top diagram is shown as a topographic map in the bottom diagram. Contour lines on the bottom diagram are drawn at intervals of 20 feet, starting with 0 at mean sea level.

SOURCE: U.S. GEOLOGICAL SURVEY.

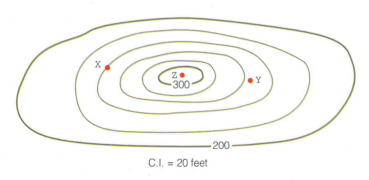

C.I. = 20 feet

FIGURE 6.8

Reading elevations from a contour map with contour interval of 20 feet. The elevation of X is 240 feet, because it falls on a contour with that elevation. Point Y falls between the 240- and 260-foot contours, so its elevation must be between those values. Its horizontal position is about three-quarters of the way between the two, and assuming a uniform slope, you might estimate an elevation of 255 feet. The half-way elevation method makes no assumptions about uniformity of slope. Point Y has a *half-way elevation* of 250 ±10 feet: 250 is half-way between 240 and 260, and ± 10 indicates that point Y falls between 240 (250–10) and 260 (250+10). Note that the error term (±10) is found by dividing the contour interval by 2. What is the half-way elevation of point Z, at the top of the hill? (310 ± 10 feet).

perpendicular to the contour line) on the downhill side. A hachured contour has the same elevation as the normal (unhachured) contour immediately downhill from it.

Contour Interval

Contour lines are drawn on a map at evenly spaced intervals of elevation. The difference in elevation between two consecutive contours on the same slope is called the **contour interval** (abbreviated *C. I.*). It is a constant for a given map, unless otherwise stated, and is usually given at the bottom of the map just below the graphic scale (see Fig. 6.2).

The choice of contour interval depends on (1) the level of detail the topographer wishes to portray, (2) the scale of the map, and (3) the range in elevation or **relief** of the area to be mapped. Obviously, Florida and the Rocky Mountains cannot be mapped at the same scale with the same contour interval. Much of the Rocky Mountain area is mapped with a 100-foot interval, whereas a 2- or 5-foot interval is common for Florida.

Index Contour

As a general rule, every fifth contour, starting from sea level, is an **index contour,** which is drawn as a heavy line and labeled with its elevation (see Fig. 6.2). Contours between index contours are usually not labeled.

Depression Contours

Depression contours are closed contours with hachures (short lines perpendicular to the contour) directed inward; they encircle a depression. All elevations within a depression contour are lower than the contour.

Reading Elevations

The elevations of points that fall on a contour line are the same as the elevation represented by that contour line. On most maps, only index contours are labeled. To find the values of contours in between, count the number of contours above (or below) an index contour, multiply that number by the contour interval, and add (or subtract) that product from the value of the index contour. An example is given in Figure 6.8.

The elevation of a point that does not fall on a contour must be estimated. An

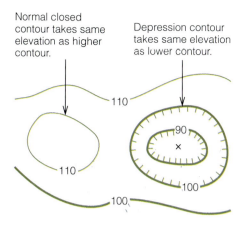

Depression contour takes same elevation as lower contour.

110

90
×

110

100

100

FIGURE 6.9

Elevations of normal and depression contours lying between two other contours. *C. I.* = 10 ft. The elevation of the closed contour on the left is 110 feet, as it takes the elevation of the higher one. The elevation of the outer *depression* contour, on the right, is the same as the *normal* contour downslope, namely 100 feet. The elevation of the inner depression contour is 10 feet lower, or 90 feet. The bottom of the depression is less than 90 feet, but more than 80 feet, because there is no 80-foot contour. The half-way elevation at *X* in the bottom is 85 ± 5 feet.

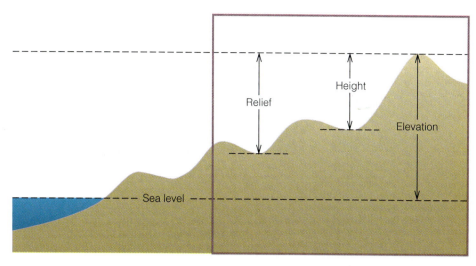

Height

Relief

Elevation

Sea level

FIGURE 6.10

This profile view shows that the *elevation* of the hill on the right is measured from sea level, whereas its *height* is the difference in elevation between the top and bottom of the hill. *Relief* is the difference in elevation between the highest and lowest points in a specified area, such as the one that is outlined.

estimate can be made by interpolation, assuming the slope between adjacent contours is uniform. For example, a point one-quarter of the way between contours with elevations of 200 and 220 feet (*C. I.* = 20 feet) would have an elevation of about 205 feet. However, slopes are often not uniform, so another approach is to give the *half-way elevation* between the two contours. A **half-way elevation** is the elevation half-way between the values of adjacent contours; thus, the elevation of a point between contours can be stated as the half-way elevation plus or minus one-half the contour interval. Figure 6.8 provides examples.

A normal closed contour that lies between a higher and a lower contour always takes the same elevation as the higher one. A depression contour in the same situation takes the same elevation as the lower one, as illustrated in Figure 6.9.

Height and Relief

If someone asked you what your height was, you would say something like 5 feet 9 inches. This is the distance from the floor to the top of your head. You can also talk

about the **height** of a hill, which is the difference in elevation between the top of the hill and the bottom.

A related but different term, **relief,** refers to the difference between the highest and lowest elevations in a given area. For example, in Winnebago County, Wisconsin, the highest elevation is about 920 feet, and the lowest is about 745 feet. Therefore, the relief of the county is 175 feet (920 − 745 = 175 ft). In Jefferson County, Colorado, immediately west of Denver, the highest and lowest elevations are approximately 11,700 feet and 5100 feet; the relief is 6600 feet. Relief is also used in a relative sense; a mountainous area has a high relief, whereas a plain has a low relief; Jefferson County has a high relief; Winnebago County has a low relief.

Figure 6.10 illustrates the differences between elevation, height, and relief.

Bench Marks and Spot Elevations

A **bench mark** is a point whose elevation and location have been precisely determined by government surveyors; its location is marked by a small brass plate. Bench marks are designated on maps by the symbol *B. M. Spot elevations* are also shown for many section corners, bridges, road intersections, hilltops, and the like (see Fig. 6.2). Bench marks

and spot elevations are used in conjunction with aerial photographs to construct topographic maps. Two aerial photos, taken from different points but overlapping the same area, will provide a three-dimensional view of the land surface when viewed through a stereoscopic viewer. By orienting the photos so that they are in exact perspective, two beams of light from different sources can be focused at any elevation. If the superimposed beams are moved around a landform, they will trace a line at a precise elevation that is determined from bench marks plotted on the photographs. Aerial photographs are discussed further in Chapter 7.

Making a Topographic Map

Given a number of elevations on a planimetric map (Fig. 6.11), a topographic map can be constructed as follows:

A. Select a contour interval that will illustrate the topography of the area. Choice of *C. I.* will depend on the total relief of the area. The *C. I.* used in Figure 6.11 is 10 feet. This means that contours will be drawn through all those points with elevations divisible by 10. From the elevation values shown in Figure 6.11A, you can see that contours should be drawn representing elevations of 100, 110, 120, and 130 feet.

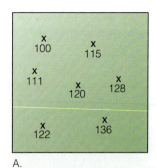
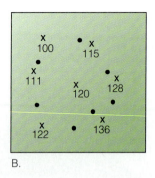
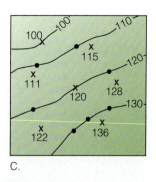

A.   B.   C.

FIGURE 6.11

Sketches illustrate how a contour map is made. A. Elevations are given for points (X's) on map, and a contour interval is chosen (10 feet in this case). B. Points with elevations divisible by the contour interval are located and marked (dots). C. Smooth contour lines are drawn and labeled as appropriate.

B. Mark the position of each contour line by interpolating between adjacent points of known elevation. The simplest way to do this is to assume a constant or even slope between points, then estimate the position between points at which the elevation represented by the contour line falls. For example, somewhere between the points in Figure 6.11B marked 122 and 136, there will be an elevation of 130 feet. There will also be an elevation of 130 feet between points marked 120 and 136. The locations of these 130-foot-elevation points shown on Figure 6.11B are rough estimates only; more information would be needed to locate them more accurately.

C. Start with any elevation, and connect points of the same elevation to make contour lines. Smooth out the lines and keep them reasonably parallel. Label each of the contour lines with the elevation that it represents (as in Fig. 6.11C), or label only index contours. Once you understand the process, you probably will be able to draw contour lines without first marking the points between known elevations, as outlined in step B.

Topographic Profiles

A topographic profile shows the shape of the land surface as it would appear in a cross section; it is like a side view of the land surface. Topographic profiles portray the shape of the land surface along a par-

ticular *line of profile.* They are useful for many practical purposes, such as planning roads, railroads, pipelines, canals, and the like, or for estimating the volume of material that will need to be excavated or filled during construction. Profiles are most easily made along straight lines, but also can follow curved paths, such as a road or a stream.

A topographic profile is made from a contour map using the following procedure (Fig. 6.12):

A. Select the line or path along which the profile is to be made, such as line A-B in Figure 6.12A.

B. Record the elevations along the line as shown in Figure 6.12B. To do this, lay the straight edge of a piece of paper along the line of profile. You can use the graph paper on which you will make your profile, or a separate piece of paper altogether. Mark on the paper the ends of the profile line and the exact place where each contour line meets the edge of the paper. If you are doing this on graph paper, use the top or bottom of the graph paper to make the marks. Label each mark on the paper with the elevation of the corresponding contour. Also mark the positions of any streams that cross the line of profile, because they will be low points on the profile.

C. Set up the graph on which the profile will be drawn (Fig. 6.12C). First note the differences in elevation between the highest and lowest points along

the line of profile; this will determine the range of elevations on your profile. You will also need to choose a vertical scale. It is common to use different vertical and horizontal scales in order to exaggerate the topography, as discussed in the next section. Once you have chosen the appropriate vertical scale, label elevations on the vertical axis of either printed graph paper or on a graph you have constructed yourself. Each horizontal line on this graph will represent a specific elevation, and the range of elevations must include the range encountered along the profile.

D. Transfer each of the marks made along the profile to the appropriate place on the graph paper (Fig. 6.12C). If you made marks on a separate piece of paper, place that paper along the bottom of your graph paper. Mark the ends of the profile on the graph paper. Then mark the contour and stream points on the graph paper at their appropriate elevations. This is done by going straight up from the mark on the paper (or at the bottom of the graph paper if you made marks directly on it) to the horizontal line representing the same elevation; make a small dot on the paper at this point.

E. Connect the points on the graph paper with a smooth line representing the topography (Fig. 6.12C). When crossing a valley or a hilltop, there will be adjacent marks with the same elevation. Instead of connecting them with a straight line, draw your profile line so it goes up over a hilltop or down into a valley. In the case of a stream valley, the low point in the valley will be where the stream crosses the line of profile.

Vertical Exaggeration of Topographic Profiles

Profiles commonly are drawn with a vertical scale that is larger than the horizontal scale. This **vertical exaggeration** emphasizes topographic features that otherwise might not show up on the profile. The amount of vertical exaggeration is determined by the ratio of the horizontal map

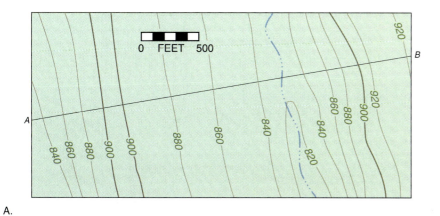

A.

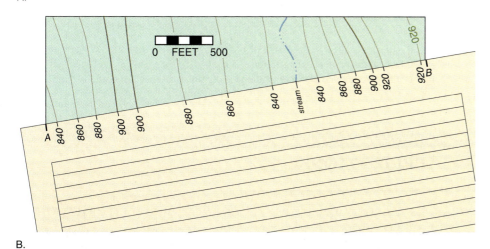

B.

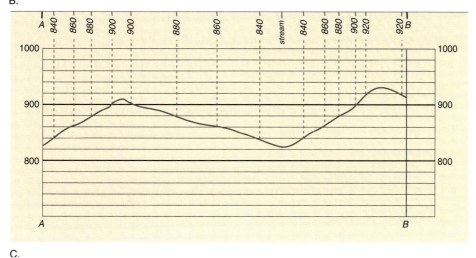

C.

FIGURE 6.12

Construction of a topographic profile. A. A line of profile (A-B) is chosen. B. Intersections of contours and the stream are marked, and elevations are noted on paper laid along the profile line. C. A vertical scale is chosen, and the points from the previous step are transferred to the appropriate elevation. A smooth line is drawn to connect the points and complete the profile.

scale (for example, 1″ to 1 mile) to the vertical scale on the profile (for example, ⅛″ to 20′).

To calculate the vertical exaggeration of a profile, first convert the horizontal scale and the vertical scale of the profile to the same units. For example,

horizontal scale is 1″ to 1 mile
convert to 1″ to 5280′
(1 mile = 5280′)
vertical scale is ⅛″ to 20′
convert to 1″ to 160′
(8 times 20′ = 160′)

Next, divide the number of feet per inch in the horizontal scale by the number of feet per inch in the vertical scale:

One inch on horizontal scale/one inch on vertical scale = 5280′/160′ = 33

The vertical exaggeration is 33× (33 times). For example, the distance representing a vertical difference in elevation of 25 feet on the profile would represent a horizontal distance of 25′ × 33 = 825′ on the horizontal scale.

Figure 6.13 shows the profile from Figure 6.12C exaggerated (A and B) and non-exaggerated (C). Note that an exaggeration of 5× was chosen for the profile in Figure 6.12.

Gradient

Gradient represents the change in elevation over a specified distance and often is expressed as feet (or meters) per mile (or kilometers). A gradient of 10 feet/mile means that the elevation of a given point is 10 feet higher than it is a mile away downhill. On a contour map, gradient is determined along a line or stream course by (1) using contour lines to determine the difference in elevation between two points, (2) using the horizontal scale to determine the distance between the same two points, and (3) dividing the vertical difference by the horizontal distance. For example, if the elevation along a stream changes 60 ft in a distance of 7.6 miles, the gradient is 7.9 feet/mile (60 feet divided by 7.6 miles).

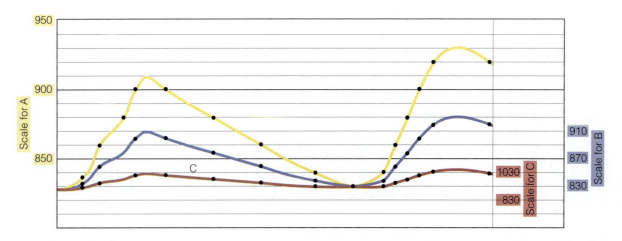

FIGURE 6.13

The profile from Figure 6.12 is shown using three different vertical scales. Assume a horizontal scale of 1 inch to 500 feet. In profile A, the vertical scale, shown in yellow on the left side of the profile, is 1 inch to 50 feet, so the vertical exaggeration is 500/50 = 10 times. In profile B, the vertical scale, shown in purple on the right side of the profile, is 1 inch to 100 feet, so the vertical exaggeration is 500/100 = 5 times. In profile C, the vertical scale, in red, is 1 inch to 500 feet, so the vertical exaggeration is 500/500 = 1 times—there is no vertical exaggeration.

APPLICATIONS

GEOLOGY AND OTHER SCIENCES DEPEND ON MAPS TO CONVEY BASIC INFORMATION, BECAUSE THEY PROVIDE A WAY TO VISUALIZE AT A CONVENIENT SCALE. CONTOUR MAPS ARE ESSENTIAL FOR RECOGNIZING AND UNDERSTANDING THE CHARACTER AND ORIGIN OF MANY LANDFORMS. GEOLOGICAL DATA FROM A LARGE AREA, WHEN PLOTTED ON A MAP, MAY PRESENT A PICTURE THAT COULD NOT BE SEEN OR UNDERSTOOD FROM THE PERSPECTIVE OF A HILLTOP IN THE FIELD. ALTHOUGH MAPS ARE USED MOST TO DISPLAY INFORMATION, THEY ARE OFTEN HELPFUL IN BASIC SCIENTIFIC ENDEAVORS SUCH AS ESTABLISHING RELATIONS BETWEEN POINTS OR FEATURES, MEASURING VARIATIONS, OR CLASSIFYING ENTITIES.

OBJECTIVES

If you complete all the problems, you should be able to:

1. Define latitude and longitude.

2. Describe the boundaries of a quadrangle map in terms of latitude and longitude, and locate a point on a map using these coordinates.

3. Locate a point using the Universal Transverse Mercator (UTM) system.

4. Locate or describe a parcel of land using the U.S. Public Land Survey System, and give its area in acres.

5. Give the dimensions and area of a section and township (in miles and square miles).

6. Number the sections of a township if they are not already numbered on the map.

7. Determine the scale of a map and use it to measure distances.

8. Convert among verbal, fractional, and graphic scales.

9. Give the magnetic declination of a map (assuming it is printed on the map) and explain what it means.

10. Determine what the various symbols used on a map mean (symbols for streams, roads, houses, etc.).

11. Use a contour map to determine elevation, height, and relief.

12. Use the characteristics of contours to determine steepness of slope, direction of stream flow, and locations of hills and valleys.

13. Determine the contour interval of a map.

14. Starting with a map showing elevations, make a topographic map by drawing contours.

15. Construct a topographic profile and determine its vertical exaggeration.

16. Determine the gradient of a stream using a topographic map.

PROBLEMS

1. Convert the following fractional scales to verbal scales; show your calculations:
    a. 1:13,226. One inch to _____ feet.

    b. 1:88,000. One inch to _____ mile(s).

    c. 1:125,000. One centimeter to _____ kilometers.

2. Convert the following verbal scales to fractional scales; show your calculations:
    a. One inch to 2,000 feet.

    b. One inch to 4 miles.

    c. One centimeter to 15 kilometers.

3. Use the UTM method to describe the location of the Radio Tower in Figure 6.4 to the nearest 10 m.

4. Use the township-range method to describe the locations of the areas labeled A, B, C, D, E, F, G, and H in Figure 6.5C.

5. One section (one square mile) contains 640 acres.
    a. What is the area, in acres, of a half-section? A quarter-section?

    b. What is the area, in acres, of A, B, C, D, E, F, G, and H in Figure 6.5C?

6. Figure 6.14 is a map showing several streams and the elevations of selected points. Draw contours on the map using a *C. I.* of 20 feet. Label index contours and make them darker than intermediate contours.

7. Using the topographic map in Figure 6.15:
    a. Determine the contour interval.

    b. Give the elevations of points A, B, C, D, E, and F.

    c. On a separate piece of paper, draw a topographic profile along the line A-B using a vertical scale of 1 inch to 80 feet.

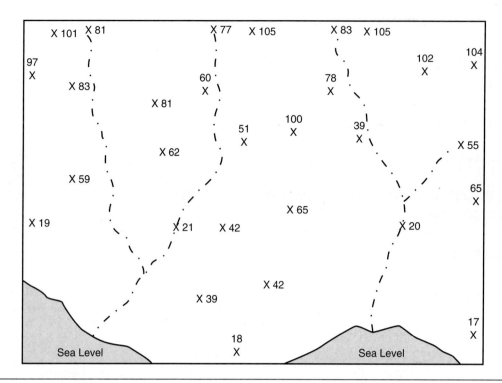

FIGURE 6.14

*X*s mark locations with indicated elevations. Dot-dash lines are streams that flow into the ocean. Draw a contour map using a *C. I.* = 20 feet.

    d.   If the horizontal scale of the map is 1 inch to ¼ mile, what is the vertical exaggeration of the profile? Show your calculations.

    e.   Determine the gradient of the westernmost (or left-hand) stream branch between the southern and northern boundaries of the map (points *X* and *Y*). Show your calculations.

Use the map provided by your instructor for problems 8 through 27.

  8.   What is the name of the quadrangle?

       What year was it published?

  9.   What is the name of the quadrangle to the east?

       The southwest?

       The north?

10.   What is the latitude of the southern boundary?

11.   What is the latitude of the northern boundary?

12.   What is the longitude of the eastern boundary?

13.   What is the longitude of the western boundary?

14.   What is the size of the quadrangle in angular units?

15.   Locate the feature designated by your instructor in terms of latitude and longitude, to the nearest minute.

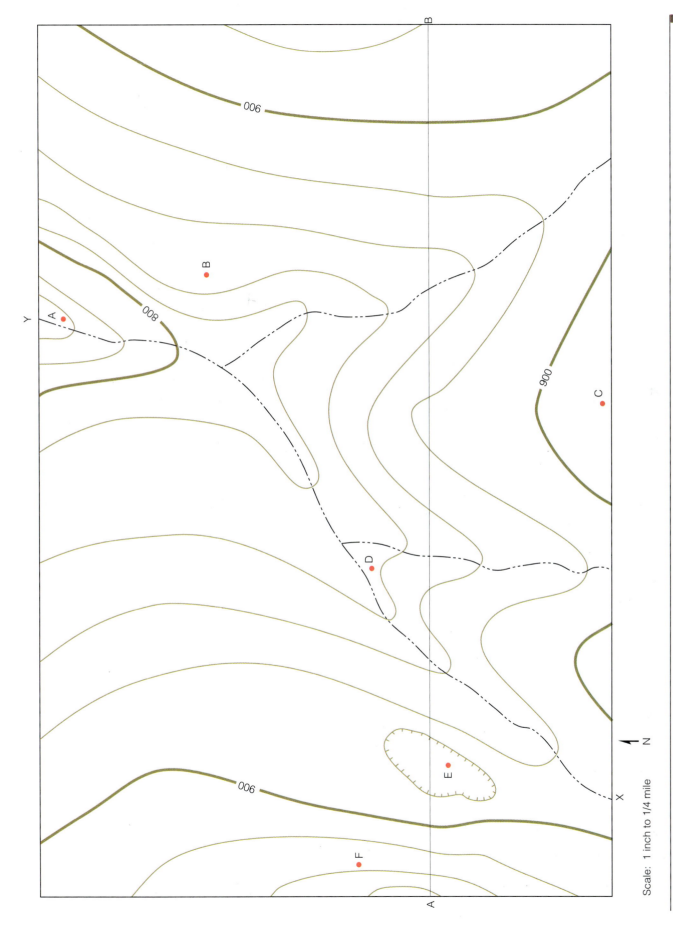

Scale: 1 inch to 1/4 mile

FIGURE **6.15**
Topographic map for Problem 7.

16. Locate the feature designated by your instructor in terms of UTM coordinates, to the nearest 100 m.

17. If the map is subdivided by the Township-Range method, locate the feature designated by your instructor to the nearest ¹⁄₁₆th of a section.

18. What is the approximate size of the area designated by your instructor (in acres, if the map is subdivided by the Township-Range method; in square miles, if not)?

19. What is the fractional scale of the map?

20. Use the graphic scale to determine the distance in miles and kilometers between the features designated by your instructor.

21. What is the approximate verbal scale for this map in terms of inches and miles?

22. If you wanted to enlarge part of the map to a scale of 1 inch to 1000 feet, by what factor would it have to be enlarged? Explain your answer.

What would the enlargement factor be if you wanted a scale of 1 cm to 100 m?

23. What magnetic declination (in degrees) is indicated on the map, and for what year?

24. What is the contour interval for this map?

25. What is the highest elevation within the area designated by your instructor?

What is the lowest elevation in that area?

What is the relief in that area?

What is the height of the feature designated by your instructor in the same area?

26. Find the building designated by your instructor (black square), and give its elevation.

27. In what direction does the water flow in the stream designated by your instructor?

28. Go to *http://www.lib.utexas.edu/Libs/PCL/Map_collection/National_parks/,* or link to author's *homepage* (see Preface). Select Devils Tower National Monument, Wyoming, and answer the following:

   a. What is the elevation of Devils Tower?

   b. What is the contour interval of the map?

   c. What is the approximate height of Devils Tower?

   d. What is the top of Devils Tower like? Is it jagged, flat, dome-like?

   e. What does the dashed line that more or less circles Devils Tower represent?

   f. How steep is the north side of Devils Tower, as measured by the angle between horizontal and the side of the Tower? To figure this out, you need to know the map scale, and none is given. However, from another source, it was found that the maximum distance from the west to east side of the 5100-foot contour that encircles the top of the tower is about 180 feet. You can see from the map that the horizontal distance between the 5100- and 4600-foot contours on the north side of Devils Tower is also about 180 feet. Thus, the elevation changes about 500 feet over a horizontal distance of 180 feet. Use this information to determine the average slope, in degrees, by either (1) making a sketch and measuring the angle with a protractor, or (2) using a trigonometry. (It's about 70°)

   For a picture and more information about Devils Tower, try going to *http://dci-server.dcomp.com/Sundance/dtower.htm* (or link to author's homepage).

In Greater Depth

29. Figure 6.16 is part of a Massachusetts road map. The distances between towns or intersections are given in miles.

   a. Determine the verbal and fractional scales.

   b. Draw graphic scales in both miles and kilometers.

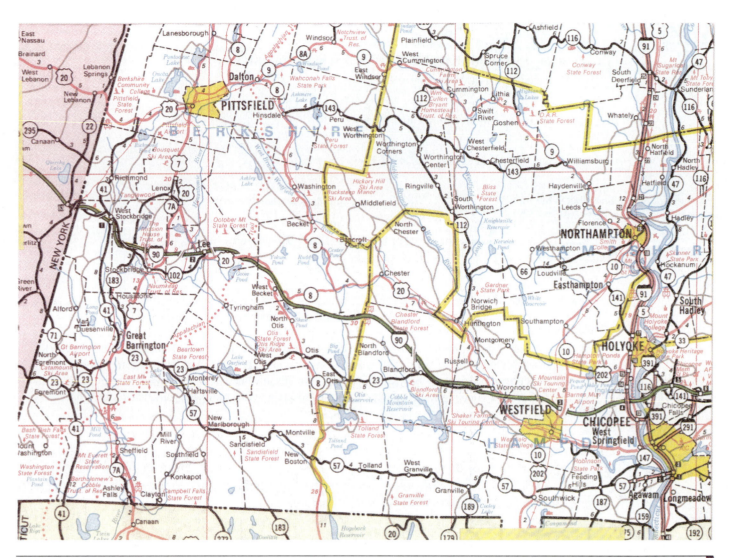

FIGURE 6.16
Portion of road map of Massachusetts for Problem 29.

    c.   Use your scales to measure the direct-line distance between Pittsfield and Northampton.

    d.   Use the mileages given on the map to determine the shortest road mileage between the two towns.

7

Aerial Photographs and Satellite Images

Overview

Views of Earth from the perspective of an airplane or a satellite are helpful for many geological purposes. The view is recorded directly by photographs or indirectly by electronic scanning of preselected wavelengths of the electromagnetic spectrum. Aerial photographs—in black and white, natural color, or infrared color—taken from airplanes can have scales similar to 7½- or 15-minute quadrangle maps. Pairs of overlapping aerial photographs viewed with a stereoscope provide a three-dimensional view. Landsat satellite images, obtained by electronic scanning from polar orbits at altitudes of 200 to 1000 km, are available in black and white or false color. The purpose of this lab is to introduce the various kinds of images so they can be used to advantage in subsequent lab exercises. You will learn to view aerial photographs stereoscopically and determine their scale, and you will learn how to interpret the colors of false-color satellite images, and use a series of images taken from a geostationary orbit at 35,900 km to monitor an ongoing process.

Materials Needed

Pencil and eraser • Ruler • Calculator • Stereoscope (provided by instructor)

INTRODUCTION

As you know or can imagine, the view you see from an airplane looking down is very different from that on the ground. You are able to see things you may not have noticed before, or see familiar things from a new perspective. This kind of perspective is invaluable in geological studies and can be preserved by means of photographs or other kinds of images.

Electromagnetic Spectrum

What people see, or what various kinds of sensing instruments detect, is **electromagnetic energy,** in the form of waves, which are either radiated or reflected from objects on the ground. Figure 7.1 shows that waves are characterized by a specific **wavelength** (the distance between crests of waves) and **frequency** (the number of complete vibrations per second); the height of the wave, or

amplitude, is a measure of the intensity of the radiation.

Different types of electromagnetic energy have different wavelengths, as summarized in Figure 7.2, an illustration of part of the **electromagnetic spectrum.** Our eyes detect visible light, which is energy of a specified range of wavelengths (or frequencies). Special films detect energy we cannot see, such as infrared, and special instruments detect other energy, such as microwaves.

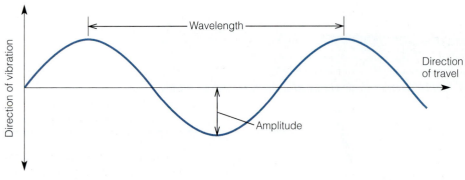

A. Longer wavelength, lower frequency

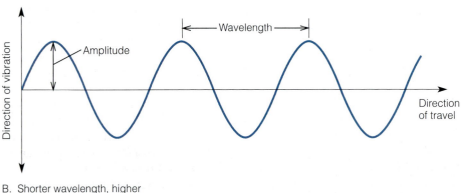

B. Shorter wavelength, higher

FIGURE 7.1
Examples of two waves. A. Wave with longer wavelength undergoes fewer complete vibrations per second, so has a lower frequency. B. Wave with shorter wavelength has higher frequency.

TYPES OF IMAGERY

Photography

Photographs are taken with a camera and recorded on film. Types include black and white, infrared black and white, natural color, infrared color, and various combinations of film and filters. Some examples are shown in Figure 7.3. Black-and-white and true-color films are sensitive to visible light and a bit of the ultraviolet (wavelengths of 0.3 to 0.7 μm—see Fig. 7.2), whereas infrared black-and-white and infrared color films are sensitive to the near-infrared (0.7 to 0.9 μm), as well as the visible spectrum. Camera filters generally are used with infrared films to cut out all or part of the visible spectrum. The result is that colors recorded by infrared color film are not true colors but **false colors.** For example, healthy, green plants appear red, not green, and clear, clean water appears black.

Electronic Scanning

Scanners are detectors that electronically record selected parts of the electromagnetic spectrum. These data can be transmitted from space and converted to a television or photograph-like image, or processed in other ways. An example is the weather-satellite image used by television meteorologists. The most commonly used scanners detect *natural* visible and infrared wavelengths reflected from the surface.

In another method, called *Side Looking Airborne Radar (SLAR),* microwaves (wavelengths of 1 to 30 cm) are beamed down from one side of an airplane, reflected from the ground back to the airplane, and recorded by a scanner. This is the essence of the type of radar that may be familiar to you speeders. One advantage of this method is that microwaves pass through clouds, so cloudy weather is not a problem.

Examples of images produced from data collected from scanners are shown in Figure 7.3.

AERIAL PHOTOGRAPHS

Most aerial photographs are taken from airplanes, although some of the more spectacular ones were taken by astronauts from space. Because of their low distortion, the most useful for extracting data are **vertical photographs,** those taken with the camera pointing straight down. **Oblique photographs,** taken at other angles, are excellent for illustrations, but features become distorted. Black-and-white photos are the most common, and such photographs of most of the United States have been available since the early 1950s. The U.S. Geological Survey provides high-altitude (40,000 feet), cloud-free, black-and-white and color-infrared photographs of the 48 conterminous states at scales of 1:80,000 and 1:58,000, respectively. It also offers high-quality color-infrared photos taken from 20,000 feet (scale 1:40,000); black-and-white photos can be made from these if desired.

Vertical photos are taken at regular intervals as an airplane flies along a pre-determined flight line at a specific altitude. The intervals are such that photographs along the flight line overlap each other by about 60 percent, and adjacent flight lines overlap by about 30 percent. A common photograph size is a square 230 mm (approximately 9 inches) on a side.

Scale

If the distances between the same two points on both the ground and on the photograph are known, the average fractional scale of the photo can be determined by multiplying their ratio by the fractional scale of the map:

Photo scale ≈ (photo distance/map distance)× fractional scale

For example, assume two road intersections are 31 mm apart on the photo; the *photo distance* is 31 mm. The same two intersections are 25 mm apart on a map; the *map distance* is

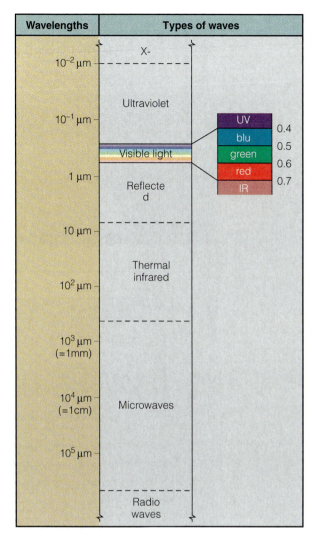

Wavelengths	Types of waves

X-

10^{-2} μm

Ultraviolet

10^{-1} μm

UV — 0.4
blu — 0.5
green —
red — 0.6
IR — 0.7

Visible light

1 μm

Reflected

10 μm

Thermal infrared

10^2 μm

10^3 μm (=1mm)

10^4 μm (=1cm)

Microwaves

10^5 μm

Radio waves

FIGURE 7.2

A portion of the electromagnetic spectrum. Wavelengths are in micrometers (10^{-6} meter or one millionth of a meter; symbol, μm); boundaries between types of waves are approximate.

25 mm. If the *fractional scale* of the map is 1:50,000:

Photo scale ≈ (31 mm/25 mm)× (1/50,000) ≈ 0.000025 = 1/40,000

Distortion

The *approximately equals* symbol (≈) is used in the scale equation because, unless the terrain is perfectly flat, the scale of most aerial photographs varies from one part of the photo to another. This *distortion* is greatest near the edges of the photo and most severe where relief is high. When compared to points at intermediate elevations, points at high elevations are shifted away from the center of the photo, and points at low elevations are shifted toward the center. In addition, the flying altitude

varies as hills and valleys are crossed. The altitude above hills or mountains is less than above valleys, so the photo scale is larger over hills and mountains and smaller over valleys. (Larger scale means that the quotient of the fractional scale is a larger number; the fractional scale 1:50,000, with quotient 0.00002, is a larger scale than 1:62,500, with quotient 0.000016.)

Stereoscopic Viewing

Overlapping aerial photographs can be viewed with a **stereoscope** to see the image *stereoscopically,* that is, in three dimensions (Fig. 7.4). The slight difference in perspective of the two photos resembles the slightly different perspectives from which our two eyes see objects;

our brains process this information to give us three-dimensional viewing.

To use a stereoscope:

1. Place the photos so that identical points on each photo are directly below each lens of the stereoscope. The photos in this manual are already in position.

2. Select an obvious point on the photos and place the stereoscope over them. Adjust the interpupillary distance of the stereoscope for your eyes.

3. As you look through the stereoscope, let your right eye look at the right-hand photo and your left eye look at the left-hand photo. If the three-dimensional scene doesn't pop out at you, rotate the stereoscope slightly about a vertical axis. The stereoscope is properly aligned when an imaginary line connecting the centers of the two lenses is parallel to a line connecting equivalent points on the two photos. If you are having difficulty seeing three dimensions, draw a light line on the lab manual photos between equivalent points to guide you.

You may be able to view a stereopair (a pair of overlapping aerial photographs) stereoscopically without a stereoscope. Use the same approach recommended to see those strange-looking images in the comic section of your Sunday newspaper. Put the two photographic images close to your face with your nose on the line separating them. Stare straight ahead while gradually moving them away until the three-dimensional image pops into focus.

The height of features is exaggerated three or four times in the typical stereoscopic view. One of the most misleading effects of this is that slopes are also exaggerated. For example, a 15° slope appears to be about 40°, and a 30° slope appears to be about 60°. This will not cause you problems in this class, but it is something that you should realize.

Comparison with Topographic Maps

Figure 7.5 is a stereographic pair of aerial photos of a volcanic cone near Idaho Falls, Idaho, and Figure 7.6 is a topographic map of the same area. Look at the photos with

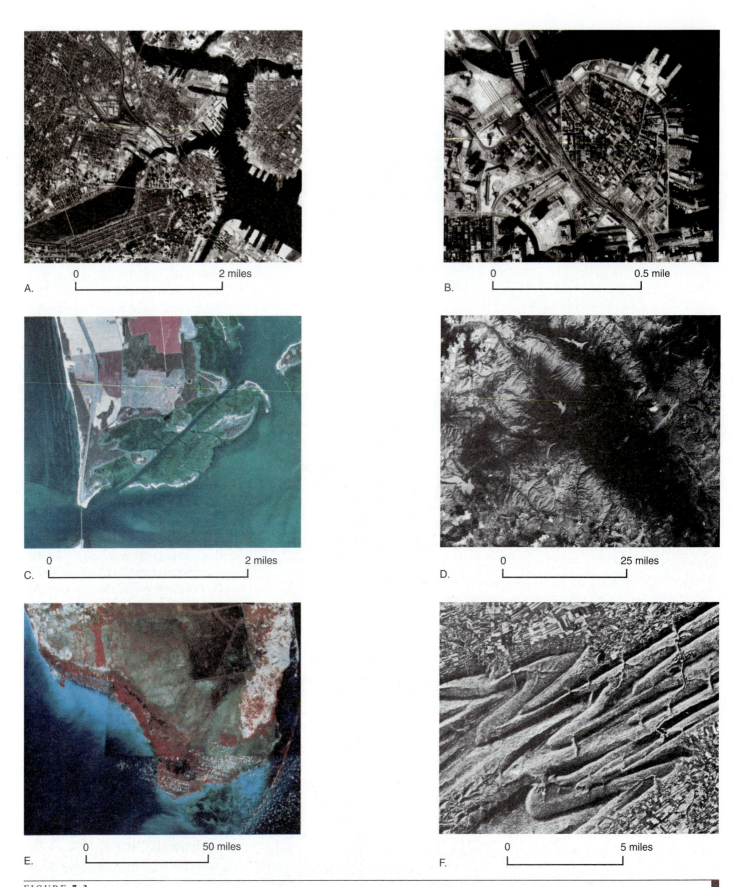

A.

0   2 miles

B.

0   0.5 mile

C.

0   2 miles

D.

0   25 miles

E.

0   50 miles

F.

0   5 miles

FIGURE **7.3**

Examples of different aerial and satellite images. A. High-altitude, black-and-white photograph of Boston, Massachusetts (scale 1:80,000). B. Low-altitude, black-and-white photograph of Boston (scale 1:20,000). C. High-altitude, color-infrared photograph of Cape Charles, Virginia (scale 1:58,000). D. Single-band, Landsat-satellite image of Boulder, Colorado, area (scale 1:1,200,000). E. Color-infrared Landsat image of the southern tip of Florida (scale approximately 1:2,400,000). F. Side-Looking Airborne Radar (SLAR) image of Appalachian Mountains, central Pennsylvania (scale 1:250,000).

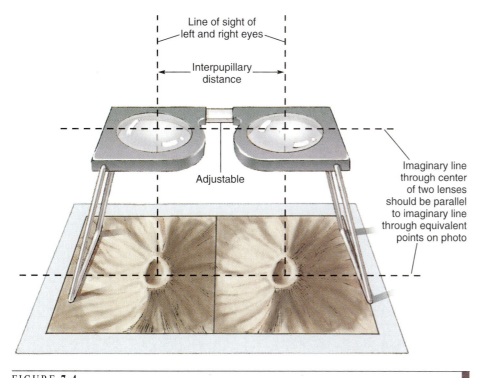

a stereoscope and compare them with the map. Note how the cone in the stereo pair is exaggerated vertically and how pronounced the crater in the center appears.

SATELLITE IMAGES

Most remote sensing from satellites is done from one of two types of orbits, polar or geostationary (Fig. 7.7). **Polar orbits** take satellites over or near the north and south poles at altitudes of 200 to 1000 km. As the satellite orbits, the Earth spins below it, so that, with time, the satellite will pass over all or most of the Earth. Satellites designed to study the Earth's surface fly in polar orbits. **Geostationary orbits** (Fig. 7.7) are much higher (35,900 km) and follow the equator. The orbital velocity is such that the satellite remains in the same apparent position above the Earth. Communication and weather satellites use geostationary orbits.

FIGURE 7.4
Stereoscope in position for three-dimensional viewing of a stereopair of aerial photographs.

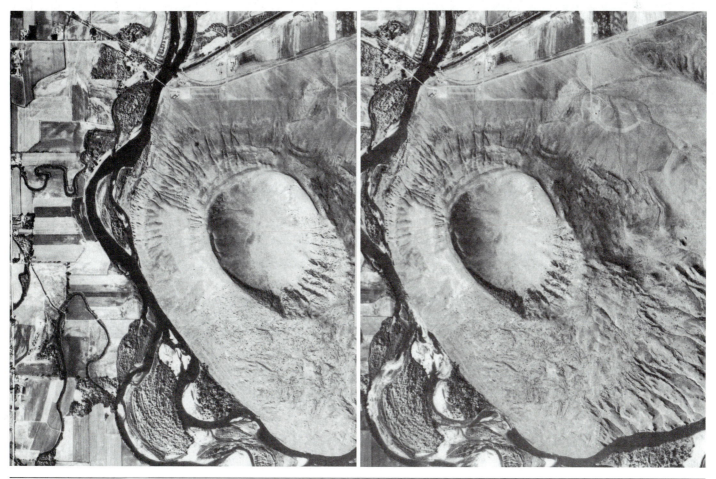

FIGURE 7.5
Stereographic pair of one of the Menan Buttes near Idaho Falls, Idaho. The Menan Buttes are volcanic cones made of basalt ash. Normally, basalt cones are made of walnut-size cinders, but here, rising basaltic magma encountered abundant groundwater in the gravel along the Snake River and erupted explosively, forming fine ash-size particles.

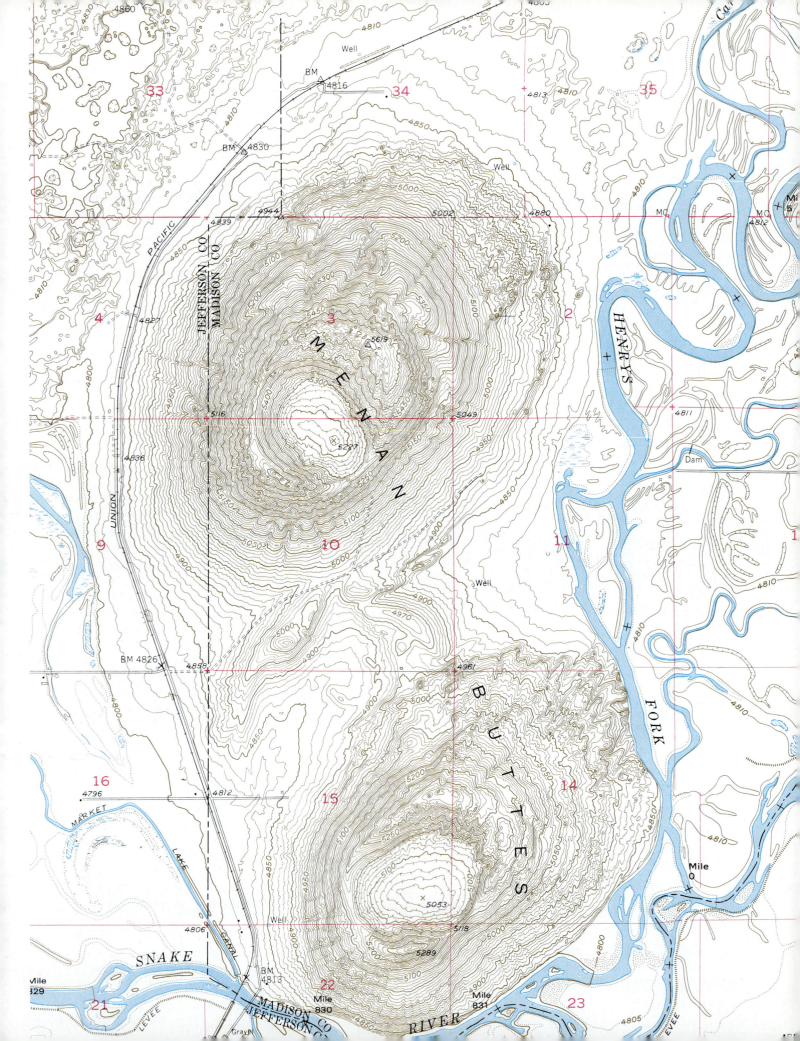

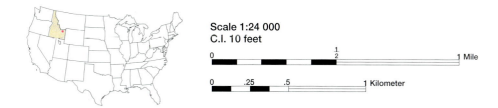

Scale 1:24 000
C.I. 10 feet

0   1/2   1 Mile

0   .25   .5   1 Kilometer

FIGURE 7.6
Portion of topographic map of Menan Buttes from Menan Buttes, Idaho, 7½ minute quadrangle.

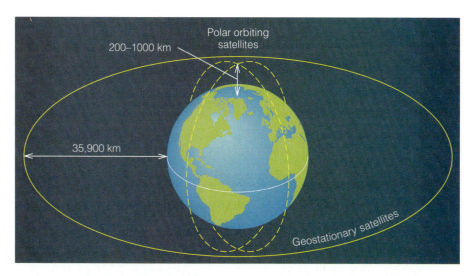

Polar orbiting satellites

200–1000 km

35,900 km

Geostationary satellites

FIGURE 7.7
Polar and geostationary orbits of satellites. *Landsat* satellites are in polar orbit.
SOURCE: DATA FROM RAY HARRIS, *SATELLITE REMOTE SENSING: AN INTRODUCTION*, 1987 ROUTLEDGE & KEGAN PAUL, LONDON & NEW YORK.

The most useful information for geologists has been gathered by the *Landsat* satellites in polar orbits. *Landsat 1* was launched in 1972; *Landsat 7* is scheduled for launch in 1998. All carry a *Multispectral Scanner (MSS)* capable of scanning a 185-km-wide swath in four separate wavelength bands (or spectra), corresponding to green, red, and two *reflected infrared* bands. *Landsats 4 and 5* also carry a *Thematic Mapper,* a seven-band scanner that includes a band for heat or *thermal infrared* (10.4 to 12.5 μm). Data from the scanners are either transmitted directly to receiving stations on Earth or stored for later transmittal. The data generally are "processed" or "enhanced" by computer before a visual image is produced. Enhancement involves geometric corrections and improvements in the visual quality of the image.

The most dramatic Landsat satellite images are those produced in color infrared (false colors), as shown in Figure 7.8. False colors are obtained by projecting the four wavelength bands (green, red, and two near-infrared) through filters and combining them to form a false-color composite image. In false-color images, reds (the false color assigned to infrared) signify growing vegetation; the more vigorous and dense the growth, the more intense the red color. As vigor and density decrease, pale red is replaced by shades of white, green, or blue, depending on type and moisture content of the underlying soil or rock; geologic features may be dramatically highlighted by these differences in shading and color. Water is black if clear, but with an increase in sediment load, it becomes pale blue; very shallow water takes on the color of the bottom sediment. Human-made features have tones that relate to the material from which they are made. Asphalt roads are dark, concrete roads light, and buildings are various colors.

FIGURE 7.8
False-color *Landsat* image of the Great Salt Lake, Utah.

APPLICATIONS

LIKE MAPS, AERIAL PHOTOS AND SATELLITE IMAGES PROVIDE A WAY TO VIEW THE EARTH AT A CONVENIENT SCALE. UNLIKE MAPS, THEY ARE DIRECT IMAGES AND CONVEY FIRST-HAND INFORMATION, NOT INTERPRETED INFORMATION. GEOLOGISTS USE AERIAL PHOTOS TO RECORD DATA, HELP INTERPRET RELATIONS, AND GENERALLY GUIDE THEM IN THE FIELD. CAREFUL STUDY OF AIR PHOTOS AND SATELLITE IMAGES, IN CONJUNCTION WITH FIELD WORK, ENABLES HYPOTHESES TO BE FORMULATED AND TESTED, AND AERIAL IMAGES THUS ARE AN INTEGRAL PART OF MANY ASPECTS OF GEOLOGICAL SCIENCE.

OBJECTIVES

If you complete all the problems, you should be able to:

1. Identify common features on an aerial photograph or a false-color satellite image.

2. Determine the scale of an aerial photograph or satellite image if you know the actual distance between two points on the photo.

3. View a pair of suitable aerial photos stereoscopically (some may be unable to do this).

4. Use a sequence of satellite images to monitor a volcanic eruption.

PROBLEMS

1.   To help you get started using aerial photos, refer to Figures 7.5 and 7.6 and answer the following:

    a.   Which of the Menan Buttes is shown in Figure 7.5? How do you know?

    b.   Draw an arrow pointing north on Figure 7.5.

    c.   Outline on Figure 7.6 the area covered by the right-hand aerial photograph in Figure 7.5.

    d.   From what direction was the sunlight coming in Figure 7.5? How do you know?

    e.   Determine the approximate scale of Figure 7.5; show your work. Locate the features on the map and the photo that you used for your measurements.

    f.   The Menan Buttes formed from the accumulation of basaltic ash during a small but unusually violent volcanic eruption. In what direction was the wind blowing when the southern Menan Butte formed? The northern butte? Explain your answer.

g.   During what time of year were the photographs taken? What is the evidence?

2.   To illustrate how familiar things appear from the perspective of an airplane, identify the labeled features in Figure 7.9.

A.   E.   H.

B.   F.   I.

C.   G.   J.

D.

Which way is north? (Hint: Use the shadows. The photographs were taken about noon in late April.)

FIGURE 7.9
Aerial photographs (stereographic pair) of Granite Falls, Minnesota (scale 1:17,996), for use with Problem 2.

101

3. To help you learn to interpret satellite images, refer to Figure 7.8, a composite, false-color, *Landsat* image of the Great Salt Lake, Utah, and answer the following (north is toward the top of the image):

a. In what part of the area is vegetation the most abundant? How do you know?

Notice that the vegetation patterns differ; nearer the lake, the pattern is splotchy, whereas in the mountains (known as the Wasatch Range), the color is more even. Explain.

b. On the east side of the image, in the red area, there is a discontinuous white line trending north-northwest. What is the white?

Some white patches can also be seen in the red areas on the south side of the image. What are they?

c. Two large areas of white occur southwest and west of the lake, and some smaller ones on the north. Knowing the name of the lake and realizing that it has been larger in the past, what might these white areas be?

d. Much of the land in the center and western part of the image, including the islands, is a greenish tan color. What would you expect to find in those areas?

e. Locate Salt Lake City on the southeast end of the Great Salt Lake. Interstate 80, a light gray line on the image, goes directly west from downtown Salt Lake City to the lake, where the road bends southwest. Immediately southeast of the bend, you can see an evaporating pond with a shape that points west. What might the evaporating pond be used for?

f. The lake is very dark blue on the south end, and light blue on the north end and to the west of the small island on the southwest. The colors are separated by straight lines, an indication that the cause is human activity. The line across the middle of the lake marks the position of the causeway for the Southern Pacific Railway, and levees block the area on the southwest. In light of this information, suggest a reason for the color differences in the lake.

g. The largest island (Antelope Island), in the southeastern part of the lake, is approximately 25 km long. What is the approximate scale of the image?

h. The world's largest open-pit mine, at Bingham Canyon, Utah, appears on this image. It is the light blue area near the south edge of the image; Antelope Island "points" at it. It is principally a copper mine, but when copper prices are low, recovery of minor, but more valuable, elements such as gold and silver enable the mine to continue operating. What are the length and width of the area disturbed by this mining operation?

To help you gain some perspective into the amount of land devoted to the mining operation, compare it to the visible area of the Tooele Army Depot to the west. The army depot is directly west of the mine in the valley between the two red mountain ranges; it is marked by the area of squares and rectangles that are aligned northwest-southeast. This is where 40 percent of the United States' chemical weapons—enough to kill everything on the planet (National Geographic, v. 189, no. 1, 1996)—are stored, inside igloo-shaped warehouses. In response to a Congressional order to destroy all of the United States' toxic weapons by 2004, an incinerator will be built at Tooele, at a cost of half a billion dollars.

4. a. Go to *http://southport.jpl.nasa.gov:80/scienceapps/instruments.html* (or link to it through the author's homepage—see *Preface*), and select *SIR-C/X-SAR* to learn about one of the newest remote sensing systems.

   (1) What do the letters S, I, and R represent?

   (2) What will the images be used to measure?

   (3) What three wavelengths of the electromagnetic spectrum will be used? What general name (from Fig. 7.2 in this lab manual) is given to radiation of these wavelengths?

   b. Next go to *http://satftp.soest.hawaii.edu:80/space/hawaii/index.html* (or link to it through the author's homepage—see *Preface*), and click on *Remote Image Navigator.* Below the map are four buttons for the selection of four types of images: Shuttle photographs, *Landsat,* SIR-C, and Aerial Photographs.

   (1) To see how these images differ, choose each type and click on the outlined area that best covers the island of Maui. Describe each and indicate what kinds of things that image shows better than the others.

   (2) Next choose *Aerial Photographs,* and click on the easternmost outlined area on the big island of Hawaii. This will give you an image of the Royal Gardens area, which was overrun by basaltic lava flows emanating from the Pu`u `O`o crater in the early 1980s. You may want to visit *Go to Virtual Field Trips* for a tour of the Big Island or one of the other Hawaiian Islands.

   c. Go to *http://www.winona.msus.edu/academicdeptsfolder/geology/imagearchive/remoteimages/remoteimages.html* (or link to it through the author's homepage—see *Preface*), and select the *Wyoming state relief map from radar imagery.* Mountain ranges show very well on this map. Use a state road map of Wyoming that gives the names of mountain ranges, and locate the following in terms of latitude and longitude:

   (1) Bighorn Mountains

   (2) Absaroka Mountains

   (3) Teton Mountains

IN GREATER DEPTH

5.   Figure 7.10 shows Geostationary Operational Environmental Satellite (GOES) images taken from 35,900 km above the Earth following the May 18, 1980, eruption of Mount St. Helens. The times (Pacific Daylight Time) at which the images were taken are given, and the state boundaries are superimposed on the images.

    a.   Transfer the boundaries of the ash cloud from each image onto the map in Figure 7.11 using the state boundaries as guidelines. Label each ash-cloud boundary with the time of the image. Write the time of each image in Table 7.1

    b.   Using Figure 7.11, measure the *maximum* distance from Mount St. Helens to the leading edge of the ash cloud for each image, and enter your results in Table 7.1.

    c.   Calculate the travel time represented in each image by subtracting the time of the eruption (8:32) from each image-time, and enter the result in Table 7.1.

    d.   Prepare a graph of the data in Table 7.1, using Figure 7.12, by plotting the travel time on the y (vertical) axis and distance traveled, in kilometers, on the x (horizontal) axis for each image. Connect the data points. When the points are connected, do they form a straight line, a smooth curve, or an irregular line?

    e.   The velocity at which the ash plume moved is represented by the slope of the lines connecting the points; the steeper the slope, the slower the velocity. What does the graph tell you about the wind velocity—was it constant, did it gradually increase or decrease, or was it variable?

    f.   The velocity at which the ash plume traveled can be calculated by dividing the distance traveled during a particular time interval by the time interval. For example, if the ash cloud traveled 425 km in 200 minutes, velocity, $v = 425$ km/200 min. $= 2.13$ km/min. Calculate the overall, or *average,* velocity between 8:32 and 16:15. Give the answer in (1) km/min, (2) km/hour (multiply (1) by 60), and (3) miles/hour (1 km $= 0.62$ miles).

    g.   In what general direction was the prevailing wind (see Fig. 7.10)?

    Was the direction the same along the entire path, or did it change; if so, how?

    h.   Based on the color-shading of the ash plume during the period of the images, predict where the thickest deposits of airfall ash were likely to have accumulated, and outline them on Figure 7.11.

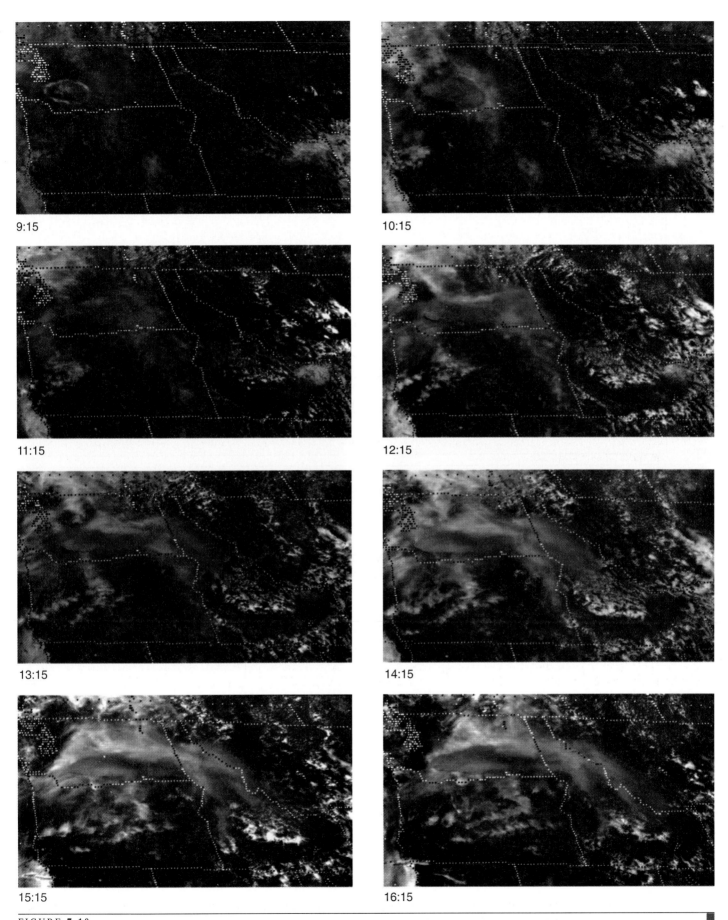

9:15

10:15

11:15

12:15

13:15

14:15

15:15

16:15

FIGURE 7.10

Satellite images taken by GOES from a geostationary orbit illustrate the movement of the ash cloud following the May 18, 1980, Mount St. Helens eruption. For use with Problem 5.

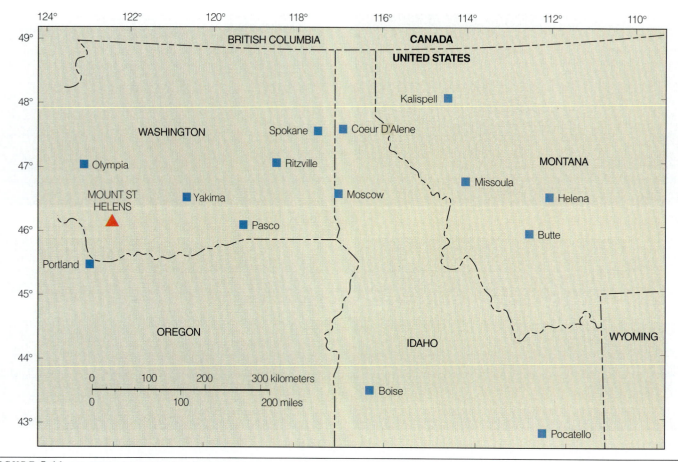

FIGURE **7.11**
Outline map of area affected by ash from Mount St. Helens. For use with Problem 5.
SOURCE: U.S. GEOLOGICAL SURVEY.

T a b l e   7 . 1

GOES DATA, MOUNT ST. HELENS, 1980

TIME OF IMAGE	MAXIMUM DISTANCE FROM MOUNT ST. HELENS	TRAVEL TIME

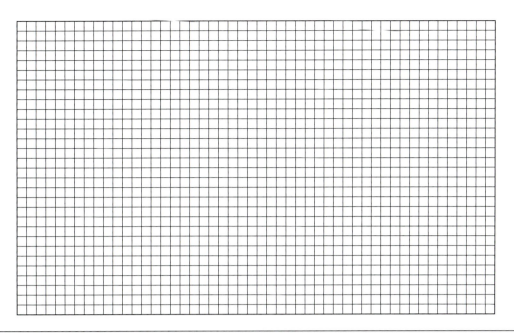

FIGURE 7.12
Graph for use with Problem 5.

Part 4

SURFACE PROCESSES

chapter

8

Streams and Humid-Climate Landscapes

Overview

Running water is the most important agent of erosion, even in most arid regions, and many common landforms owe their origin to streams. The characteristics of individual stream valleys and surrounding landscapes are determined by the material on which they are developed, the climate, time, and changes in elevation relative to base level. The volume of water that passes a given location on the stream during a particular time interval can be measured, and this information used to predict the size and frequency of floods. In this lab, you will learn how to use data from gaging stations to predict the extent of floods with various recurrence intervals, how to determine the size and shape of a drainage basin, and how to recognize and interpret common erosional and depositional features of landscapes formed by running water.

Materials Needed

Pencil, eraser • Colored pencils • Calculator • Ruler

INTRODUCTION

Most of Earth's landscapes are shaped by weathering, mass wasting, and erosion by running water. Weathering causes surface materials to disintegrate or decompose, mass wasting moves them downhill, and streams erode and carry them away. Running water is an important agent of erosion: even in the driest deserts, stream channels, though generally dry, and other evidence of running water are abundant.

Human lives are affected by rivers in many ways. Many cities are located on rivers and draw their water supplies from them. Rivers provide transportation and recreation. Rivers also flood, causing millions of dollars in damage annually.

In this lab, you will learn about streams and their valleys, and the landscapes they produce in humid regions—those with annual precipitation of more than about 50 cm (Fig. 8.1).

RUNOFF AND DRAINAGE BASINS

Streams are part of the hydrologic cycle, as shown in Figure 8.2. They carry water precipitated on the surface back to the oceans as surface runoff. Some of the precipitation returns to the atmosphere by evaporation and transpiration (emission through the leaves of plants), and some soaks into the ground and is carried away below the surface (the *infiltration loss* in the equation below). Water that enters the streams and makes up the *runoff* comes from direct overland flow, and from water that moved underground before being discharged into the streams. Runoff due to overland flow can be expressed as:

runoff = precipitation − (infiltration loss + evaporation + transpiration).

Anything affecting one of these terms will affect runoff, so the percentage of

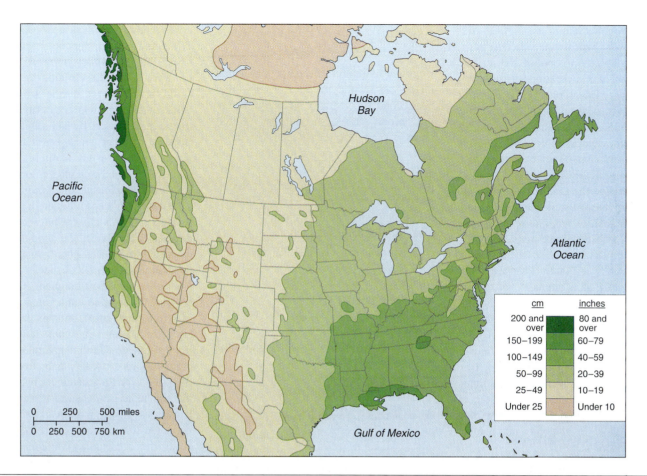

FIGURE 8.1

Average annual precipitation in the United States and Canada. Areas with more than 50 cm per year are considered humid. From *Geosystems,* 2nd edition by Robert W. Christopherson. Copyright © 1994. Reprinted by permission of Prentice-Hall, Inc. Upper Saddle River, N.J.

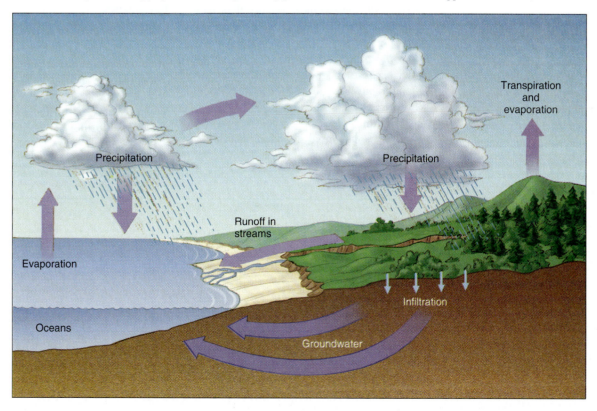

FIGURE 8.2

The hydrologic cycle. This chapter emphasizes runoff.

precipitation that naturally runs off varies considerably worldwide. In Arizona, where precipitation is about 25 cm/year, evaporation and infiltration are high, and runoff is very low, typically 2 cm/year or less. In contrast, precipitation in Alabama is about 125 cm/year, and because infiltration and evaporation are low, runoff is about 75 cm/year. Human activity also affects runoff. For example, runoff increases when the land surface is paved so that rain doesn't soak into the ground, or when vegetation is removed so that the amount of transpiration is reduced.

When water runs off on the surface, it goes downhill and into a stream. The stream is part of a drainage network in which smaller streams feed larger streams. As shown in Figure 8.3, each stream, no matter how large or small, has an area that it drains called a **drainage basin.** Drainage basins are separated by **divides,** which are higher than the land on either side of them. Rain falling on one side of a divide goes downhill into one drainage basin; that falling on the other side goes into another drainage basin.

STREAM CHANNELS AND VALLEYS

Figure 8.4 shows a contour map of a stream and a topographic profile drawn along its length. The **longitudinal profile** illustrates how the **gradient** (see Chapter 6) gradually decreases from the **head** to the **mouth** of the stream. The mouth is the **base level** for a stream: it limits the depth to which that stream can erode. A stream adjusts its channel and longitudinal profile in response to changes in discharge, base level, and erodibility of the rock or sediment over which it flows, and with the passage of time. Ideally, the adjustments lead to a near *balance* between erosion and deposition along the course of a stream and produce a smooth longitudinal profile, as shown in Figure 8.4. A stream that does not have a smooth profile erodes or deposits so as to attain one: waterfalls and rapids are eroded, lakes or ponds along streams are filled.

The size of a stream channel and the velocity and volume of water all increase downstream. The volume of water per unit of time is the **discharge** and is given by this equation.

$$discharge = velocity \times cross\text{-}sectional$$
$$area\ of\ channel$$

Common units for discharge are cubic meters per second (m^3/sec) or cubic feet per second (ft^3/sec); for velocity, meters per second (m/sec) or feet per second (ft/sec); and for cross-sectional area, square meters (m^2) or square feet (ft^2). As discharge increases, so do all of its components. Thus, during flooding, velocity and channel size increase as the volume of water increases. Stream velocity, channel area, and discharge are recorded at *gaging stations* on many streams throughout the United States. For example, during the summer flood of 1993, the gaging station on the Mississippi River at St. Louis, Missouri, recorded a peak discharge of 1,070,000 ft^3/sec on August 1. This is more than eight times the average August discharge of 133,000 ft^3/sec.

Features of Streams and Their Valleys

Streams vary, from turbulent mountain streams rushing down narrow valleys to great rivers with wide valleys flowing across a nearly flat landscape. As streams vary, so do their characteristic features.

Streams with steep gradients tend to erode downward more rapidly than they erode laterally. Therefore, they typically have narrow valleys with V-shaped cross-profiles. Longitudinal profiles are irregular because of the presence of waterfalls and rapids. Figure 8.5 illustrates these features.

With decrease in gradient, lateral erosion becomes more important, and broad valleys develop. Such streams have a vari-ety of features, as illustrated in Figure 8.6. The actual stream channel is much narrower than the valley, most of which is occupied by the **floodplain,** the area that could be submerged during a flood. Just adjacent to the channel are **natural levees,** low ridges formed by deposition during flooding. A river channel is not straight, but meanders or winds about in the floodplain; a bend in the channel is a **meander.** Erosion on the outside of a meander forms a **cutbank,** and deposition on the inside of a meander forms a **point bar.** The **meander belt** is the zone in the floodplain within which meanders occur. Erosion and deposition cause the channel to change position within the floodplain. The channel may take a short cut across a meander loop to form a **cutoff,** or abandon the loop altogether to form an **oxbow lake.** Where floodplains are wide and natural levees are high, tributary streams may flow in the floodplain for long distances before joining the main river; such tributaries are known as **yazoo streams. Stream terraces** are step-like benches above the level of the present-day floodplain. They represent the remnants of pre-existing floodplains or valley floors.

Floods and Recurrence Intervals

A flood occurs when a stream overflows its channel. The size of a flood, as measured by maximum discharge, or by **stage** (elevation of water surface), varies from year to year. By analyzing the frequency of floods of various sizes, a *recurrence interval* can be developed for a river at a particular locality. The **recurrence interval,** usually measured in years, is the

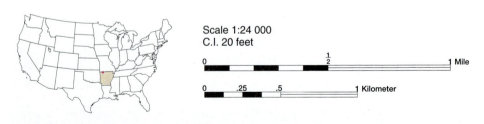

Scale 1:24 000
C.I. 20 feet

▶ FIGURE 8.3
The drainage basin of Clayborn Creek and its tributaries is outlined on this portion of the Beaver, Arkansas/Missouri, quadrangle. Note that the divides connect the highest elevations between adjacent drainage basins. Water falling within the drainage basin moves downhill toward Clayborn Creek and eventually makes its way into the White River.

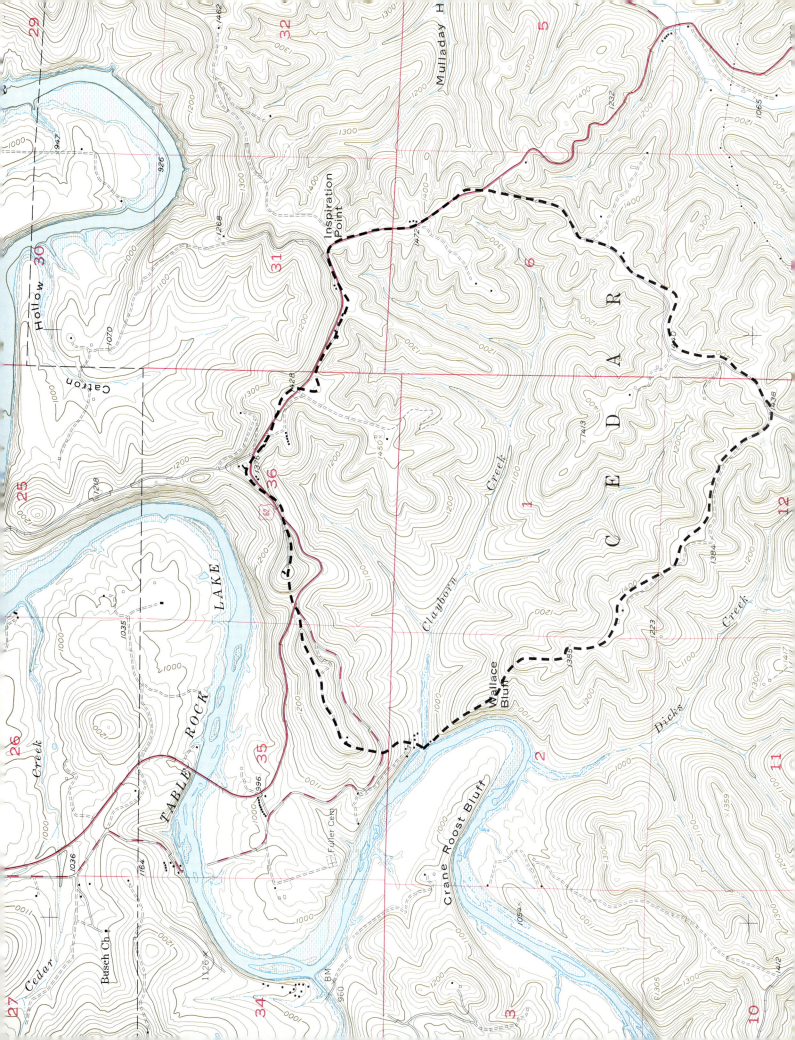

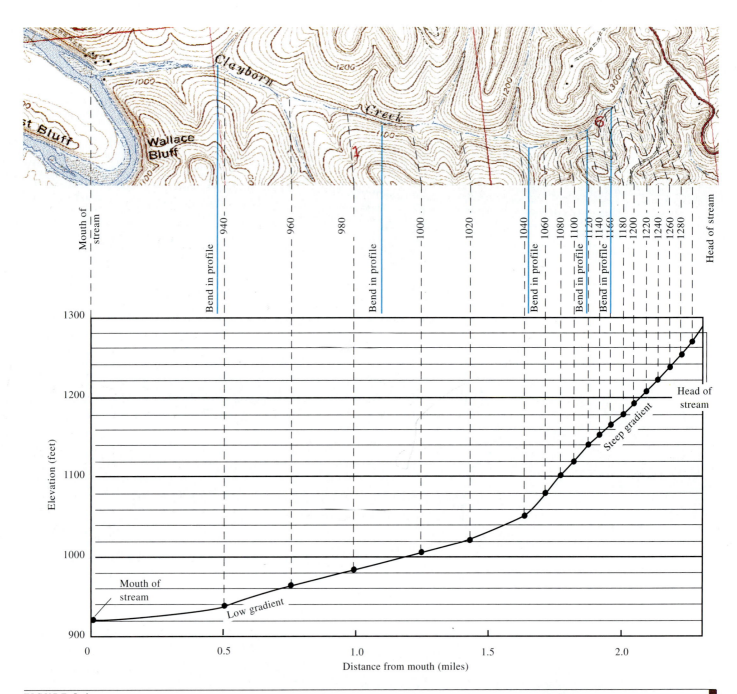

FIGURE 8.4

The longitudinal profile along Clayborn Creek was drawn by marking, on the top of the profile paper, points where contours cross the stream and labeling these points with their elevations. The profile follows the bends in the stream, not the straight-line distance between the head and the mouth. An easy way to follow along the stream when marking the contour-crossing points is to hold the paper down with a pencil where the stream bends, then rotate the paper to follow along the next stream segment. The locations of the bends are shown here for illustrative purposes only. Note that the gradient decreases from the head to the mouth of the stream. The vertical exaggeration is 16.7 times. (The horizontal scale is 1:24,000, and the vertical scale is 1 inch to 120 feet).

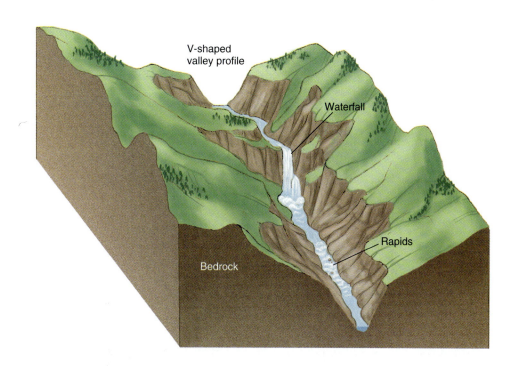

FIGURE **8.5**
Features of a stream with a steep gradient.

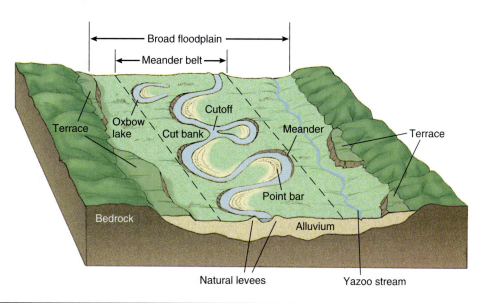

FIGURE **8.6**
Features of a stream with a gentle gradient.

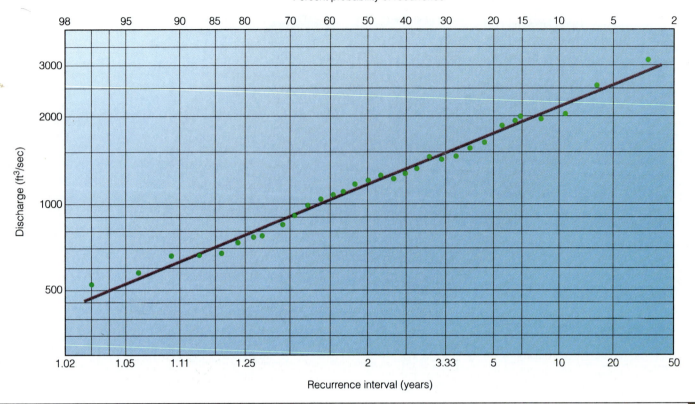

Percent probability of recurrence

FIGURE 8.7
Flood-frequency curve for Rock Creek near Red Lodge, Montana, 1932 to 1963, using maximum annual discharge data from the USGS. From the *Recurrence Interval (RI)* scale (bottom), it can be predicted that, on average, a discharge of 2600 ft³/sec will be attained every 20 years. Using the *Percent Probability of Recurrence (P)* scale on top, the same thing can be said in a different way: there is a 5% probability that a discharge of 2600 ft³/sec will be reached in a given year. $P = 100 \ (1/RI)$.

average interval between floods of a particular size. Thus, *on average,* a 100-year flood will recur at intervals of 100 years. That does not mean that a flood that size could not occur two years in a row; it means that the chance of it occurring in *any* year is 1 in 100. A *flood-frequency curve* plots discharge, or in some cases, stage, against recurrence interval, as illustrated in Figure 8.7.

Floodplain zoning is based on recurrence intervals. If planners know what stage a stream will reach during, say, a 50-year flood, they can determine what part of the floodplain will be underwater, and they can zone accordingly. The data necessary to determine the size of the 50-year floodplain are available from flood-frequency curves and from gaging-station records.

LANDSCAPES IN HUMID AREAS

Controlling Factors

Landscapes in humid regions tend to have rounded, soil-covered slopes, ridges and valleys, and extensive stream deposits. Landscapes reflect not only the climate in which they developed, but also the processes by which they were eroded, the characteristics of the underlying materials, and changes in position relative to base level. Landscapes developed on homogeneous rock or unconsolidated sediment will differ from those developed on rocks of varying susceptibility to weathering and erosion. The erosional patterns offer clues as to what is underneath, and this is particularly evident with stream drainage patterns, as

shown in Figure 8.8. If the area is uplifted, or base level drops, erosion by downcutting will increase; if base level rises or the land subsides, deposition will occur.

Evolution of Stream Valleys

Stream valleys change with time. If the controlling factors remain constant, the changes are more or less predictable, and a stream may evolve through early, middle, and late stages. The changes are similar to those seen along the length of a large, well-developed river, from its head to its mouth. Because early-stage streams, or segments of streams, are well above base level, they erode principally by downcutting. Such streams have narrow valleys, commonly with rapids and waterfalls. With time, the gradient decreases, *headward erosion* lengthens the stream,

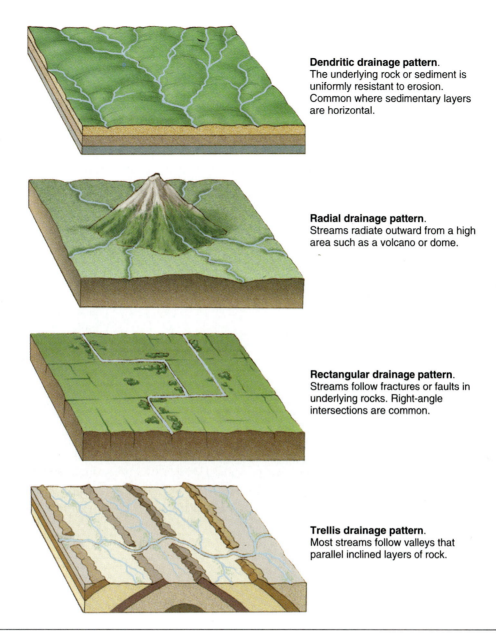

Dendritic drainage pattern.
The underlying rock or sediment is
uniformly resistant to erosion.
Common where sedimentary layers
are horizontal.

Radial drainage pattern.
Streams radiate outward from a high
area such as a volcano or dome.

Rectangular drainage pattern.
Streams follow fractures or faults in
underlying rocks. Right-angle
intersections are common.

Trellis drainage pattern.
Most streams follow valleys that
parallel inclined layers of rock.

FIGURE **8.8**
Four common drainage patterns.

and tributaries become longer and more
significant. Rapid headward erosion may
cause one stream to intersect and divert
the flow of another stream in a process
known as **stream capture** or **stream
piracy.** During the middle stage of development, lateral erosion increases and
greater meandering becomes possible,
valleys broaden and develop floodplains,
while waterfalls and rapids disappear. By
the late stage of stream evolution, rivers
have wide floodplains with oxbow lakes
and yazoo rivers.

When you have completed Problem 1,
Table 8.1 will contain a list of some of the
characteristics of early-, middle-, and late-
stage streams.

Evolution of Landscapes

Just as stream valleys change with time or
position, so does the general landscape.
With increased development of tributaries
and lengthening of streams, more of the
surrounding land becomes part of the
valley system, hilliness increases, and
drainage divides become more prominent.
As time goes on, gradients of stream segments decrease and hills are worn away.
Early geologists conceived an *erosion
cycle,* in which landscapes passed through

early, middle, and late stages, each with its
own list of characteristics. The erosion-
cycle idea *per se* probably is too simplis-
tic, because the controlling factors are
rarely constant. For example, climates differ from place to place and change with
time, base level changes as sea level rises
or falls, and tectonic activity may have a
profound influence in determining how
landscapes develop. Nevertheless, most
stream valleys and landscapes have *characteristics of either early, middle, or late
stages,* which serve to distinguish them,
even though they may not have resulted
from an ideal erosion cycle.

APPLICATIONS

GEOLOGISTS HAVE APPROACHED THE STUDY OF STREAMS AND ASSOCIATED LANDSCAPES IN SEVERAL WAYS. EARLY APPROACHES WERE DESCRIPTIVE SYSTEMS OF CLASSIFICATION OF NATURAL PHENOMENA, WHICH ASSUMED CAUSE-AND-EFFECT RELATIONS AMONG VARIOUS FEATURES. WORKING MODELS WERE CONSTRUCTED TO HELP UNDERSTAND RIVERS AND RIVER SYSTEMS. EVENTUALLY, ENOUGH NUMERICAL DATA BECAME AVAILABLE FROM BOTH NATURAL AND MODEL SYSTEMS TO DEVELOP MATHEMATICAL MODELS, AND QUANTITATIVE PREDICTIONS COULD BE MADE. THE PROBLEMS IN THIS LAB ILLUSTRATE BOTH QUALITATIVE AND QUANTITATIVE APPROACHES.

OBJECTIVES

If you complete all the problems, you should be able to:

1. Identify the following features on a map or aerial photograph: floodplain, meander, meander belt, oxbow lake, cut bank, point bar, natural levee, cutoff.

2. Determine whether the characteristics of a stream or landscape are those of early, middle, or late stage.

3. Identify dendritic, radial, trellis, and rectangular drainage patterns, and use them to predict the general structure of the underlying rocks.

4. Use data from a gaging station to calculate the *percent probability of recurrence* or the *recurrence interval* of a particular discharge or river stage and to construct a flood-frequency curve.

5. Outline a drainage basin on a topographic map, estimate its area, and calculate its annual runoff from precipitation, evaporation, transpiration, and infiltration data.

6. Draw both cross and longitudinal profiles of a stream valley.

PROBLEMS

1. Figures 8.9, 8.10, and 8.11 illustrate segments of three major U.S. rivers that show different characteristics. Examine the maps and complete Table 8.1; note that the maps are at different scales. If you are unable to complete some parts, explain why. The purpose of this problem is to learn to recognize certain stream features on a map and to learn what kinds of stream features typically are associated with one another.

    The three rivers in Figures 8.9, 8.10, and 8.11 are examples of streams with characteristics of early, middle, or late stages. Based on the information you have recorded in Table 8.1, which is which?

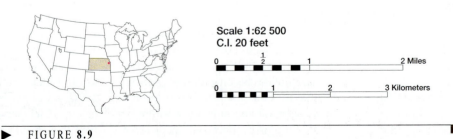

▶ FIGURE 8.9
Portion of topographic map of Leavenworth, Kansas, 15-minute quadrangle, for Problem 1.

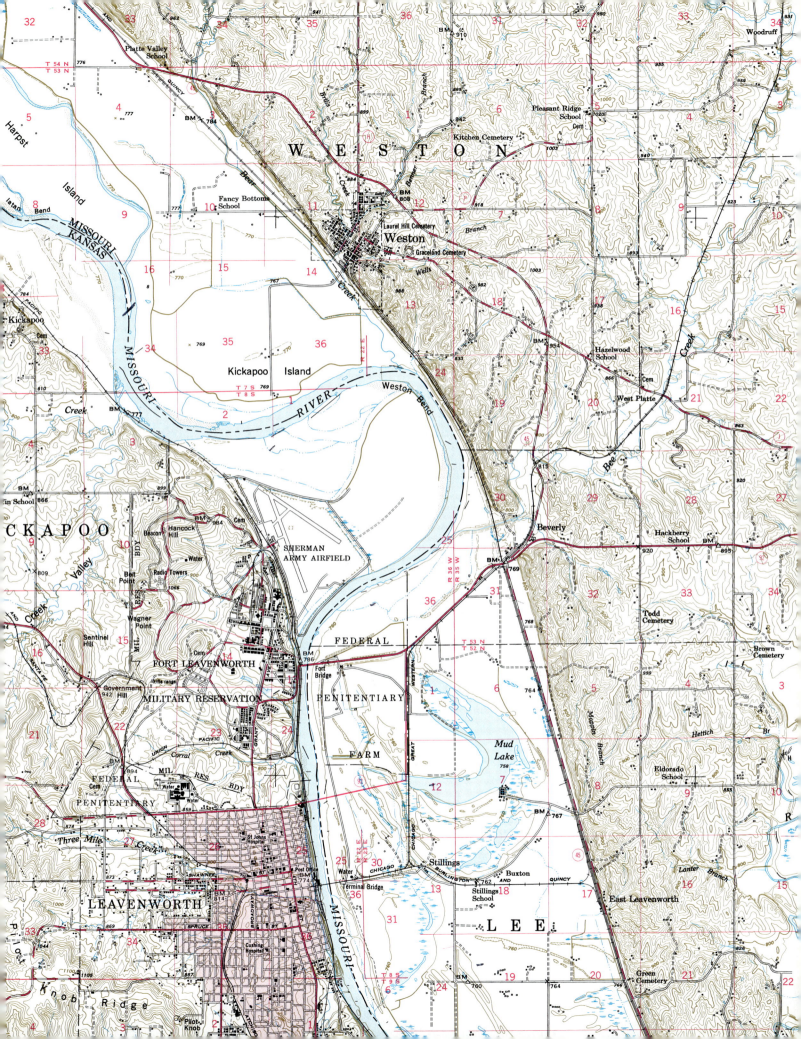

Table 8.1

CHARACTERISTICS OF THREE STREAM VALLEYS

STREAM CHARACTERISTICS	MISSOURI RIVER	ARKANSAS RIVER	MISSISSIPPI RIVER
Presence of:			
floodplain			
meanders			
natural levees			
back swamps			
yazoo streams			
oxbow lakes			
cutoffs			
filled former channels			
Gradient within map area (in feet per mile)			
Width of floodplain (if present and measurable)			
Width of meander belt			
Floodplain width compared to meander-belt width (divide width of floodplain by width of meander belt)			
Sketch of valley profile, including channel			
Probable importance of: vertical erosion (downcutting) lateral erosion			

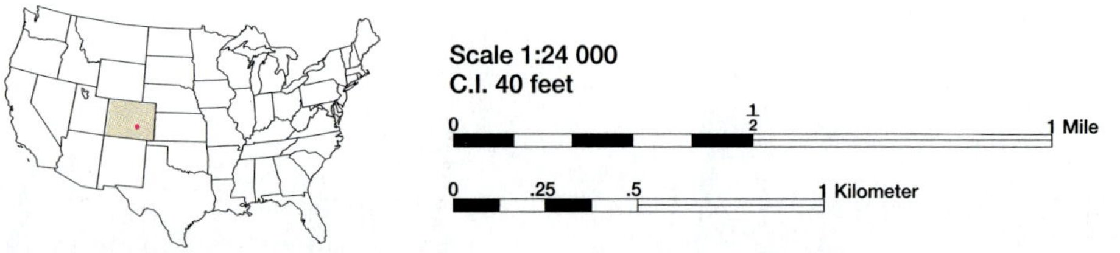

Scale 1:24 000
C.I. 40 feet

0   1/2   1 Mile

0   .25   .5   1 Kilometer

▶ FIGURE 8.10
Portion of topographic map of Royal Gorge, Colorado, 7½-minute quadrangle, for Problem 1.

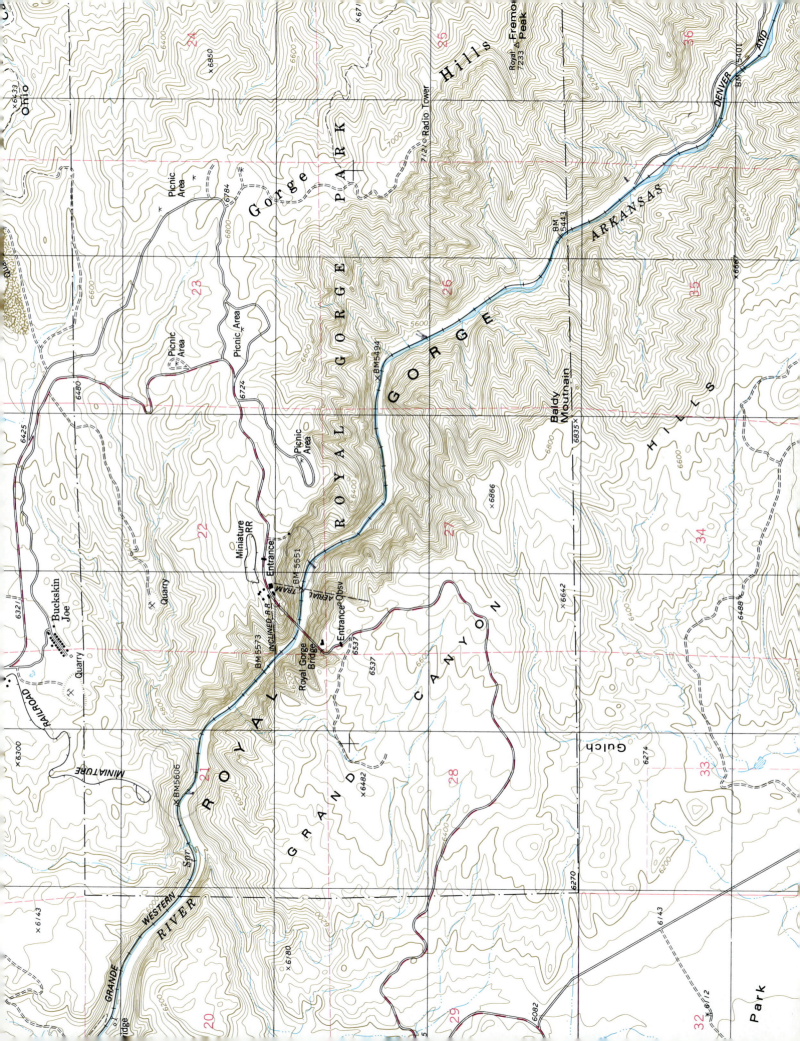

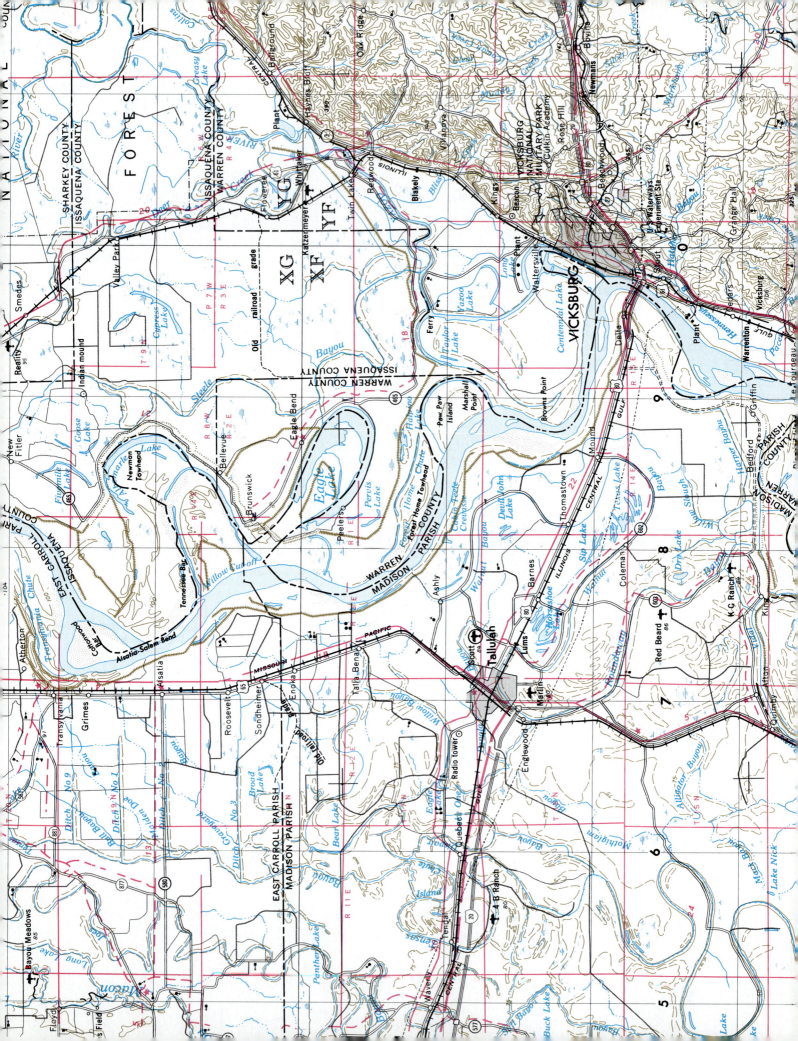

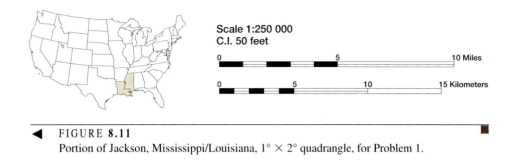

Scale 1:250 000
C.I. 50 feet

0	5		10 Miles

0	5	10	15 Kilometers

◄   FIGURE **8.11**
Portion of Jackson, Mississippi/Louisiana, 1° × 2° quadrangle, for Problem 1.

2.   Refer to Figure 8.12, a portion of the Kaaterskill, New York, quadrangle. This area, in the northeast corner of the Catskill Mountains of southeastern New York, shows a contrast in topography between the east and west sides of the map. The western part of the area is underlain by sedimentary rocks with nearly horizontal layers, and the eastern part is underlain by sedimentary rocks with steeply inclined layers.

a.   How do the drainage patterns in the eastern and western parts of the area differ?

b.   Are rocks in the eastern or western part of the area most resistant to erosion? How do you know?

c.   Which of the two creeks, Schoharie Creek or Plattekill Creek (between its head and West Saugerties), has the steepest gradient?

Which do you think is eroding headward most actively, and why?

What is the likely future for the headwater tributaries of Schoharie Creek?

d.   Kaaterskill Creek has committed stream piracy and "beheaded" Gooseberry Creek. Use a colored pencil to indicate on the map the probable course of Gooseberry Creek before this tyrannous event.

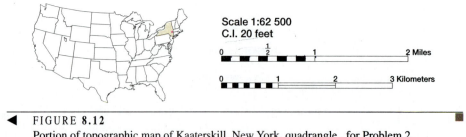

Portion of topographic map of Kaaterskill, New York, quadrangle , for Problem 2.

3.   Figure 8.13 is a map showing the Minnesota River at Jordan, Minnesota. The obvious broad valley on the map was carved when much more water was coming down the river than at present. This took place about 10,000 years ago during the melting of continental glaciers and the draining of a vast lake (Glacial Lake Agassiz—see Fig. 10.6).

   a.   To help visualize the past and present Minnesota River valley, draw a topographic profile along line A-A', Figure 8.13. Use Figure 8.14 for your profile. Based on the profile shape and the contours shown on the map, locate and mark the position of the *present-day* floodplain on your profile. What evidence is there on your profile and on the map for an older floodplain or floodplains?

   b.   The gaging station at the bridge north of Jordan monitors the *stage,* or level, of the river, as well as discharge and water quality. The stage here is the number of feet above the gaging-station elevation of 690'. Table 8.2 shows the maximum stage reached during selected years from 1935 through 1994. A relative magnitude, *m,* has been assigned to each stage and is recorded in column 3 of Table 8.2. The highest maximum stage was reached in 1965, so its magnitude, *m,* is 1; the lowest maximum stage was attained in 1940 (not shown in the table), and its magnitude is 60 (the table summarizes 60 years of records).

   Construct a **flood-frequency curve** as follows:

   (1)   Calculate the *percent probability of recurrence, P,* for each of the events listed in Table 8.2, and record it in column 4 of that table. *P* is the percent probability that a stage of a particular height will recur—be reached or exceeded—in a given year. It is determined by:

$$P = 100 \times m/(n + 1)$$

   where *n* is the number of years for which records were kept, and *m* is the relative magnitude of the event. For example, for a stage that ranked 30th in magnitude for a period of 60 years, $P = 100 \times 30/(60 + 1) = 49.2\%$.

   (2)   Plot the data in Table 8.2 on the graph paper in Figure 8.15. Be sure to use the scale on the top of the graph, labeled *Percent probability of recurrence (P)* when plotting your numbers. Then draw a best-fit straight line through the data points to complete the flood-frequency curve. The graph paper is called probability paper, because it can be used to determine the probability, or likelihood, that the event plotted on it will occur. In the example above, there is a 49.2% probability that a stage equal to that reached during the $m = 30$ event will be reached in any given year.

   c.   The **recurrence interval,** or average *time* interval between similar-sized floods, is another way of describing recurrence. The recurrence interval, *RI,* is given by:

$$RI = (n + 1)/m$$

   and *RI = 100/P.* Values for the recurrence interval are shown on the lower part of the graph in Figure 8.15. Note that a 49.2% probability of recurrence also means that a stage of that value would be reached, on average, about every two years. The probability of a 50-year flood (one with a recurrence interval of 50 years) in any given year is 0.02, or 2%. Thus, there is always a probability of a 50-year flood, even if one had occurred just the previous year. Conversely, there is no guarantee that a 50-year flood will occur every 50 years.

   The maximum stage of the summer of 1993 flood was 33.52 feet. Using your flood-frequency curve and best-fit straight line, determine what the recurrence interval was for that flood at this location.

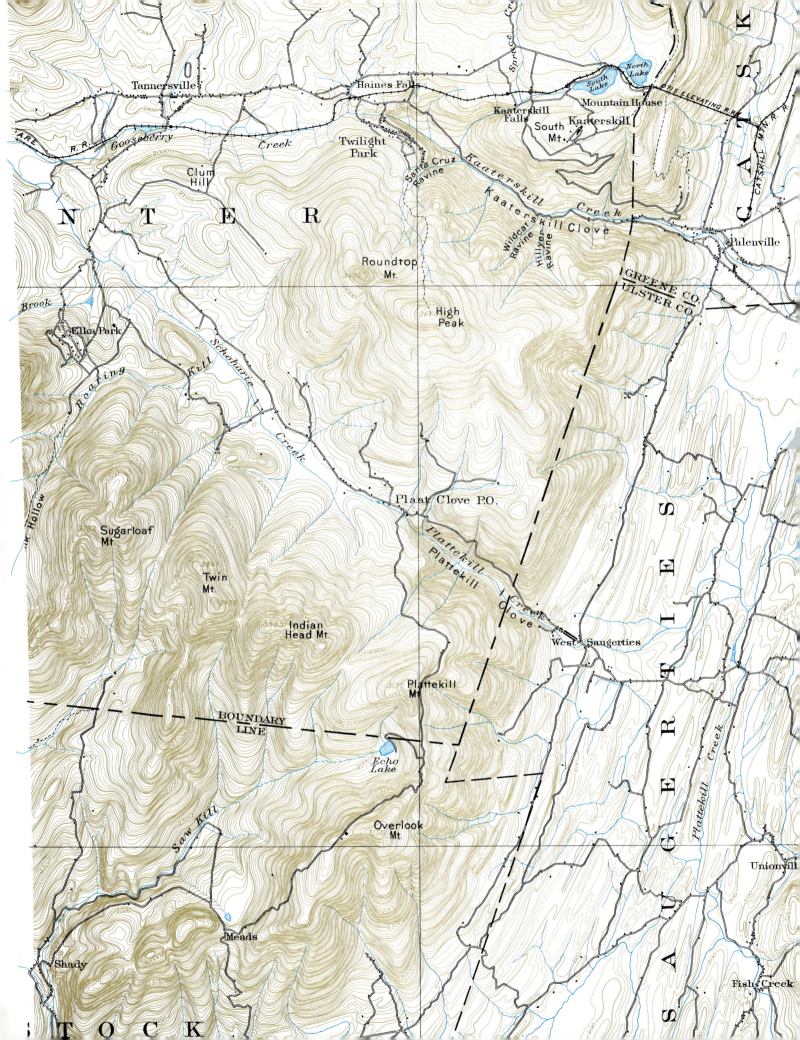

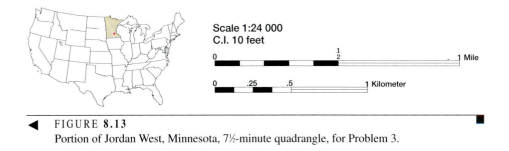

Scale 1:24 000
C.I. 10 feet

0   1/2   . 1 Mile

0   .25   .5   1 Kilometer

◀ FIGURE **8.13**
Portion of Jordan West, Minnesota, 7½-minute quadrangle, for Problem 3.

d. Use the flood-frequency curve you have constructed (Fig. 8.15) to determine the maximum stages for 2-, 20-, and 100-year floods (those with recurrence intervals of 2, 20, and 100 years). From the stage values, calculate the elevation of each of these floods (by adding stage to 690 feet), and use different-color pencils to outline the 2-, 20-, and 100-year floodplains on the topographic map.

    The floodplains you have outlined can be used by government agencies for land-use planning, or by insurance companies to set their flood insurance rates. For example, 2-year floodplains should not be used for housing or other permanent types of development, and flood insurance rates would be prohibitive. People building in areas of the floodplain with longer recurrence intervals should know they can expect to be flooded sometime, and should be prepared to take that risk.

A   **A'**

FIGURE **8.14**
Graph for profile along line A-A¢, Figure 8.13. For Problem 3a.

Table 8.2

SELECTED DATA FROM THE GAGING STATION ON THE MINNESOTA RIVER NEAR JORDAN, MINNESOTA, 1935–1994
(SOURCE: USGS)

YEAR	MAXIMUM STAGE, FEET	MAGNITUDE, M	P
1936	22.43	24	
1942	14.36	47	
1946	18.22	39	
1952	28.31	4	
1958	12.30	54	
1959	8.37	58	
1964	16.15	45	
1965	34.37	1	
1969	32.85	3	
1975	23.71	17	
1976	11.19	56	
1984	27.54	6	
1985	25.05	14	
1986	26.30	7	
1990	20.23	31	

4. Go to *http://water.usgs.gov/* (or link to it through the author's homepage—see *Preface*) and link to *United States Surface-Water Data Retrieval* through *National Water Information System (NWIS)*. Click on a state of your choice, then locate a gaging station of interest by clicking on *a map of (state name)* or *a list of counties in (state name)*. If you retrieve the map of the state, choose the county you want and click within it. The outlined areas are major drainage basins. If you choose the county list, click on the county name. Either way, you will receive a list of gaging stations in the county and can choose the one you want by clicking on it. Note that there is a *help* function at the bottom of the page to help you with terms. Click on the gaging station you want and answer the following questions:

a. What is the name of the stream? In what county is the gaging station? What is the name of the drainage basin, and what is its area?

b. Click *current conditions data*, if that is an option: What is the current stream flow (discharge)? How does it differ from the long-term average on this date? How much has the stream flow varied within the past week? What is the current gage height, and how much has that varied in the past week? What is the current velocity?

c. Click on *map of region surrounding station*, and describe the location of the gaging station as best as you can so that someone could drive to it, or print a *Zoom In* and *Zoom Out* version of the map.

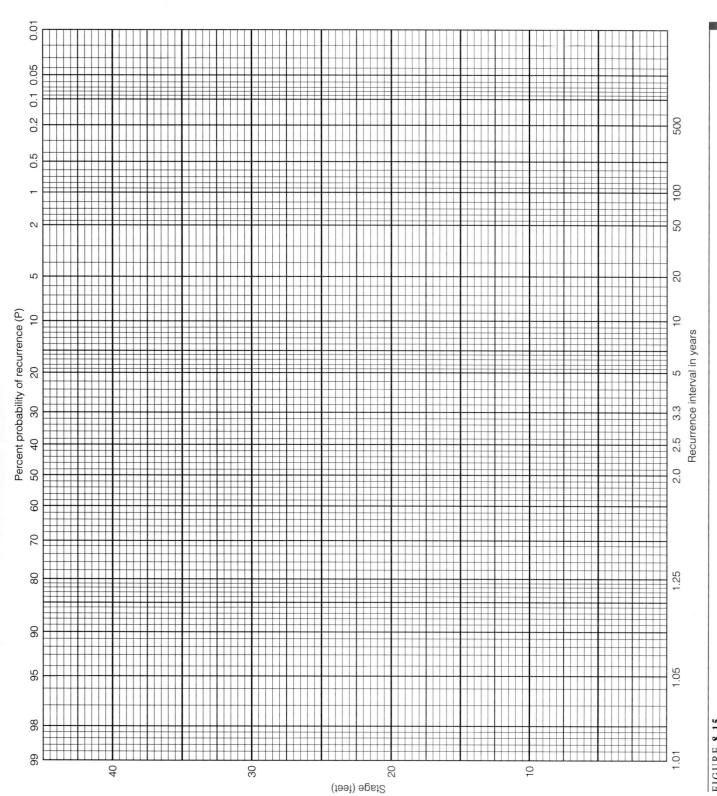

FIGURE **8.15**
Graph paper for Problems 3b and c.

d. Click on *Historical streamflow daily values,* choose *Graph,* then *Small Graph,* then *Retrieve Data.* It may take several minutes for the graph to appear. In which year or years was the discharge the highest? The lowest? Does discharge seem to vary from year to year in the same way (for example, is it always highest in the spring and lowest in the winter, or doesn't there seem to be a regular pattern)?

e. Click on *peak flow data,* then, under "data retrieved," select *only annual peaks,* and under "output format," select *tab delimited text data file,* and choose the *MM/DD/YYYY* "Date format." Click on *Retrieve Data.* This is the basic data from which *flood-frequency curves* are constructed and from which the *percent probability of recurrence* and *recurrence intervals* are calculated. Unless directed to do more by your instructor, you only need to use this list to find the year in which the peak discharge was the highest and the year in which it was the lowest.

IN GREATER DEPTH

5. Figure 8.16 is a portion of the Urne, Wisconsin, quadrangle to be used for the following problems. The Urne quadrangle is in the unglaciated or driftless area of Wisconsin. Note that little flat or gently sloping land can be found in the Alma quadrangle, except in the floodplain of the Mississippi River. This is one indication that the area has not been glaciated and that streams have been the principal eroding agents. If glaciers had covered the area, the topography would be more subdued; hilltops would have been eroded, valleys would have been wholly or partly filled with glacial drift, and surface drainage would not be as well developed.

a. On a separate piece of paper, draw a longitudinal profile from the mouth of Deer Creek to its head at the pond near the center of Section 3. First, find the length of the stream (or the profile) by following all the curves and measuring the actual distance along the channel; mark and label those points where an index contour crosses the stream. This can be done most easily by using the edge of a piece of paper. Use a vertical scale of 1 inch to 100 feet for the profile.

What is the vertical exaggeration? Show your calculations.

In what part of the stream is the gradient greatest? How do you know?

What is the shape of the profile? Is it concave or convex upward? Is it smooth, or are there abrupt changes in elevation such as might be caused by a waterfall?

Does the shape of the profile suggest that the stream has or has not attained a near balance between erosion and deposition, and how did you tell?

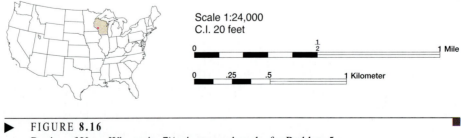

Scale 1:24,000
C.I. 20 feet

▶ FIGURE **8.16**
Portion of Urne, Wisconsin, 7½ minute quadrangle, for Problem 5.

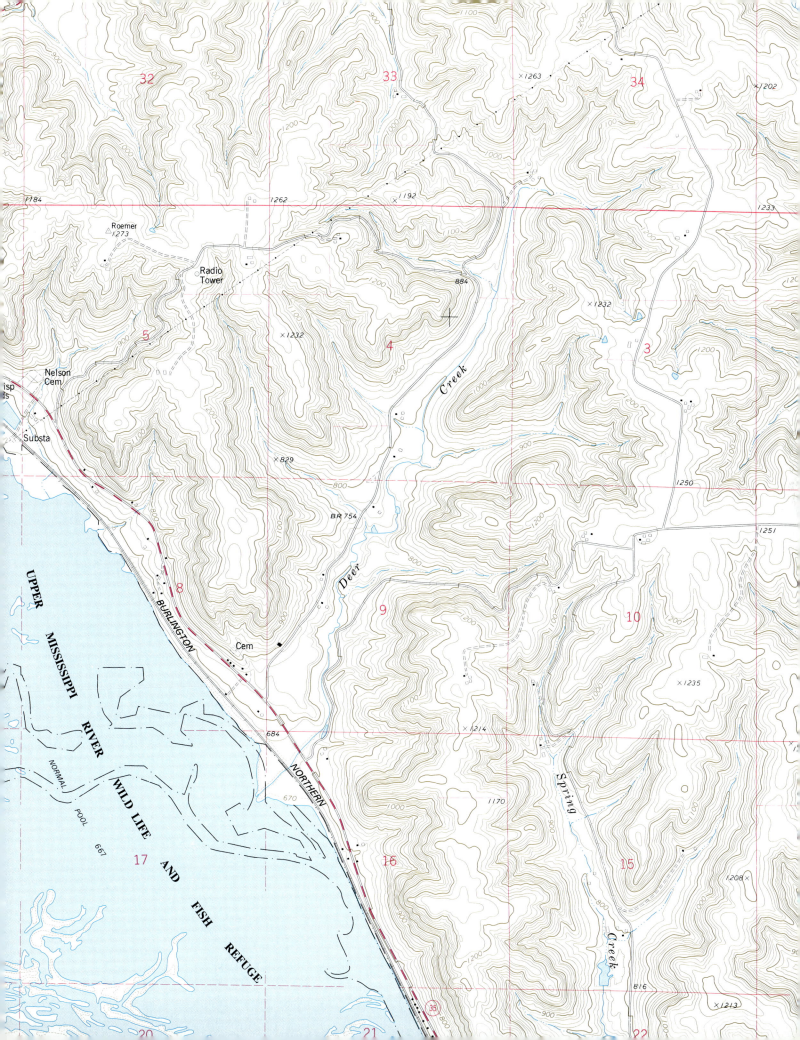

b. Outline the stream channels and drainage basin of Deer Creek and its unnamed tributaries. The boundary of a drainage basin is a drainage divide, the line of separation between adjacent drainage basins. It can be a ridge, summit, or tract of high ground. Water falling within the drainage basin will flow toward the mouth of the stream; water falling outside the drainage will flow into another stream system.

c. Estimate the area of the drainage basin in square miles by adding the number of whole sections and fractions of sections that fall between the divides (one section = 1 square mile).

Using your estimate, calculate the area of the drainage basin in square feet (1 mile = 5280 feet). Show your work.

d. The average annual precipitation for this area is 29.1 inches (2.4 feet). Calculate the average volume of water that falls on the drainage basin per year in cubic feet (volume = area × annual precipitation). Show your work.

e. The average volume of water that runs off and flows to the mouth of Deer Creek per year is the average annual discharge of Deer Creek. Calculate the average annual discharge, assuming the different conditions specified below:

(1) Infiltration loss, evaporation, and transpiration are all zero, and no groundwater is contributed from outside the drainage basin.

(2) Infiltration loss = 2 inches per year, evaporation + transpiration = 15 inches per year, and no groundwater comes from outside the drainage basin. Hint: What percent of the annual precipitation runs off? Show your calculations.

chapter

9

Groundwater

Overview

Groundwater is water beneath Earth's solid surface that occupies open spaces in rock or sediment. The permeability of the rock or sediment determines how easily water can move through it. The flow of groundwater enables us to continue to pump water from a hole in the ground without the well immediately running dry. It also enables us to contaminate our groundwater by carelessly disposing of wastes. In most rocks, water occupies cracks and tiny open spaces between individual grains or crystals. However, in carbonate rocks, the slightly acidic groundwater can dissolve the rock, forming large, water-filled openings. Problems for this chapter show how the shape of the water table can be determined in areas where lakes are abundant, how the flow direction of groundwater can be determined and used to predict paths of pollutants, how confined and unconfined aquifers differ, and how the surface topography in areas underlain by carbonate rocks is shaped by the action of groundwater.

Materials Needed

Pencil, eraser • Calculator • Ruler

INTRODUCTION

Water below the surface of the Earth is called groundwater. It infiltrates from the surface, and is part of the water cycle, as illustrated in Chapter 8. Typically, surface water percolates downward through unconsolidated soil and sediment until it enters a zone in which all of the interconnected spaces between sediment grains are filled with water, as illustrated in Figure 9.1. Near the surface, in the **unsaturated zone,** or **zone of aeration,** these *pore spaces* are mostly filled with air. Below the **water table,** interconnected pore spaces are filled with water in the **zone of saturation.** In humid regions, the elevation of the water table generally rises below hills and falls below valleys; its shape is commonly a subdued replica of the land surface.

OCCURRENCE AND MOVEMENT OF GROUNDWATER

Porosity

The **porosity** (P), or percentage of void space in the rock or sediment, determines the volume of water in the zone of saturation. In most rocks or sediments, the only pore spaces are those between adjoining particles or the narrow openings along fractures. In carbonate rocks and some lava flows, however, there may be large open spaces, and water may actually flow as underground rivers.

Permeability and Flow

The rate or velocity at which groundwater moves or flows is determined by the **permeability** (ease of conducting water) and the **hydraulic gradient,** or the slope down

133

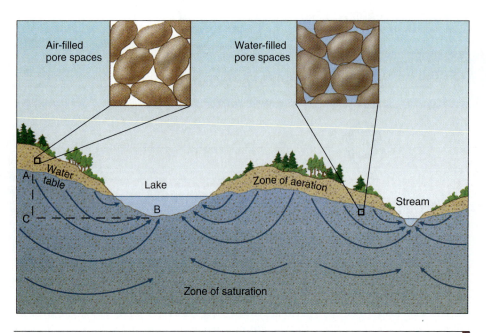

Cross section illustrating zones of aeration and saturation, the water table, and flow paths of groundwater in an unconfined aquifer. Water going from *A* to *B* has a hydraulic gradient equal to *A-C/C-B*. Exploded views show that pores in the zone of aeration are filled with air, whereas those in the zone of saturation are filled with water.

which it flows. Rock or sediment with large, interconnected pores has high permeability; if few pores are connected, permeability is low. Groundwater will flow faster if it moves down a steeper slope through a highly permeable material with a low porosity, a relation expressed by a form of Darcy's law:

$$v = \left(\frac{K}{P}\right)\left(\frac{h}{l}\right)$$

where *v* is flow velocity (also known as *seepage velocity*); *K* is the *hydraulic conductivity,* which depends on the nature of the medium and is directly related to permeability; *P* is the effective porosity; *h* is the head, or the difference in elevation between the top and bottom of the flow path; and *l* is the length or horizontal distance of flow. The *hydraulic gradient* is *h/l*. Figure 9.1 shows a common situation in a humid area in which water flows underground, following a curved path, and discharges into a stream or lake. Water going from point A to point B, Figure 9.1, drops a distance *A-C* and travels a horizontal distance *C-B: A-C* is the head, *h*, and *C-B* is the length, *l*, in Darcy's law.

Aquifers

An **aquifer** is a porous and permeable sediment or rock from which a useful amount of groundwater can be obtained. Sand and gravel, sandstone, limestone, dolomite, basalt flows, and some fractured granites and metamorphic rocks are good aquifers.

An **unconfined aquifer,** like the one in Figure 9.1, is one in which movement of water is unrestricted, and the aquifer is recharged by water percolating downward from the overlying surface. Sand and gravel make excellent unconfined aquifers.

A **confined aquifer** is overlain by, or sandwiched between, comparatively impermeable **confining layers,** as shown in Figure 9.2. Common confining layers are clay and shale. Confined aquifers are recharged where they are exposed at the surface (the *recharge area*), and by whatever water can leak through the confining beds.

Springs

Springs occur where groundwater comes to the surface. This may happen where fractures, faults, caves, or a contact between a permeable layer and underlying impermeable layer intersect the surface.

EXAMPLE

If, in Figure 9.1, the head (*A-C*) is 75 m, the horizontal distance traveled (*C-B*) is 1.2 km, the hydraulic conductivity of the sediment through which the water moved is 3.0×10^{-3} cm/sec, and the porosity is 20 percent, calculate the velocity of the water discharging into the stream in cm/sec.

Because the answer should be in cm/sec, first convert *A-C* and *C-B* to centimeters: 75 m = 75 m × 100 cm/m = 7500 cm = 7.5×10^3 cm; 1.2 km = 1.2 km × 1000 m/km × 100 cm/m = 120,000 cm = 1.2×10^5 cm. Then use Darcy's Law, where *A-C* = *h*, *C-B* = *l*, the hydraulic conductivity is *K*, porosity is *P*, and the velocity is *v*:

$$v = (K/P)\,(h/l)$$
$$v = (3 \times 10^{-3} \text{ cm/sec}/0.2)(7.5 \times 10^3 \text{ cm}/1.2 \text{ cm} \times 10^5 \text{ cm})$$
$$v = 9.4 \times 10^{-4} \text{ cm/sec}$$

What would the velocity be in centimeters per day? First determine the number of seconds in a day. One day has 24 hours, each hour has 60 minutes, and each minute has 60 seconds. Therefore, there are 24 hours/day × 60 minutes/hour × 60 sec/minute = 86,400 sec/day = 8.64×10^4 sec/day. The velocity per day is, $v = 9.4 \times 10^{-4}$ cm/sec × 8.64×10^4 sec/day = 81 cm/day.

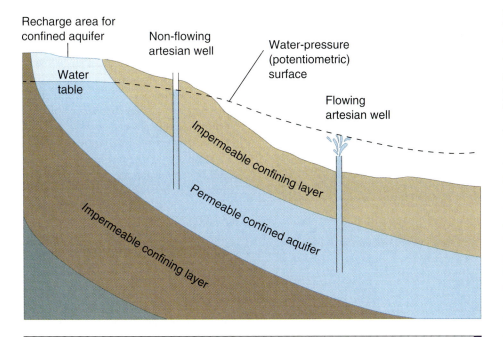

FIGURE 9.2
Cross section illustrating a confined aquifer system with recharge area, water-pressure surface, and flowing and non-flowing artesian wells.

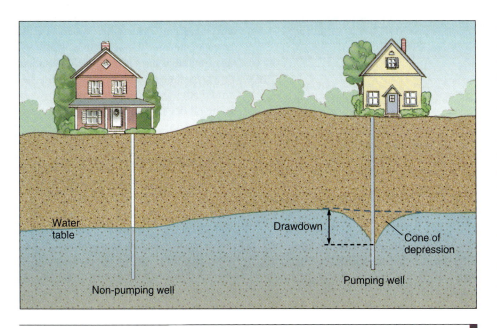

FIGURE 9.3
Cross section illustrating a non-pumping and pumping well in a simple unconfined aquifer. Pumping causes the water table to be drawn downward, forming a cone of depression.

HUMAN USE

We make use of groundwater by drilling wells and pumping it to the surface. We also contaminate groundwater by carelessly disposing of contaminants in a way that allows them to enter the groundwater system.

Wells

A well is a hole dug or drilled into the ground for the purpose of obtaining groundwater (Fig. 9.3). Wells may penetrate unconfined or confined aquifers. In either case, water from the aquifer flows toward the well, an area of low relative pressure. If the aquifer is unconfined, the hole will fill with water up to the level of the water table. If the well is in a confined aquifer, water will generally rise above the level of the aquifer, forming an **artesian well** (see Fig. 9.2).

Water is extracted from most wells by pumping. Pumping causes the level of the water and the pressure in the well to drop, allowing more water to flow into the well. If the aquifer is unconfined, continuous pumping causes a **cone of depression** to develop on the water table around the well, as shown in Figure 9.3.

If the amount of water withdrawn from an unconfined aquifer exceeds the amount added by recharge, the water table throughout the region will drop. Excessive withdrawal from confined aquifers will lower the **potentiometric,** or **water-pressure, surface,** the level to which water will rise in the well without pumping (Fig. 9.2).

In extreme cases, extraction of water from pore spaces in the aquifer may cause the pore spaces to collapse. As a result, the aquifer may be permanently damaged. In some cases, porosity in the aquifer or in confining beds decreases so much that the land surface subsides. For example, so much water has been withdrawn from the lake sediment on which Mexico City is built that the opera house has settled 3 m, and parts of the city have subsided as much as 8 m.

Fluids with Different Densities

Lighter, less dense fluids tend to float on heavier, more dense fluids, whether in a mixing bowl in the kitchen or in a permeable rock below the surface. Four natural

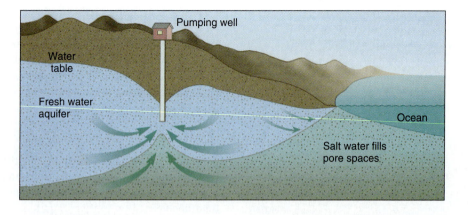

FIGURE 9.4
Cross section illustrating saltwater encroachment caused by excessive withdrawal of freshwater.

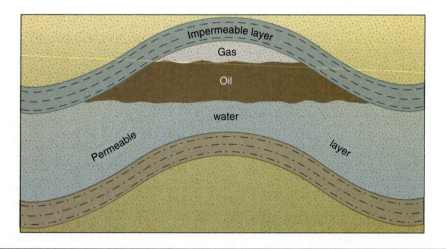

FIGURE 9.5
Oil and gas rise in permeable rocks until they encounter an impermeable layer and are trapped, as illustrated in this cross section.

fluids, listed in order of decreasing density, are saltwater, freshwater, oil, and natural gas—if all four are present, saltwater will be on the bottom, gas on the top.

Saltwater fills the pore spaces in sediment below the oceans, and some continental aquifers contain saline water. In coastal areas, saline and fresh groundwater come in contact, as shown in Figure 9.4. Because saltwater is more dense, it underlies freshwater in the groundwater system. Active removal of freshwater by pumping will change the position of the saltwater interface and may cause infiltration of salt-water into some wells, a problem known as saltwater encroachment.

Oil and natural gas, being insoluble in water and lighter than either saline or fresh groundwater, rise through permeable rocks until they reach an impermeable barrier. If that barrier cannot be surmounted, the gas and/or oil is trapped, and an oil or gas pool will form. As shown in a simple example in Figure 9.5, when water, oil, and gas are all present, gas, being the lightest, rises to the top, while water sinks to the bottom and oil settles in the middle.

Contamination

Contamination of groundwater can be a severe problem because it is generally not detected until it affects a well. Once contaminated, an aquifer is difficult and expensive to purify. Figure 9.6 illustrates some of the ways in which groundwater can be contaminated by human activities. Many contaminants are somewhat soluble in water, allowing them to disperse and making it difficult or impossible to recover them. In some cases, it is possible to dispose of such materials by injecting them into deep, permeable rocks containing dense saltwater that is isolated from freshwater aquifers. Insoluble contaminants may segregate within the system on the basis of density. For example, waste oil disposed of at the surface would percolate down to the top of the water table, but would not mix with underlying freshwater. If the oil does not migrate very far laterally, it may be possible to reclaim it.

GROUNDWATER IN CARBONATE ROCKS

Rocks made of carbonate minerals are soluble in acid. Limestone and calcite marble are particularly susceptible to dissolution. Groundwater is slightly acidic, because it contains dissolved carbon dioxide derived from the atmosphere or from decay of

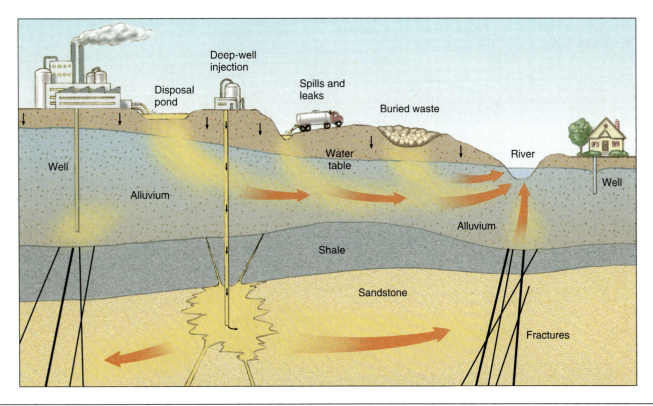

FIGURE 9.6

This cross section illustrates a number of ways in which groundwater can be contaminated by human activity.

SOURCE: DATA FROM USGS CIRC. 875.

organic matter in soil. Water combined with dissolved carbon dioxide makes carbonic acid (H_2CO_3), which then acts on carbonate minerals. The overall reaction is:

$$H_2O + CO_2 + CaCO_3 = Ca^{+2} + 2HCO_3^{-1}$$

As a result, calcium (Ca^{+2}) and bicarbonate (HCO_3^{-1}) ions are carried away in groundwater, and cavities form in carbonate rocks.

Caves, Disappearing Streams, and Sinkholes

The largest subsurface cavities are **caves,** which commonly develop as groundwater flows along sedimentary layers or fractures. Where dissolution occurs along horizontal surfaces, such as sedimentary layers, caves have dendritic or braided patterns when viewed on a map. Where fractures control water flow, a more regular pattern is common, with chambers intersecting at 90°. Most caves form at or below the water table, but they can also form above it. Caves serve as channels for groundwater flow and act as underground rivers. **Disappearing streams** are surface streams that plunge underground to become underground rivers. **Sinkholes** are depressions formed by dissolution of carbonate rock beneath the overlying soil or by the collapse of portions of near-surface caves.

Karst Topography

Areas underlain by carbonate rocks, in which dissolution has been active, have a distinctive surface appearance known as **karst topography.** Figure 9.7 shows features of karst topography in various stages of development. In the early stages, scattered sinkholes appear, and the surface drainage is disrupted as some streams disappear underground. Continued cave collapse leads to the presence of valleys, which form as sinkholes enlarge and join. Eventually, all that may remain are haystack-shaped hills consisting of rock that has not dissolved.

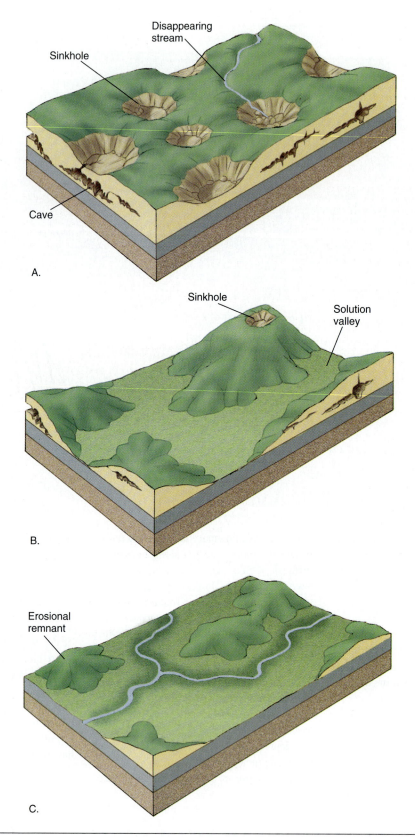

Disappearing
stream

Sinkhole

Cave

A.

Sinkhole

Solution
valley

B.

Erosional
remnant

C.

FIGURE 9.7

The evolution of karst topography. A. Scattered sinkholes and disappearing streams characterize the early stage. B. Continued dissolution and collapse of caves leads to the formation of valleys. C. Eventually, all that remains are scattered limestone hills.

BECAUSE GROUNDWATER AND THE ROCK OR SEDIMENT THROUGH WHICH IT MOVES ARE NOT DIRECTLY VISIBLE, OUR ABILITY TO UNDERSTAND AND PREDICT ITS BEHAVIOR IS STRONGLY DEPENDENT ON MODELS OR HYPOTHESES. THESE MODELS CAN BE TESTED BY VARIOUS MEANS. WHERE CARBONATE ROCKS HAVE BEEN DISSOLVED BY GROUNDWATER, DIRECT OBSERVATIONS CAN BE MADE IN CAVES AND AT LARGE SPRINGS. IN MOST AREAS, HOWEVER, DATA ARE OBTAINED FROM WELLS, WHICH ARE SAMPLES OF A VERY SMALL VOLUME. IN THIS LAB, YOU WILL LEARN SOME OF THE METHODS USED BY HYDROGEOLOGISTS TO INTERPRET AND PREDICT THE MOVEMENT OF GROUNDWATER, AND ITS RELATION TO THE ROCK OR SEDIMENT THAT CONTAINS OR CONFINES IT.

OBJECTIVES

If you complete all the problems, you should be able to:

1. Recognize distinctive karst topography and the following features on a topographic map: sinkhole, disappearing stream, solution valley.

2. Use water-table contours and flow lines to predict groundwater flow.

3. Use water-level elevations, whether in lakes or wells, to contour the water table or draw it on a cross section.

4. Explain how water wells in confined and unconfined aquifers differ.

5. Use water-table and water-pressure-surface information to determine how deep a well must be drilled to encounter water from confined and unconfined aquifers.

6. Determine a hydraulic gradient.

7. Predict how fluids of different densities would behave in a groundwater system.

PROBLEMS

1. Mammoth Cave, in Kentucky, is one of the best-known caves in the United States—and the largest, with more than 560 km (348 miles) of passageways. The entrance to the cave is near the northeast corner of the topographic map in Figure 9.8, where red lines mark roads. Three areas with different topography are evident in this map. In the northeast, a series of ridges, some with nearly flat tops, is dissected by valleys without streams. In the northwest, the valleys are not as deep, but they do contain streams. Both areas extend south to a sharp topographic break, known as the Dripping Springs Escarpment, which is just north of the Louisville and Nashville Road shown on the map. The third area, south of the escarpment, is characterized by numerous depressions.

   The area is underlain by layers of sedimentary rock that are inclined slightly to the northwest. Rocks are of two kinds: (1) sandstone (with some interbedded shale), which is essentially insoluble in groundwater; and (2) limestone, which is soluble in groundwater.

   a. Outline on the map those areas with abundant depression contours, whether they are in the north or the south. In which of the two types of rock, sandstone or limestone, do the depressions most likely occur?

   b. The two types of rock, sandstone and limestone, occur as two layers, one on top of the other. Which of the two is on top? How did you tell?

c. Why are there no streams in the northeast part of the map?

d. What evidence suggests that there once were streams in the northeast?

e. Why are there streams in the northwest?

f. In what direction does Gardner Creek (southeast corner of map) flow? What happens to it?

2. Figure 9.9 shows the flow pattern of groundwater in the Mammoth Cave area (red lines) and contours of the water-pressure or potentiometric surface (called water-level contours in the Legend).

a. What relation is there between the flow lines and the contours (that is, are they parallel, perpendicular, or is there no relation)?

b. The black dot on the northwest side of Cave City marks the location of a cyanide spill that occurred along Interstate Highway 65 (not shown on either Figure 9.8 or 9.9) in 1980. The spill was cleaned up quickly, but what if it had not been? Would the cyanide have found its way into the underground streams of Mammoth Cave? Explain. Could the cyanide have reached the Green River? If so, where?

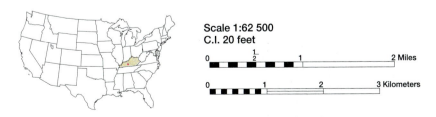

◄ FIGURE 9.8
Portion of Mammoth Cave, Kentucky, 15-minute quadrangle, for Problem 1.

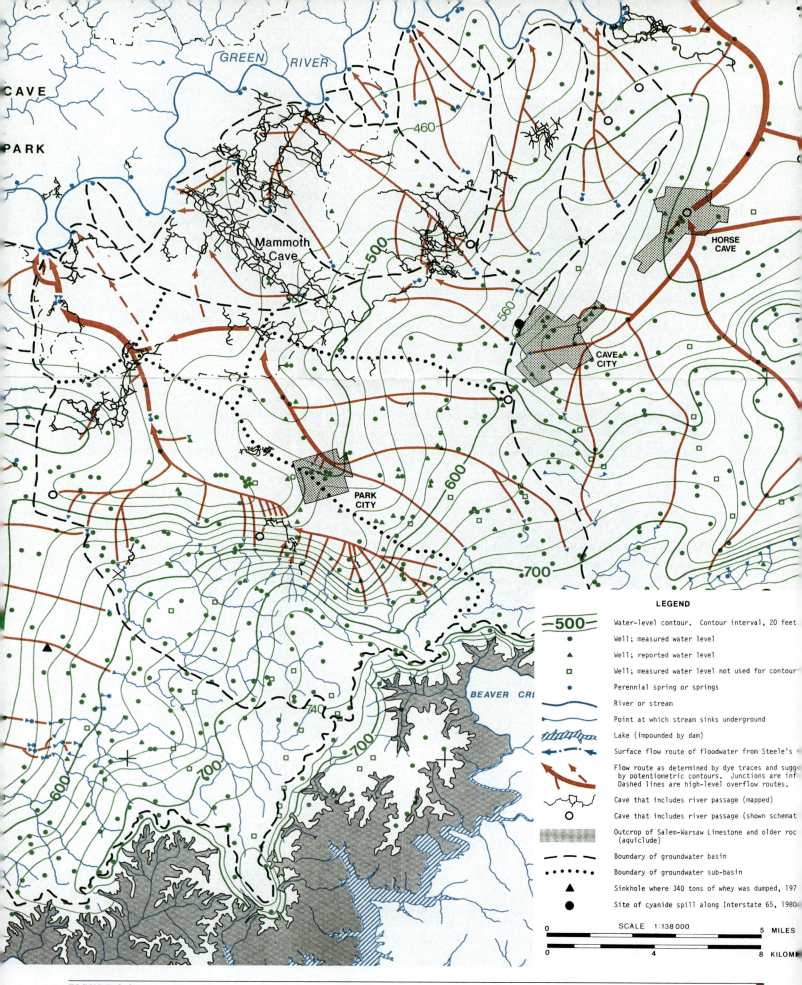

LEGEND

—500—	Water-level contour. Contour interval, 20 feet
●	Well; measured water level
▲	Well; reported water level
▢	Well; measured water level not used for contour
●	Perennial spring or springs
~~	River or stream
▷	Point at which stream sinks underground
~~	Lake (impounded by dam)
◄·—·►	Surface flow route of floodwater from Steele's
➤	Flow route as determined by dye traces and suggested by potentiometric contours. Junctions are inferred. Dashed lines are high-level overflow routes.
∿	Cave that includes river passage (mapped)
○	Cave that includes river passage (shown schematically)
▒	Outcrop of Salem-Warsaw Limestone and older rock (aquiclude)
– – –	Boundary of groundwater basin
· · · ·	Boundary of groundwater sub-basin
▲	Sinkhole where 340 tons of whey was dumped, 197
●	Site of cyanide spill along Interstate 65, 1980

SCALE 1:138 000

0 _____ 5 MILES

0 _____ 8 KILOM

FIGURE 9.9
Portion of *Groundwater Basins in the Mammoth Cave Region, Kentucky*, map for Problem 2. *Quinlan and Ray, 1981,* revised 1989.

c. A creek disappears at point X on the map (west-southwest of Park City). Dye placed in the creek just before it disappears can be used to determine where the water reappears on the surface.

    (1)   Predict where the dye would reappear at the surface, and mark those points on the map.

    (2)   What is the hydraulic gradient in feet per mile between the point where the creek disappears and the first place its waters reappear? Hint: Use the water-level contours to calculate *h* and the map scale to measure *l*. Show your calculations.

    (3)   If the dye took 152 hours to travel the distance in (2), what was its average velocity in feet/hour?

3.   Figure 9.10 is part of the Lakeside, Nebraska, quadrangle, and is part of the Sand Hills area of western Nebraska, an area composed of old sand dunes that now are mostly stabilized by vegetation. The dunes themselves have irregular surfaces with many small depressions. This makes for a complex, "busy" contour pattern. Some of the low areas between the dunes contain small lakes. Because the sand is so permeable, these lakes form where the water table intersects the surface, and their elevation, printed on most lakes, is that of the water table.

    a.   Using the lake elevations in the same way you used spot elevations in Chapter 6, Problem 5, draw a contour map of the water table. Use a contour interval of 5 feet, starting with the 3900-foot contour on the east side of the map.

    b.   In what direction does the water table slope, and what is the average gradient in feet per mile?

    c.   Assuming the gradient just determined is the hydraulic gradient for groundwater entering the system at Ellsworth and discharging near Lakeside, calculate the velocity, in cm/sec, at which it would flow between these points. Assume a *hydraulic conductivity, K,* of 1×10^{-3} cm/sec and a *porosity, P,* of 20%. Hint: Use Darcy's Law, but first convert your hydraulic gradient from units of feet/mile to feet/foot; once this is done, the hydraulic gradient will be a unitless number.

    d.   If the distance between Ellsworth and Lakeside is 7 miles (11.3 km = 1.13×10^5 cm), how long would it take for the groundwater to make this journey?

    e.   Note the windmill in the SW1/4, Sec.22, T24N, R44W (southeast of Lakeside) and its map symbol; the surface elevation is also given. Many other windmills are shown on the map by symbol only. The windmills are connected to water wells and used to pump groundwater to fill water tanks for cattle. Locate the windmill in the NW1/4, Sec.19, T24N, R43W. How high must the water be raised by the pump to fill the stock tank? Explain your answers.

4. Go to *http://water.usgs.gov/public/wrd002.html* (or link to it through the author's home page—see Preface), and select *Wisconsin* (at the time of writing, it was the only state that had the information needed for this question; try other states if you wish, but the path to the well data may be different than described here) by clicking on it in the list of states. Choose *Groundwater Data*, then select one of the wells, such as *MT-0007* in Marinette County.(north-eastern Wisconsin).

a. How deep is the well?

What is the elevation at the top of the well, at the land surface?

b. What type of rock or sediment makes up the aquifer?

c. What is the current depth to water in the well?

Is this the same as the depth to the top of the water table?

d. How much has the depth to water varied in the past year?

e. For how many years has a record been kept of the depth to water in the well?

What is the highest level for the period of record? The lowest?

What is the range in depth to water in the well for the period of record?

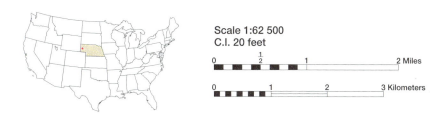

Scale 1:62 500
C.I. 20 feet

◄ FIGURE **9.10**
Portion of Lakeside, Nebraska, 15-minute quadrangle, for Problem 3.

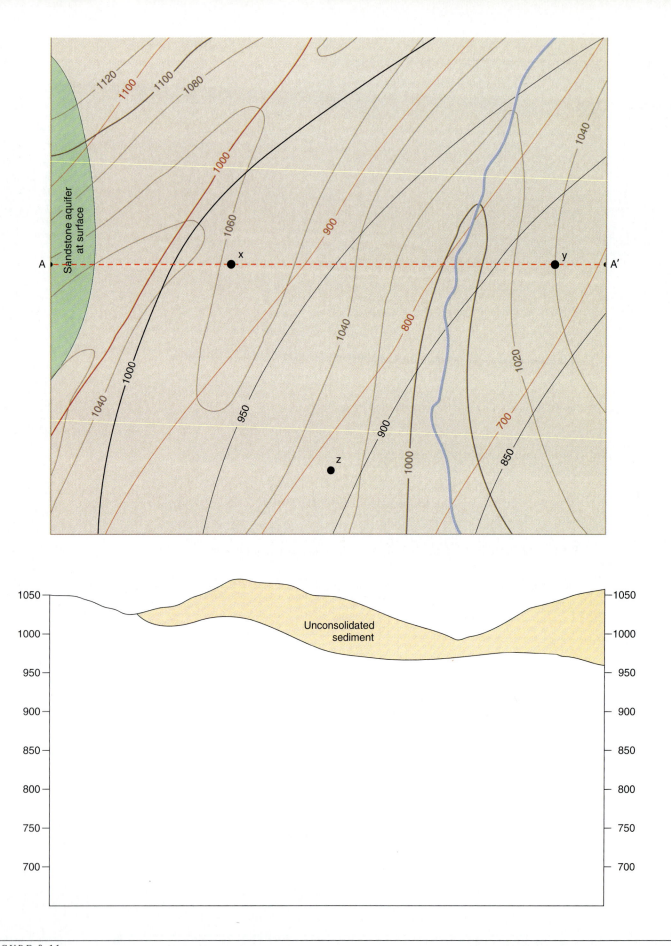

FIGURE 9.11

Top: Map for Problem 5 shows surface contours in brown, contours on the top of a sandstone aquifer in red, and contours of the water-pressure surface of the sandstone aquifer in black. *Bottom:* Cross section along line A-A′.

IN GREATER DEPTH

5. Refer to the top of Figure 9.11, a map with three types of contours. Regular surface contours are shown in brown. The top of an important sandstone aquifer is contoured with red lines. The water-pressure surface (potentiometric surface) is shown in black. The stream (in blue) flows throughout the year and is recharged from both surface and groundwater.

   a. At the bottom of Figure 9.11 is a cross section from points A to A′ on the map. It shows the surface topography along that line and the unconsolidated sediment that covers most of the area. Although not shown on the cross section, the bedrock beneath the unconsolidated sediment is shale, and below the shale is the sandstone aquifer. Complete the cross section by drawing the top of the sandstone aquifer and the water-pressure surface. To do this, use the intersections between the contours and line A-A′ in the same way you would to draw a topographic profile.

   b. Shallow wells in the unconsolidated sediment at points X and Y fill with water to within 10 feet of the surface. Using this information, mark the positions of X and Y on the cross section and sketch the probable position of the water table in the unconsolidated sediment.

   c. Assume that the shale bedrock below the unconsolidated sediment and above the sandstone aquifer is impermeable, meaning that no connection exists between the water in the sandstone aquifer and the near-surface water. How deep would a well have to be drilled (that is, how far below the surface) to intersect the sandstone aquifer at X? At Y? At Z?

   Assuming the wells were cased (lined so water cannot enter the hole) where they pass through the unconsolidated sediment, how high (to what elevation) would water from the sandstone aquifer rise in each of the three wells?

   Are these deeper wells flowing artesian wells, non-flowing artesian wells, or non-artesian wells? Explain.

   d. In the late 1940s and early 1950s, a small chemical plant was located near point X on the map. One of their waste by-products was an orange chemical. At first they disposed of it by drilling a shallow well into the unconsolidated sediment, but they were soon caught because the orange chemical reappeared. Where do you think it reappeared? Mark the general location on the map. Explain your answer.

The chemical company then drilled a 150-foot disposal well. At first they made a heavier-than-water mixture by adding a heavy liquid to the orange chemical, but after a few years, that fluid was detected. How do you think it was detected?

Next, they mixed the chemical with oil to make a lighter-than-water fluid. What do you think happened to the waste this time? Explain your answer.

chapter

10
Glaciation

Overview

Glaciers appear where temperatures are low and enough snow is accumulated and compacted over the winter that the resulting ice survives the summer. Alpine glaciers flow down mountain valleys and carve a distinctive landscape. Continental glaciers are great sheets of ice in polar regions that bury nearly the entire landscape. Glaciers erode, transport, and deposit great quantities of sediment. Continental glaciers covered nearly 30 percent of Earth's land surface during the Pleistocene Epoch, as evidenced by their deposits. In this lab, you will learn to recognize and analyze erosional and depositional features of glaciers on aerial photographs and topographic maps. You will also learn how measurements taken over a period of years can be used to monitor the activity of a glacier.

Materials Needed

Pencil and eraser • Colored pencils • Ruler • Calculator • Stereoscope (provided by instructor)

INTRODUCTION

Glaciers are slow-moving, thick masses of ice of two main types, *alpine and continental*. **Alpine glaciers** occur in mountains. They typically begin at high elevations and flow down preexisting stream valleys to lower elevations, where they melt, as shown in Figure 10.1. As ice moves downhill, it erodes the valley and imparts a character to it that is unmistakable: a glaciated valley is straighter, deeper, and wider than the original stream valley, and has a U-shaped, rather than V-shaped, cross profile. Should the ice melt and the glacier recede from the valley, evidence of its presence is etched into the landscape.

Continental glaciers are thick, broad ice sheets that spread out to cover virtually the entire landscape (Fig. 10.2). They may reach thicknesses of 4000 m or more, and are not confined by valley walls. Present-day continental glaciers cover most of Antarctica and Greenland, but during the Pleistocene Epoch of the Quaternary Period, approximately 2 million to 10,000 years ago, much of the land area in the northern hemisphere was covered.

Material eroded by glaciers is transported by the moving ice and by meltwater. Deposition occurs when the ice melts or the running water slows. The deposits of alpine glaciers, left in valleys or at valley mouths, are distinctive when young, but because they are left in areas of active erosion, they all but disappear within a few thousands of years, geologically a very short period of time. Continental glacier deposits, on the other hand, may be several hundred meters thick and impart a distinctive character to the landscape that remains for tens of thousands of years. Some glacier deposits have been converted to sedimentary rock and survived for more than two billion years.

Oblique aerial photograph of an alpine glacier near Mt. Spurr, Alaska, taken in late summer. The glacier starts in a cirque and flows down the valley to its terminus, which is marked by a prominent end moraine. The sides are marked by lateral moraines, and medial moraines occur within the glacier. The snow line is at the elevation where the medial moraines disappear, and it approximates the equilibrium line.

FIGURE **10.2**
Oblique aerial photograph of Antarctica, illustrating a continental glacier.

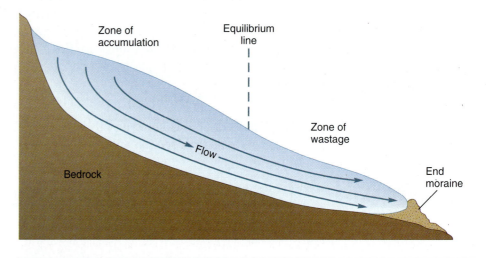

FIGURE **10.3**
Longitudinal profile of a valley glacier, illustrating zones of accumulation and wastage, equilibrium line, flow lines, and end moraine.

FORMATION, MOVEMENT, AND MASS BALANCE

If, as years pass, snow accumulates at high elevations or high latitudes, compaction and recrystallization will eventually change the lower parts of the snow mass to ice. When thick enough, the ice responds to gravity and begins to flow downhill. It continues downhill until it melts. Thus, ice accumulates at higher elevations and moves downhill until it wastes away.

Figure 10.3 is a longitudinal cross section of an alpine glacier, illustrating the behavior of an active glacier under steady climatic conditions. Snow accumulates in the **zone of accumulation,** is buried, converted to ice, and moves downhill along paths indicated by the flow lines. Buried ice eventually resurfaces in the **zone of** wastage (or **ablation**), where it evaporates or melts. The boundary between these two zones is the **equilibrium line.**

The **mass balance** of a glacier is the change in total mass of glacier ice during a year. It is the mass added in the zone of accumulation minus the mass lost in the zone of ablation. For a glacier in perfect static equilibrium, the mass balance is zero. As long as a mass balance of zero is maintained, the glacier will neither advance nor retreat. The equilibrium line of an alpine glacier will remain at an elevation about midway between that of the upper and lower end of the glacier. A *positive mass balance* would cause the equilibrium line to move to a lower elevation and an alpine glacier to **advance** down the valley. A continental glacier would advance outward from the center of ice accumulation. A *negative mass balance* would cause the equilibrium line to retreat to a higher elevation.

Horn   Horn   Aréte   Cirque

Hanging valley

FIGURE 10.4
Oblique aerial photograph of the eastern side of the Juneau icefield, British Columbia. The U-shaped valley in the foreground was carved by an alpine glacier. Cirque and small valley glaciers are still present in the background. Other features characteristic of alpine glaciation are labeled.

An alpine glacier **retreats** up the valley, and a continental glacier retreats by shrinking in size. Alpine glaciers are very sensitive to climate, and smaller ones respond quickly to climatic change. Equilibrium lines fluctuate in response to yearly climatic differences.

Advances and retreats of continental glaciers occur in response to climatic changes of longer duration, and the response time lags behind that of the climatic change. During the Pleistocene Epoch, climatic fluctuations caused continental glaciers to advance and retreat perhaps as many as 21 times. These climatic fluctuations are recorded best in deep ocean sediments, but also can be seen in glacial sediments deposited on land.

EROSIONAL LANDFORMS

A mountainous area that is or has been occupied by alpine glaciers has distinctive erosional landforms, as shown in Figure 10.4.

A **cirque** is an amphitheater-shaped depression high on a mountain, which either is or was filled with glacial ice. If it does not contain ice, the bottom may be the site of a bedrock-basin lake, a **tarn.** The upper end, or headwall, of a cirque is very steep, and erosion by frost and glacial action causes it to migrate headward with time. *Cirque glaciers* are no larger than the cirque itself, but for many *valley glaciers,* the cirque is simply the head of the glacier; cirque and valley glaciers are types of alpine glaciers.

Headward erosion of cirques from opposite sides of a ridge may eventually cause them to join and carve a low point, or *pass,* in the ridge called a **col.** The intersection of three or more cirques may result in the formation of an isolated high peak, or **horn.**

Valley glaciers reshape preexisting stream valleys with V-shaped cross profiles into **U-shaped valleys.** Depressions left in the upper part of such valleys may be occupied by several small lakes connected by a stream; when viewed from the air or on a map, they look like a string of beads and are called **paternoster lakes.** The divides or ridges between adjacent glaciated valleys are commonly sharp, jagged features known as **arêtes.** Small tributary glaciers are unable to erode as deeply as the large glaciers they join. As a result, the junction of such former glaciers is marked by a **hanging valley,** a U-shaped valley at a higher elevation than the one it joins, commonly with a waterfall. **Fiords** (or *fjords*), which are narrow, steep-sided inlets of the sea, are submerged valleys once occupied by valley glaciers.

Erosion by continental glaciers results in a generally rounded topography, but no large, distinctive features like those formed by alpine glaciers.

DEPOSITIONAL LANDFORMS

The material deposited in association with a glacier is called **glacial drift.** Two kinds of drift are **till** (unsorted, unstratified debris deposited directly from ice) and **stratified drift** (sorted and stratified debris deposited from glacial meltwater).

Moraines are landforms composed mostly of till. They form on or within a glacier, or are left behind when the ice melts. Figure 10.1 illustrates moraines on an alpine glacier that are named on the basis of their position: **lateral moraine** (ridge at the side of the glacier), **medial moraine** (ridge at or near the middle of the glacier), and **end moraine** (ridge at the end, or terminus, of the glacier). Continental glaciers also form end moraines (Fig. 10.5). **Ground moraine** is the uneven blanket of till between the other moraines.

Stratified drift deposits are most prominent at the end of the glacier, where they consist of **outwash**—sand and gravel washed out of the glacier by running water. Outwash deposited in a valley forms a **valley train.** An **outwash plain** forms where braided meltwater streams deposit sediment over a wide area. Outwash plains are common features in areas of continental glaciation, as illustrated in Figure 10.5, but also form where alpine glaciers flow out of a valley and spread out (see Fig. 10.1). Some outwash plains, formed during glacial retreat, contain abundant, undrained depressions called **kettles** or, if filled with water, **kettle lakes,** which form by melting of buried blocks of ice (Fig. 10.5). Deposits of stratified drift that form beneath or within a glacier include **eskers** and **kames,** as shown in Figure 10.5. An esker is a long, narrow, winding ridge formed by

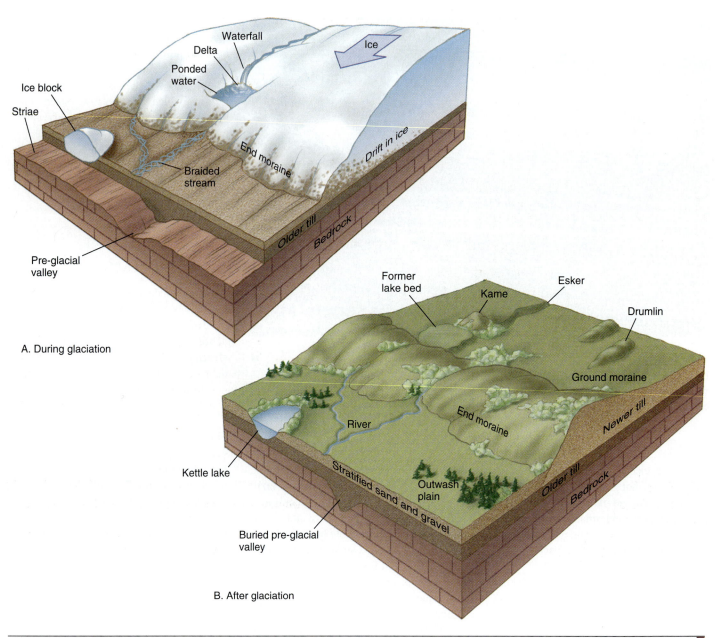

A. During glaciation

B. After glaciation

FIGURE **10.5**
A. Terminus of continental glacier. B. Same area several thousands of years after retreat.

deposition from a stream flowing within or at the base of the ice. A common type of kame is a steep-sided mound formed where meltwater flowed into a depression or hole in the ice. Eskers and kames are common features of continental glaciation in some areas, but are usually destroyed by erosion after alpine glaciation.

Figure 10.5 also illustrates **drumlins,** which are streamlined, elongated hills that are steeper on one end. Typical drumlins are 400 to 800 m long and 8 to 60 m high.

They form in swarms near the outer edge of continental glaciers and appear to have been molded from glacial drift by the advancing ice. The steeper end faces the direction from which the ice advanced.

The effects of continental glaciation extend beyond the ice itself. Abundant meltwater may pond to form large **proglacial lakes** on the perimeter of the melting glacier, as illustrated in Figure 10.6. Sedimentation in such lakes results in a flat surface, recognizable long after the water

has drained away. In some lakes, drainage occurred in stages, and *former shorelines* are left to record these stages.

Rivers draining these lakes may carve large valleys, which are later occupied by much smaller rivers. Such *underfit* streams are obviously too small to have carved the valley they now occupy. An example is the present-day Minnesota River, which occupies a broad valley cut by Glacial River Warren as it drained Glacial Lake Agassiz (Figs. 10.6 and 8.13).

FIGURE 10.6
Maximum extent of Glacial Lake Agassiz (purple area). Blue areas within Glacial Lake Agassiz are present-day lakes. Arrow indicates drainage of Glacial River Warren; the valley is now occupied by the Minnesota River.

SOURCE: DATA FROM ILLUSTRATION BY J.A. ELSON, IN MAYER-OAKES, *LIFE, LAND AND WATER*, 1967. UNIVERSITY OF MANITOBA PRESS. WINNIPEG, MANITOBA, CANADA.

APPLICATIONS

CONTINENTAL GLACIERS HAVE ONLY RECENTLY (GEOLOGICALLY SPEAKING) RETREATED FROM VAST LAND AREAS IN THE NORTHERN HEMISPHERE, AND THERE IS A POSSIBILITY THAT THEY COULD RETURN WITHIN TEN OR TWENTY THOUSAND YEARS. AN ALTERNATIVE POSSIBILITY IS THAT, WITH GLOBAL WARMING, EXISTING GLACIERS IN ANTARCTICA AND GREENLAND COULD MELT. EITHER OCCURRENCE WOULD HAVE A DEVASTATING IMPACT ON HUMANS, AND IT IS CLEARLY IN OUR BEST INTERESTS TO KNOW AS MUCH AS POSSIBLE ABOUT GLACIERS. IN THIS LAB, YOU WILL LEARN SOME OF THE SCIENTIFIC METHODS GEOLOGISTS USE TO STUDY ACTIVE GLACIERS, AND HOW IT IS POSSIBLE TO RECONSTRUCT THE HISTORY OF PAST GLACIERS.

OBJECTIVES

If you complete all the problems, you should be able to:

1. Determine from a photograph or historical measurements whether a glacier has advanced or retreated in the recent past.

2. Determine the flow direction of glacial ice from a map, photograph, or historical measurements, and given the necessary data, calculate the rate of movement or the distance moved during a given time span.

3. Recognize the following alpine-glacier erosional features on a photograph or topographic map: cirque, tarn, col, horn, U-shaped valley, paternoster lakes, arête, hanging valley, fiord.

4. Recognize the following depositional features of valley and continental glaciers on a photograph or topographic map: end moraine, lateral moraine, medial moraine, outwash plain, kettle or kettle lake, esker, kame, drumlin.

5. Interpret a map of glacier deposits.

PROBLEMS

1. Refer to the map of part of Glacier Bay National Park, southeastern Alaska, in Figure 10.7, and the stereopair of aerial photographs of North Crillon Glacier in Figure 10.8. As you can see in Figure 10.7, the two Crillon glaciers flow southwest, then bend 90° to flow in a northwest-trending valley, one to the northwest, the other to the southeast. Incidentally, the nothwest-trending valley

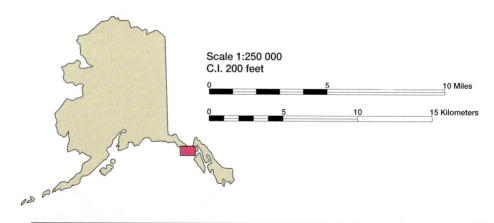

Scale 1:250 000
C.I. 200 feet

▶ FIGURE 10.7
Portion of Mt. Fairweather, Alaska/Canada, 1° × 2° quadrangle, including part of Glacier Bay National Park.

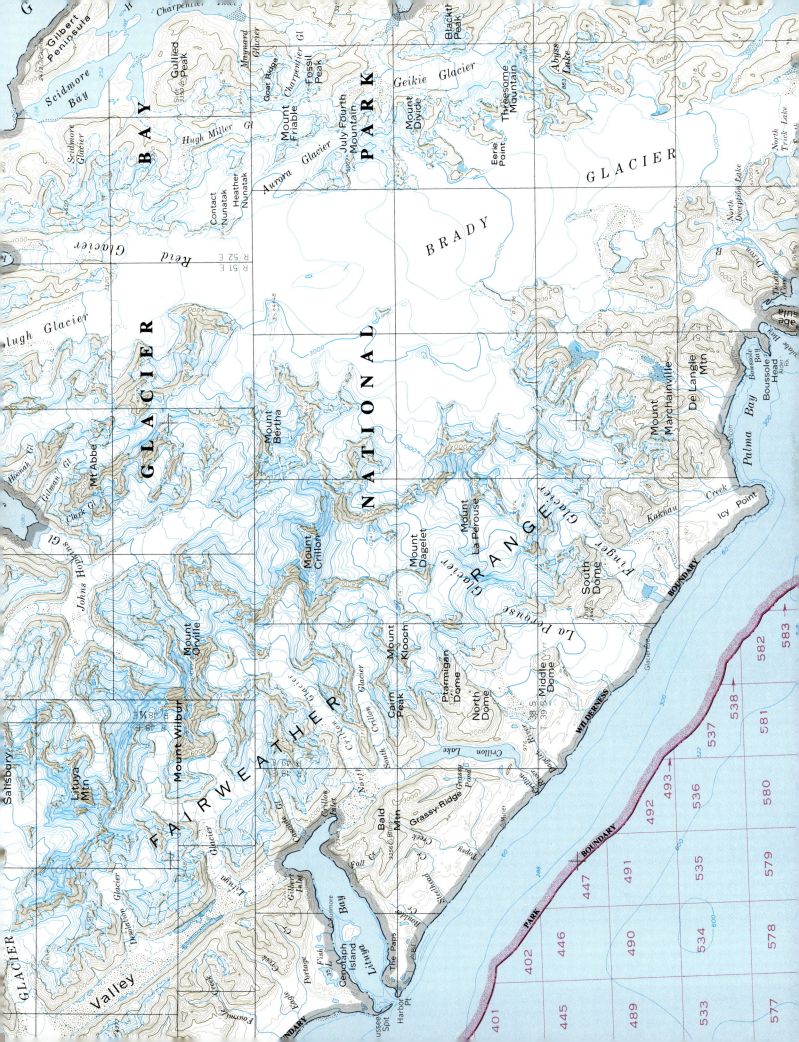

marks the position of one of the major geologic faults of southeastern Alaska, the Fairweather fault. This area is occasionally jarred by earthquakes caused by movement on the fault.

a. What evidence is there in Figure 10.7 to suggest that Lituya, North Crillon, and South Crillon glaciers were more extensive in the past?

Based on your evidence, mark their probable extent on the map.

What type of glacial feature is Lituya Bay?

What glacial feature do you think is at least partly responsible for the shallow water at the mouth of Lituya Bay?

b. The brown stippled areas on the glaciers in Figure 10.7 represent rock material on the surface of the ice. Label representative end, lateral, and medial moraines.

c. Look at Figure 10.8 with a stereoscope. North on these photos is to the upper left (compare with Fig. 10.7).
    (1) Examine the valley wall on the north side of North Crillon Glacier, and mark in yellow on the photo the height to which you think the glacier ice may once have extended.
    (2) Next, look at the small glacier north of North Crillon Glacier. What evidence is there that its ice has or has not been thicker in the past?

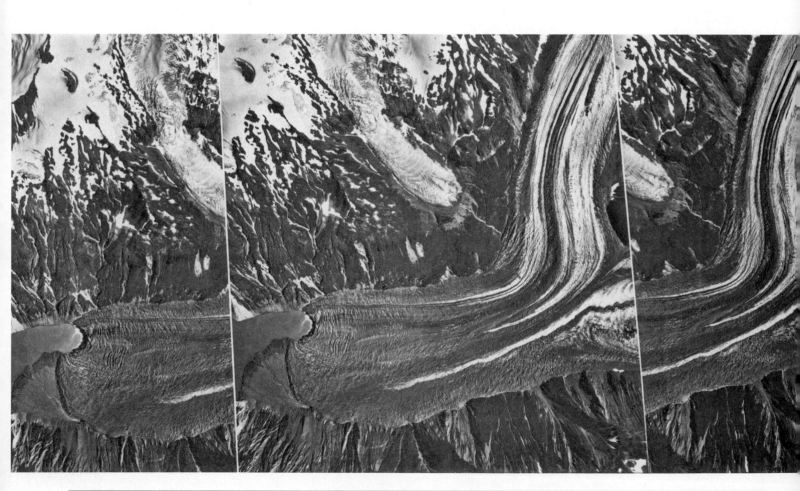

FIGURE **10.8**
Stereophotographs of North Crillon Glacier, Glacier Bay National Park (see Fig. 10.7 for location). Scale is 1:40,000.

Mark in green on the photo where you think the small glacier joined North Crillon Glacier when North Crillon was at its maximum thickness.

Note that the small glacier forms an *icefall* near the top of the photographs. Cite evidence to indicate that the small glacier may have extended farther down its valley, over an icefall, to join with North Crillon after North Crillon began to recede.

What geological term describes the valley occupied by the small glacier when compared to the valley of North Crillon Glacier?

(3)   Look at the end of the North Crillon Glacier. Explain why a big piece is missing from the end on the north side of the valley.

What feature is being formed in front of the glacier on the south side of the valley, and how does the material forming it get there?

(4)   What are the prominent dark lines on the upper end of North Crillon Glacier?

How many are there?

Assuming your number is correct, how many tributary glaciers flow into North Crillon?

(5)   Explain why the glacier is dark colored near its terminus.

(6)   Note the cracks, or *crevasses,* near the terminus of North Crillon Glacier. What is their orientation with respect to the direction of ice flow?

Do you think they form from tension (pulling apart) or compression (pushing together)? Explain your answer.

2. Glacier National Park, in northern Montana, is part of the Canadian–U.S. Waterton-Glacier International Peace Park. It sits astride the continental divide, which separates drainage to the Pacific from that to the Gulf of Mexico or to Hudson Bay. Present-day glaciers are small and occupy cirques or well-shaded areas. These particular glaciers, though always small, were probably at their maximum during the middle of the 19th century. At that time, there were some 150 small glaciers in the park area, more than twice as many as at present. During the period from about 1920 to the mid 1940s, wastage of glaciers was extensive; from that time to the present, most glaciers have retreated somewhat, although some have advanced. During the last major continental glaciation in North America, about 20,000 years ago, alpine glaciers were much more extensive in the park and occupied valleys tens of miles long. The spectacular erosional features that remain in the park reflect both earlier and recent glaciation. Figure 10.9 is a topographic map of one of the most visited parts of the park and includes the Many Glacier area.

Locate examples of the following features of alpine glaciation. Use a geographic name to describe the location (e.g., an example of a cirque glacier is Grinnell Glacier).

a. Cirque.

b. Col.

c. Horn.

d. U-shaped valley.

e. Arête.

f. Hanging valley.

g. Tarn.

h. Paternoster lakes.

Scale 1:100 000
C.I. 80 feet

▶ FIGURE 10.9
Portion of topographic map of Glacier National Park, Montana.

Devils Slide
Campground
Three Creek
Swiftcurrent
Boulder Ridge
Riser
Napi Point
Napi Rock
Singleshot Mountain
East Flattop Mountain
Going-to-the-Sun
Two Dog Flats
Mackinnon Bay
Two Dog Creek
MARY
Rising Sun
Campground
Picnic Area
Golden Stairs
Wild Goose Island
SAINT
Silver Dollar Beach
Otokomi Mtn
Rose Creek
Going-to-the-Sun Point
Cassidy Curve
LAKE SHERBURNE
Swiftcurrent Ridge Lake
Many Glacier Entrance Sta
Windy Creek
Cracker Flats
Cracker Creek
Canyon
Wynn Mtn
Allen Mtn
Cracker Mine
Cracker Lake
Rose Basin
Rose Creek
Goat Mtn
Goat Lake
Dead Horse Point
Patrol Cabin
Sunrift Gorge
Baring Creek
Baring Falls
Lost Lake
Kennedy Creek
Apikuni Mountain
Apikuni Falls
Apikuni Flat
Falling Leaf Lake
Boulder
Siyeh Pass
Matahpi Peak
Mt Siyeh
Otokomi Lake
Siyeh Glacier
Sexton Glacier
GLACIER
Siyeh Bend
Heavy Runner Mountain
Reynolds
Kennedy Lake
Crowfeet Mtn
Mt Henkel
Natahki L
Altyn Peak
Many Glacier
Swiftcurrent
Ranger Sta
Campground
Grinnell Pt Mine
Snow Moon L
Oastler Shelter
Hidden Falls
Stump L
Allen Creek
Cataract Mtn
Piegan Pass
Morning Eagle Falls
Siyeh Glacier
Piegan Mtn
Piegan Falls
Piegan Glacier
Pollock Mtn
Bishops Cap
Reynolds Mtn
Ptarmigan Tunnel
Ptarmigan Lake
Ptarmigan Falls
Ptarmigan Cr
Redrock Lake
Redrock Falls
Bullhead Lake
Fishercap Lake
Josephine Lake
Grinnell Lake
Gaging Sta
Grinnell Lake
Upper Grinnell L
Grinnell Falls
Grinnell Glacier
Angel Wing
The Salamander
Grinnell Glacier
Mt Gould
Feather Plume Falls
Cataract Creek
Haystack Butte
HIGHLINE TRAIL
THE GARDEN WALL
Big Drift
Logan Pass
Visitors Center
Triple Arches
Weeping Wall
Oberlin Falls
Hanging Gardens
Bird Woman Falls
Mt Oberlin
Logan Creek
Clements Mtn
Wall
Crystal Pt
Granite Park
THE CONTINENTAL DIVIDE
Mt Wilbur
Iceberg Lake
Iceberg Notch
Iceberg Peak
Swiftcurrent Mountain
Swiftcurrent Pass
Swiftcurrent Gl
Lookout
Chalet
Swiftcurrent Creek
North Swiftcurrent Gl
Ahern Peak
Ahern Pass
Ahern Glacier
Ahern Falls
Helen Lake
Ipasha Pk
Ipasha Lake
Ipasha Falls
Kennedy
Cosley
Wilbur
LEWIS
Bear Creek
Mt Cannon
Hidden Lake
Hidden Creek
Bearhat
Hidden Lake
HIGHLINE
The Loop
Packers Roost
Patrol Cabin
Red Rock Point
Red Rock Falls
Avalanche
Avalanche Campground
Logan
McDonald Creek
GOING-TO-THE-SUN ROAD
Mineral Creek
Ahern Creek

3. Figure 10.10 is a portion of the Glacial Map of the United States East of the Rocky Mountains, centering around Lake Michigan. It shows the kinds of glacial deposits present at the surface and, therefore, is a record of the *most recent glacial event* to have affected each area. Two major periods of continental glaciation, both within the Pleistocene Epoch, are well represented on this map. Most of the map shows, in green, deposits of Wisconsinan age, because it is the most recent glaciation. The pink areas in southern and southwestern Illinois were not affected by Wisconsinan-age glaciers, but were by glaciers of the earlier Illinoian age. The position of the ice front shifted back and forth considerably during glacial times, but each glacial age ended with total retreat.

a. What glacial feature is represented by the dark green, curved bands, some of which are named?

Was the front of the ice advancing, retreating, or staying more or less in place at the time these features were being formed?

What glacial feature, shown in light green, occurs between the dark green bands in Illinois, Indiana, and Ohio?

Was the front of the ice advancing, retreating, or staying more or less in place at the time these features were formed?

Locate the named, dark-green bands in eastern Indiana. The one on the northeast is labeled *Fort Wayne,* that on the southwest *Union City.* Which is older, the Fort Wayne or Union City, and why?

b. What do the yellow areas represent?

Explain why the yellow occupies broad areas in some places, such as central Wisconsin or northwestern Indiana, but in other places, such as northeastern Illinois or eastern Indiana, it occurs in narrow strips.

c. The familiar lakes, such as Lake Michigan, are shown in light gray on the map, but other lakes, such as Lake Wauponsee or Lake Watseka (southwest of Chicago), may be less familiar. What are these lakes?

Explain how and when they came to be, and what evidence for them you might find today.

4. All but the southwestern part of Wisconsin was covered by continental glaciers several times during the Pleistocene Epoch (Fig. 10.10). The aerial photograph in Figure 10.11 includes the boundary between glaciated and unglaciated areas. The red area on the east is hilly and partially forested. The area on the west, with the circular irrigated fields, is a flatter terrain that is mostly crop land.

   a. Which side has been glaciated, east or west, and how do you know?

   b. What type of glacial landform marks the former outer edge of the continental glacier?

   c. What kind of glacial sediment would you expect to find in the eastern part of the area?

   d. What type of glacial landform occurs in the western part of the area?

   e. What kind of glacial sediment would you expect to find on the west side of the area in the photograph?

FIGURE 10.11
High-altitude, color-infrared aerial photograph of the Coloma, Wisconsin, area (scale 1:58,000) for Problem 4. The circles are irrigated fields.

FIGURE **10.10**

Portion of Glacial Map of the United States East of the Rocky
Mountains.

EXPLANATION

AREAL COLORS

Marine sediments
Pre-Wisconsin along New Jersey coast; postglacial elsewhere.

Lacustrine sediments
*Mainly sand, silt and clay, deposited in lakes dammed by glacier ice
or by outwash sediments. Includes areas of till or bedrock. Labels
show the names of lakes except ancient Great Lakes.*

Outwash sediments
*Mainly sand and gravel, deposited by preglacial streams. Includes
some alluvium; in Kansas, Nebraska, South Dakota, North Dakota,
and Montana, includes alluvium with little or no outwash. Arrow
shows direction of flow.*

Ice-coated stratified drift
*Mainly sand and gravel, occurring as kames, eskers, kame terraces,
pitted plains and collapsed stratified drift.*

End moraines of Wisconsin age
Mostly till. Local names are lettered on or beside the moraines.

Drift, other than end moraines and outwash, of Wisconsin age
Chiefly till. Includes many areas of bedrock and older drift.

End moraines of Illinoian age
Mostly till. Local names are lettered on or beside moraines.

Drift, other than end moraines and outwash, of Illinoian age
Chiefly till. Includes many areas of bedrock.

Drift of Kansan age
Chiefly till. Includes large areas of bedrock.

Drift of Nebraskan age
Patches of till and scattered erratica.

Area not glaciated

✎ Streamline features

✎ Striation direction

⟋ Strandlines

0   25   50   100 Miles

0   50   100   150 Kilometers

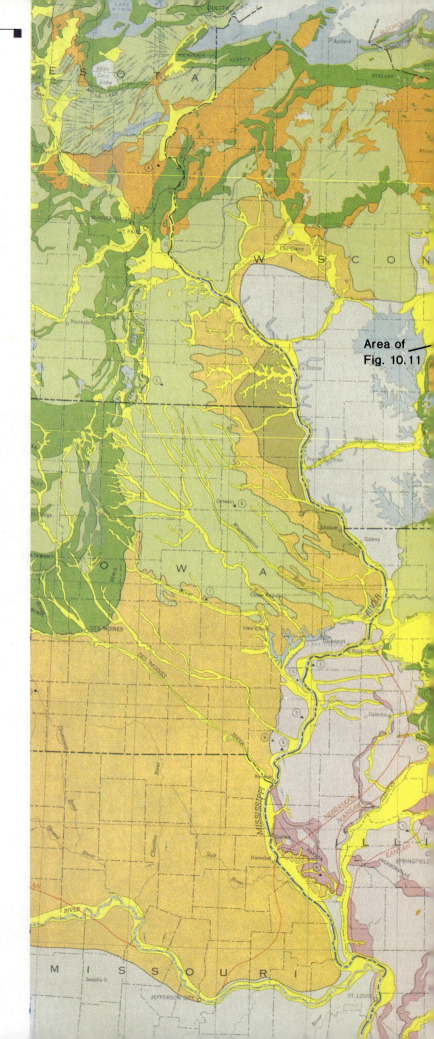

Area of
Fig. 10.11

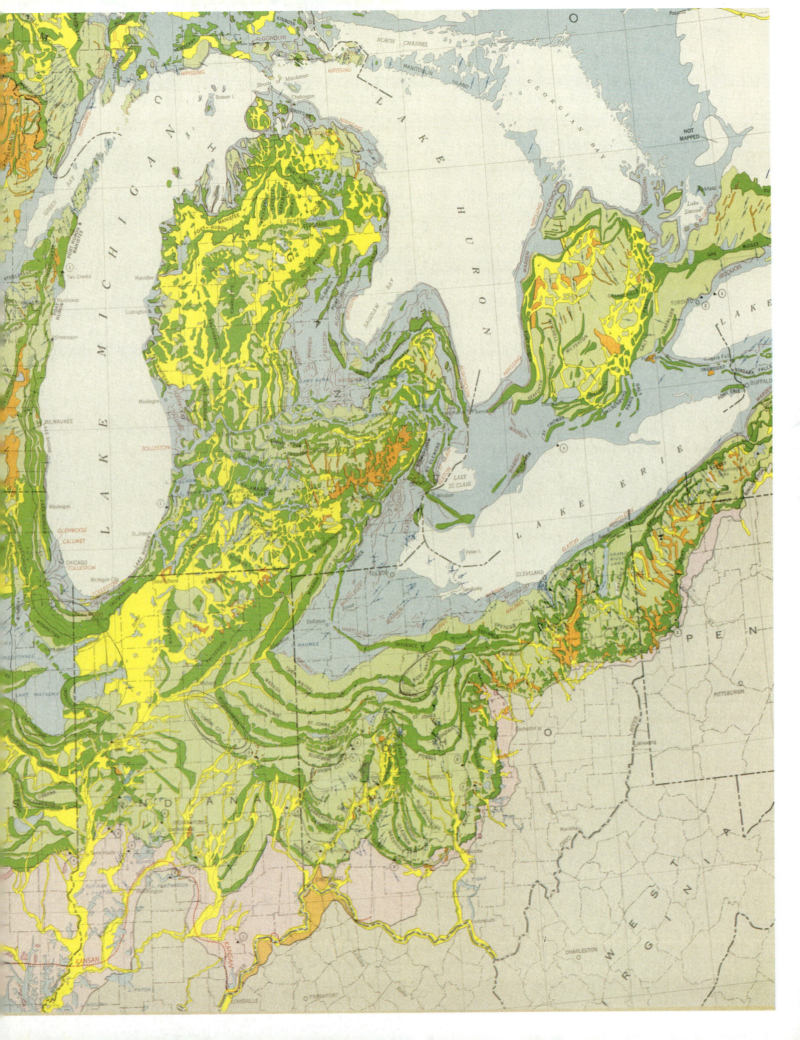

5. Figure 10.12 represents an area of southern Michigan just south of Jackson and located in one of the yellow areas of Figure 10.10. The Jackson area consists mainly of glacial deposits associated with stagnant (nonmoving) ice. These are mostly deposits from meltwater that flowed on, in, or at the base of ice that later melted away.

a. How did the several lakes—such as Mud, Crispell, Skiff, and so forth—probably form?

b. What type of glacial feature is Blue Ridge, and how did it form?

c. What type of glacial feature is Prospect Hill, and how might it have formed?

d. Newer, 1:24,000-scale maps of the north half of this area show crossed-shovel symbols in several places. What resource do you think is being extracted?

Circle several areas in the northern half of the map where you think comparable resources might be located.

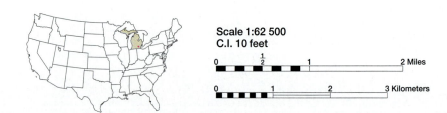

Scale 1:62 500
C.I. 10 feet

0   1/2   1   2 Miles

0   1   2   3 Kilometers

▶ FIGURE 10.12
Portion of Jackson, Michigan, 15-minute quadrangle topographic map.

6. Go to *http://www.nsidc.colorado.edu//NSIDC/EDUCATION/GLACIERS/glaciers.html* (or link to it through the author's homepage—see *Preface*), and answer the following questions.

   a. What percent of the world's land area is occupied by glaciers at present?

      During the Pleistocene?

   b. What is the approximate area covered by glaciers in the world?

      What percent of this is in Antarctica?

      In the U.S.?

      In Iceland?

      In South America?

   c. What is the difference between an *ice sheet* and an *ice cap?*

   d. What is the approximate maximum thickness of glacier ice in Antarctica?

      Approximately how thick does ice need to be before it can begin to flow as a glacier?

      How long has the Antarctic ice sheet been in existence?

   e. What is a glacial surge, and what is the maximum known rate of movement during a surge?

   f. How was the Matterhorn (in Switzerland) formed?

IN GREATER DEPTH

7. Figure 10.13 is a map that summarizes pre-1970 data for Grinnell Glacier, a cirque glacier in Glacier National Park that can be reached by a hiking trail from the Many Glacier area (see Fig. 10.9).

   a. The arrows or vectors shown on the map summarize the total movement of a number of rocks exposed on the surface of the glacier, as indicated in the *Explanation*. The lengths of the vectors depend on both the time over which observations were made and the *rate* at which the ice moved. What if we wanted to know whether the rate varied from year to year, or whether it varied from place to place in the glacier? Table 10.1 contains the data from which the length of the vectors was determined, and it shows the average annual rate of movement for one of the rocks.

      First, fill in the table by calculating the average annual rate of movement for all the rocks.

      Then separate the rates into four categories ranging from slowest to fastest (for example, if the difference between the slowest and fastest rate is 40 feet/year, and the slowest rate is 25 feet/year, the categories would be 25 to 34 feet/year, 35 to 44 feet/year, 45 to 54 feet/year, and 55 to 65 feet/year). Using colored pencils or an equivalent, highlight on the map those vectors representing the highest rates in red, second highest in yellow, third in green, and slowest in blue.

      Does the ice seem to move more rapidly in some parts of the glacier? If so, give a possible explanation.

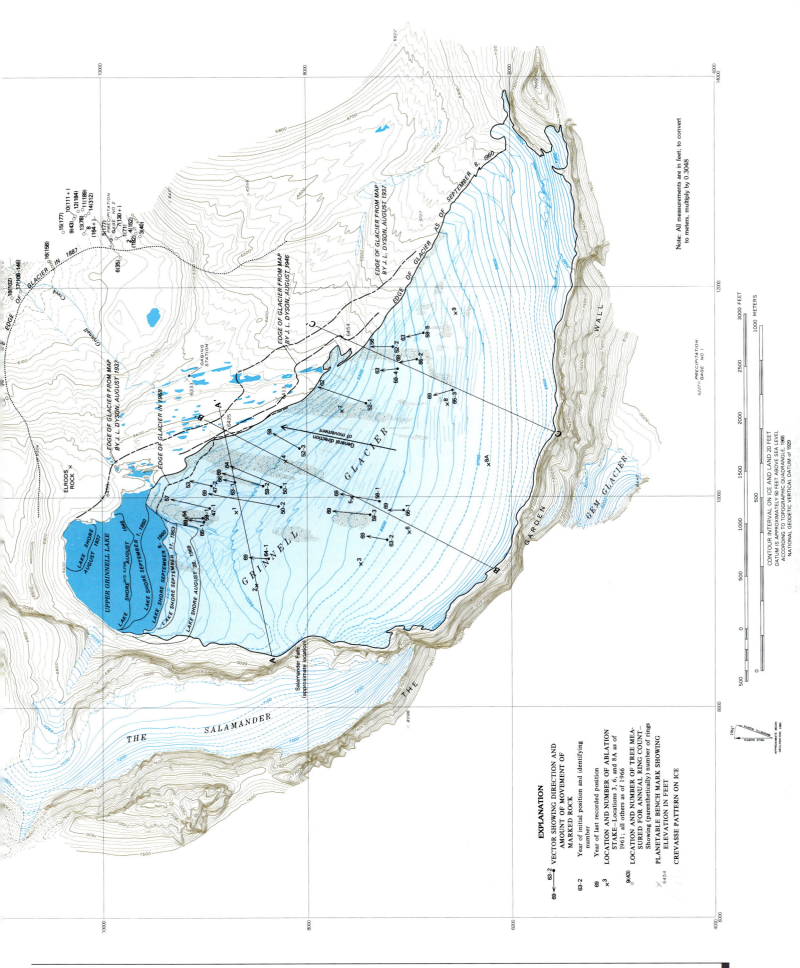

EXPLANATION

63-2 ⟵ ● VECTOR SHOWING DIRECTION AND
AMOUNT OF MOVEMENT OF
MARKED ROCK

63-2 Year of initial position and identifying
number

69 Year of last recorded position

×3 LOCATION AND NUMBER OF ABLATION
STAKE—Locations 3, 6, and 8A as of
1961; all others as of 1966

9(43) LOCATION AND NUMBER OF TREE MEA-
SURED FOR ANNUAL RING COUNT—
Showing (parenthetically) number of rings

PLANETABLE BENCH MARK SHOWING
ELEVATION IN FEET

6454 ×   CREVASSE PATTERN ON ICE

Note: All measurements are in feet; to convert
to meters, multiply by 0.3048

CONTOUR INTERVAL ON ICE AND LAND 20 FEET
DATUM IS APPROXIMATELY 50 FEET ABOVE SEA LEVEL
ACCORDING TO TOPOGRAPHIC QUADRANGLE, 1968
NATIONAL GEODETIC VERTICAL DATUM of 1929

FIGURE **10.13**
Map of Grinnell Glacier, Glacier National Park, Montana.

T a b l e   1 0 . 1

MOVEMENT OF MARKED ROCKS ON GRINNELL GLACIER
SOURCE: DATA FROM A. JOHNSON, *USGS PROFESSIONAL PAPER 1180, 1970.*

ROCK NUMBER	PERIOD	MOVEMENT (FEET)	
		Total	*Annual average*
47-1	1947–57	380	38
47-2	1947–53	215	
50-1	1950–64	530	
50-2	1950–69	700	
52-1	1952–62	485	
52-2	1952–56	210	
52-3	1952–59	285	
58-1	1958–69	385	
59-1	1959–64	170	
59-2	1959–68	355	
59-3	1959–69	355	
59-4	1959–65	275	
59-5	1959–66	320	
63-1	1963–66	110	
63-2	1963–69	190	
64-1	1964–69	185	
65-1	1965–69	170	
65-2	1965–69	180	
65-3	1965–69	185	
66-1	1966–68	80	

b. It is clear from Figure 10.13 that Grinnell Glacier has receded since 1887, when it was first described by the George B. Grinnell party. However, the position of the terminus of such a small glacier over a short time period may or may not reflect the true mass balance of the glacier. The terminus position can be strongly influenced by rate of ice movement and by bedrock topography. Another way to look at changes is by periodic observations of ice-surface elevations, such as those reported in Table 10.2.

The data in Table 10.2 are mean (average) elevations of five segments of the line B-B′ shown on the map. The segments are designated with the distance from plane table bench mark 6425, located near the intersection of lines A-A′ and B-B′ on Figure 10.13. For example, the column headed 100–500 lists the mean elevations of the segment of line B-B′ between 100 and 500 feet south-southwest of the point labeled 6425. Use these data to construct a graph on Figure 10.14 that illustrates how the average elevation of the ice surface in just the 500- to 1000-ft interval has changed with time. Plot mean elevation on the vertical *y*-axis and the date (year) on the horizontal *x*-axis. Draw lines between adjacent data points to emphasize the changes. Although these data represent one small area of the glacier, changes elsewhere on the glacier appear to be similar, so you can assume your graph represents the glacier as a whole.

How has the elevation of the glacier changed during this short time interval?

Is this generally consistent with the changes in position of the terminus shown on the map?

MEAN ICE-SURFACE ELEVATIONS OF SEGMENTS OF PROFILE B-B´, FIGURE 10.13

SOURCE: DATA FROM A. JOHNSON, *USGS PROFESSIONAL PAPER 1180*, 1970.

LINE SEGMENT (DISTANCE FROM 6425 IN FEET)	100–500´	500–1000´	1000–1500´	1500–2000´	2000–2500´
Date	Mean elevation of line segment				
1950 Sept. 14	6,460.1	6,523.3	6,564.8	—	—
1951	—	—	—	—	—
1952 Aug. 22	6,460.3	6,522.6	6,563.8	6,604.8	—
1953 Sept. 4	6,458.4	6,519.5	—	—	—
1954 Sept. 27	6,459.5	6,522.0	6,564.4	—	—
1955 Sept. 6	6,460.6	6,521.8	6,563.9	—	—
1956 Aug. 30	6,461.7	6,521.6	6,563.8	6,604.6	6,659.9
1957 Sept. 10	6,456.6	6,517.9	6,560.6	6,600.9	6,654.9
1958 Sept. 15	6,446.4	6,509.8	6,551.6	6,591.2	6,642.7
1959 Sept. 12	6,449.9	6,513.6	6,555.7	6,597.1	6,649.5
1960 Sept. 6	6,449.9	6,514.1	6,555.5	6,594.7	6,646.7
1961	—	—	—	—	—
1962 Sept. 2	6,442.8	6,506.4	6,547.4	6,596.9	6,640.7
1963 Sept. 12	6,437.5	6,499.2	6,541.8	6,582.4	6,634.4
1964 Sept. 15	6,436.5	6,499.1	6,541.1	6,581.4	—
1965 Sept. 6	6,436.4	6,499.6	6,540.9	6,580.2	6,635.0
1966 Aug. 17	6,437.6	6,501.5	6,543.1	6,582.3	6,637.7
1967	—	—	—	—	—
1968 Aug. 28	6,432.1	6,498.8	6,539.5	6,578.8	6,633.9
1969 Aug. 27	6,428.2	6,493.3	6,534.2	6,574.2	6,628.3

Does this imply that Grinnell Glacier has advanced or retreated during this period?

Has the *rate* at which elevation changed been consistent throughout the period? If not, indicate when it changed, and identify those time periods when it was generally consistent. When considering rate changes, look at rates established over periods of 5 or more years, not year-to-year changes. (Hint: Rate is expressed graphically by the slope of the line.)

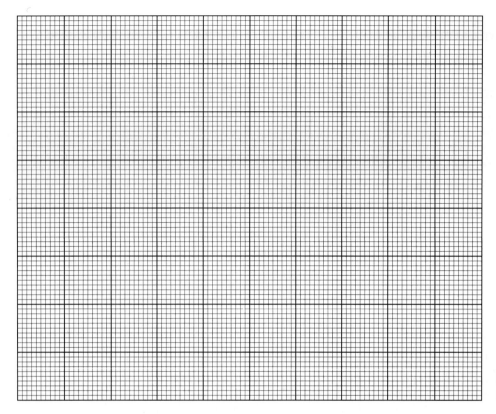

FIGURE **10.14**
Graph paper for Problem 7b.

11
Sea Coasts

Overview

Over 75 percent of the population of the United States lives near sea coasts, and their lives are affected by natural and human-caused changes that occur along the coasts. Waves striking the coasts gradually erode irregularities, and they generate longshore currents that transport sediment. Interruptions in the transportation process cause sediment to be deposited. In the long term, coastlines should become straighter, as rocky points are eroded and bays fill with sediment. But neither sea level nor continents are static, and many irregular coastlines show evidence of recent emergence or submergence. In this exercise, you will learn how the interaction between waves, longshore currents, river-mouth deposition, and changes in sea-level position affect coasts. You will learn to recognize common coastal features and use them to interpret present, past, and future coastal processes.

Materials Needed

Pencil and eraser • Ruler • Stereoscope (provided by instructor)

INTRODUCTION

Coasts are places where water and land interact. Waves, generated by wind blowing across the water, play the greatest role in this interaction. Figure 11.1 illustrates a wave and shows how its passage causes movement of water. Note that while the crest of a wave travels across the water, the water itself simply moves back and forth in a circular path. Water motion decreases to zero at the *wave base,* a depth approximately equal to half the wavelength.

As waves approach the shore, they begin to interact with the bottom, their shapes change, and the pattern of water movement changes, as shown in Figure 11.2. When the water depth becomes less than wave base, waves begin to erode and transport fine-grained bottom sediment. In the shallow *surf zone,* the wave shape breaks down, and water motion becomes turbulent; it is here that most work is accomplished. Sand and pebbles are thrown into suspension by the turbulent surf and are in near-constant motion.

Waves nearly always approach a coast with their crests at an acute angle to the shoreline. As the waves begin to interact with the bottom, they slow down and gradually bend, or **refract,** and become more parallel to the coast, as shown in Figure 11.3. In spite of refraction, however, the waves still hit the shore at an angle. This causes water within the surf zone to form a **longshore current,** which flows along the coast in the direction in which it is being pushed by the waves (Fig. 11.3). Sediment in the surf zone is carried by the current in a process known as **longshore drift.**

Wind energy is thus transferred through wave action to the surf zone, where erosion, transportation, and deposition combine to modify the coastline. We look first at common landform features that result from coastal erosion, then turn to those features formed by deposition.

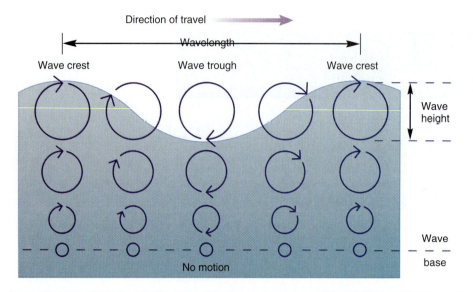

FIGURE 11.1
Cross section of wave, showing how water motion decreases to zero at wave base (a depth approximately equal to half the wavelength).

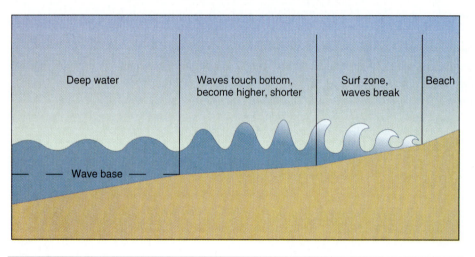

FIGURE 11.2
When waves enter water shallower than the wave base, they change form until, in the surf zone, they break, and water motion becomes turbulent.

EROSION

All coasts undergo erosion, but the intensity differs among coasts. The evidence for active erosion is seen best on rocky coasts. Figure 11.4 shows waves being refracted on approaching an irregular rocky coast. Notice that refraction causes most of the waves' energy to be focused on the **headland,** the rocky point of land between bays.

The headlands are actively eroded, and steep **wave-cut cliffs** form. Erosive activity is concentrated just above and below sea level. Undercutting of cliffs where erosion is rapid eventually causes them to collapse into the water, leaving a very gently sloping erosional surface called a **wave-cut platform.** More resistant portions of the rock temporarily left standing offshore on the wave-cut platform are **sea stacks.**

DEPOSITION

While headlands erode because wave energy focuses on them, deposition takes place in the bays because wave energy dissipates there. Deposition results in the formation of a **beach,** commonly made of sand, pebbles, or cobbles eroded from the headlands, as well as material carried to the coast by rivers. In time, the irregularity of the coastline diminishes.

Longshore drift aids in straightening out coastlines. When a longshore current passes a point and enters deeper water, it slows, and some of its load is deposited. A common feature formed by this process is a **spit,** an emergent ridge of sand extending from the point in the direction of longshore drift, as shown on Figure 11.5. A spit extending completely across the mouth of a bay is a **baymouth bar. Tombolos** are sand ridges connecting islands to the shore. They develop because the island blocks and refracts the waves and locally redirects the longshore current or deprives it of the energy needed to transport its load.

Rivers provide most of the sediment for beaches and longshore drift. If longshore currents are strong, most of the sediment at the mouth of a river is swept away. If longshore currents are weak, or if the sediment load of a river is excessive, the sediment is deposited at the mouth of the river as a **delta** (Fig. 11.5).

Low-lying coasts, such as the southeastern U.S. coast from New York to Mexico, may have long, narrow, sandbar islands paralleling the coast. These **barrier islands** (Fig. 11.6) are 2 to 5 km wide and 10 to 100 km long. They are separated from the mainland by a *lagoon* some 3 to 30 km wide. Gaps between islands, known as **tidal inlets,** allow strong tidal currents to develop as the tide rises and falls. Sediment transported by these currents is deposited as **tidal deltas** both landward and seaward of the inlets. Atlantic City, New Jersey; Miami Beach, Florida; and Galveston, Texas are on barrier islands. Barrier islands are fragile coastline features, and dramatic changes in them can be seen in the course of a human lifetime.

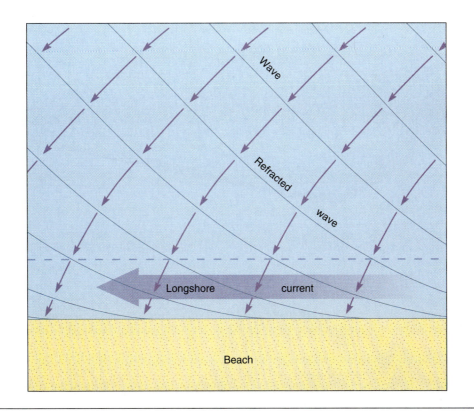

FIGURE **11.3**

Map view showing refraction (bending) of waves as they approach shore. Solid lines represent wave crests, and arrows indicate wave movement. As the water is pushed against the shore, a longshore current develops in the surf zone.

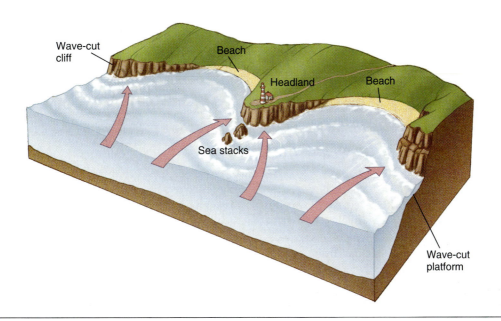

FIGURE **11.4**

Waves approaching an irregular coastline are refracted so that energy is concentrated at headlands. Erosion results in the formation of wave-cut cliffs, sea stacks, and a wave-cut platform.

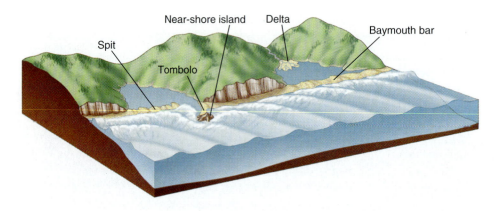

FIGURE 11.5
Longshore drift is from left to right, as indicated by the spit. Wave refraction around the nearshore island causes the tombolo to form.

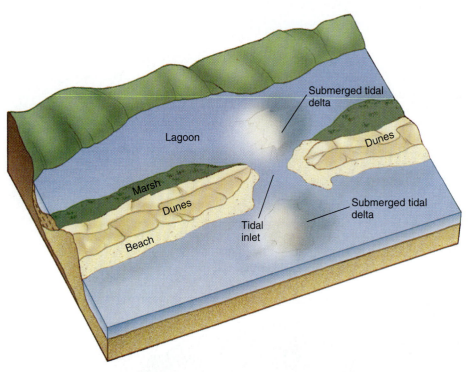

FIGURE 11.6

Features of barrier islands include beach, dunes, marsh, lagoon, tidal inlet, and associated tidal delta (usually submerged).

SUBMERGENT COASTS

The position of shorelines has fluctuated throughout geologic time. Not only does the amount of water in the oceans fluctuate, but tectonic and other forces have caused continents to rise and fall relative to sea level. These changes have been dramatic in the past two million years because of the rise and fall of sea level during the Pleistocene continental glaciations. The result is that coastlines are transitory features. There are many examples in the world of coasts that have been drowned (submerged) as sea level rose following the melting of continental glaciers.

The characteristics of a **submergent coast** depend on the nature of the landscape before submergence. If the topography was rugged, the coastline is irregular, with features such as **estuaries** (drowned river valleys) or *fiords* (drowned glacial valleys) and abundant islands. Submerged landscapes of low relief may have large estuaries, extensive tidal flats and salt marshes, and barrier islands.

HUMAN INTERACTION

Commonly, people living along coasts do not understand the power of the natural processes of coastal erosion and deposition, and they try to control them artificially. Easily eroded shores can be protected against normal storm erosion with a seawall. However, reflection of wave energy enhances erosion of the beach in front of the wall. Once that beach is gone, the longshore current is deprived of its supply of sediment, and beaches downstream may disappear. Another approach to protect the shoreline from erosion is to build *groins,* which are low walls built outward from the shore (Fig. 11.7). Other forms of breakwaters have been used to protect harbors, but where longshore drift is important, these have met with varying degrees of success, as illustrated in Figure 11.7.

EMERGENT COASTS

Emergent or uplifted coasts are found in many places around the tectonically active margin of the Pacific Ocean. Their most distinctive features are uplifted **marine terraces** (Fig. 11.8). Most of these were formed just below sea level as **wave-cut platforms.** The terraces are the result of continuing uplift, coupled with sea-level fluctuations during the Pleistocene Epoch.

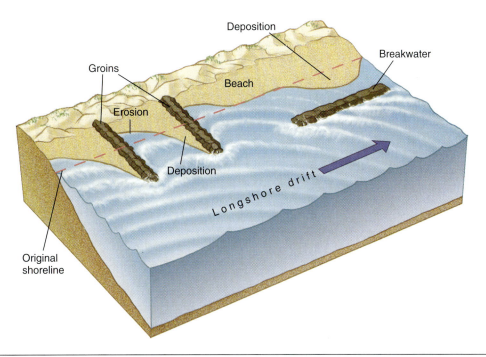

FIGURE 11.7
Groins and breakwaters interrupt longshore drift, causing erosion or deposition.

FIGURE 11.8
The gently sloping surface near Pacific Valley, California, was a submerged wave-cut platform that has been uplifted to form a marine terrace. Old sea stacks are visible on the terrace.

APPLICATIONS

COASTLINES ARE AMONG THE MOST DYNAMIC ENVIRONMENTS IN WHICH WE LIVE, SO IT IS IMPORTANT THAT WE THOROUGHLY UNDERSTAND COASTAL PROCESSES. GEOLOGISTS LEARN TO USE THE PAST AND THE PRESENT TO PREDICT THE FUTURE. THIS LAB DEMONSTRATES HOW CAREFUL OBSERVATION OF PAST RESULTS AND PRESENT PROCESSES, TOGETHER WITH THE APPLICATION OF GENERAL PRINCIPLES, LEADS TO AN UNDERSTANDING OF THE COASTAL ENVIRONMENT.

OBJECTIVES

If you complete all the problems, you should be able to:

1. Illustrate the refraction of waves as they strike a smooth or irregular coastline.

2. Explain the origin of longshore currents and longshore drift.

3. Explain how wave refraction may determine where erosion and deposition take place along a coast.

4. Identify the following erosional features on a map or photograph: headland, wave-cut cliff, wave-cut platform, sea stack.

5. Identify the following depositional features on a map or photograph: beach, spit, baymouth bar, tombolo, delta, barrier island, and associated features such as lagoon, tidal inlet, tidal delta, tidal flat.

6. Determine the dominant direction of the longshore current from a map or photograph showing features such as spits or groins.

7. Based on a topographic map, identify a coast as rocky or sandy, and use evidence on the map to determine whether erosion or deposition is the dominant process.

8. Recognize submergent and emergent coasts and their distinctive features such as estuaries, fiords, and marine terraces.

PROBLEMS

1. Figure 11.9 illustrates the fragility of barrier islands with a sequence of aerial photographs. Shown in the photographs is part of Matagorda Island, a barrier off the Texas coast in the Gulf of Mexico. Hint: Figures 11.5 and 11.6 may help you answer the following questions.

   a. Photo A, from 1943, shows a tidal inlet called Greens Bayou. Note the road to the left (southwest) of Greens Bayou. What is the minimum width of Greens Bayou? (The fractional scale is 1:10,200.)

      Based on the orientation of the waves in Photo A, determine the direction of the longshore current. What, if any, evidence is there to indicate whether this is the usual direction of longshore current?

   b. Photo B shows the same area in 1957. Note the road now. What is the minimum width of Greens Bayou now? (The fractional scale is 1:25,400.)

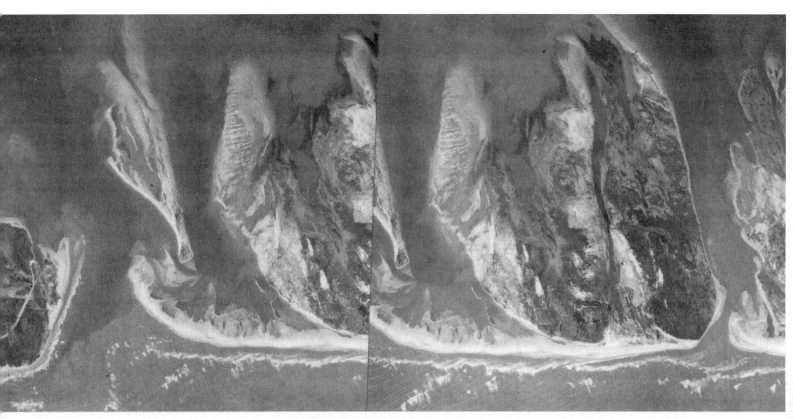

A.

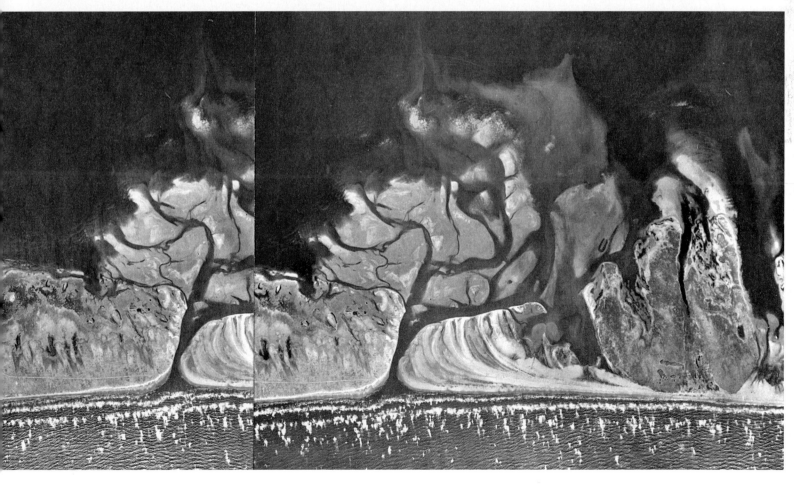

B.

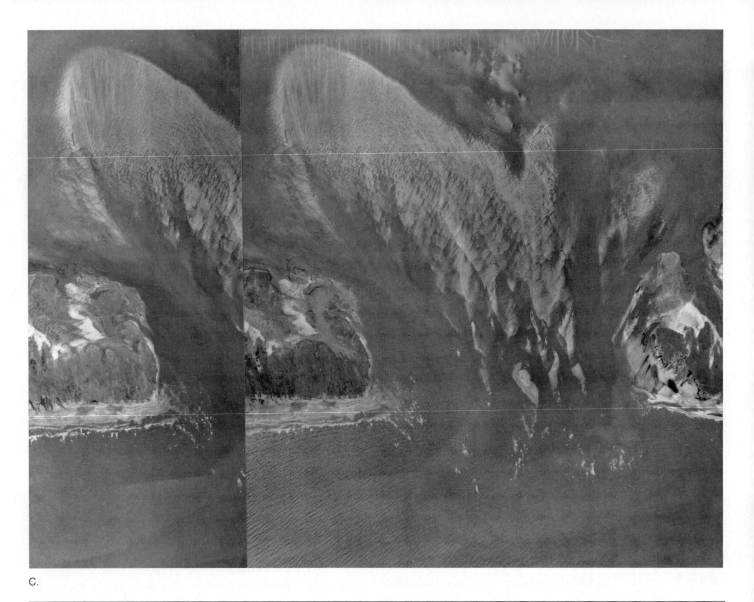

C.

FIGURE 11.9

Sequence of three sets of stereophotographs of Greens Bayou, Matagorda Island, Texas, taken in: A. 1943 (scale 1:10,200); B. 1957 (scale 1:25,400); and C. 1961, after Hurricane Carla (scale 1:18,500).

Where has erosion occurred?

Where has deposition occurred?

Explain the origin of the curved lines where Greens Bayou once was.

What feature has formed in the lagoon behind what's left of Greens Bayou? Explain how conditions have changed since 1943 to allow the formation of this feature.

c. Photo C shows the same area in 1961, after Hurricane Carla hit the Texas coast. The road is barely visible, but can be seen just behind the beach.

What is the minimum width of Greens Bayou in this photo? (The fractional scale is 1:18,500.)

Explain what happened in the vicinity of Greens Bayou during the hurricane.

Is a barrier island a good place to build a condo?

2. Figure 11.10 is a portion of the Cayucos, California, 1:62,500 quadrangle, which is about 15 miles northwest of San Luis Obispo.
   a. Is this a rocky or sandy coast?

What coastal landform is represented by the small circular land areas just offshore 2 to 6 km northwest of Morro Beach (for example, Whale Rock)?

b. What is the landform on the west side of Morro Bay, labeled Morro Bay State Park?

Morro Rock is shown as an island on older maps. What is the landform that now connects Morro Rock to the mainland?

What happens to Morro Creek as it approaches the coast? Explain. How might this situation change during a period of abundant rainfall?

c. What does the pattern of dots in Morro Bay indicate? Hint: See the table of topographic map symbols in Chapter 6.

    What is the swampy area on the east side of Morro Bay?

    What does the future hold for Morro Bay?

d. In what direction is the longshore drift in the area between the town of Morro Beach and the bottom of the map, as indicat-
    ed by (1) the breakwater and (2) natural landforms? What is the evidence?

Scale 1:62 500
C.I. 50 feet

0   ½   1   2 Miles

0   1   2   3 Kilometers

◄ FIGURE 11.10
   Portion of the Cayucos, California, 15-minute quadrangle, for Problem 2.

3. Figure 11.11 is a portion of the Boothbay, Maine, 1:62,500 quadrangle. This area, about 40 miles northeast of Portland, is made of metamorphic rocks whose resistant layers intersect the surface in generally north-south bands. The striking topography and extremely irregular coastline result from differential erosion of these rocks. Blue contours show water depth in feet.

    a. Is this a rocky coast or one dominated by sandy beaches? Cite evidence for your answer.

    b. Is erosion or deposition the dominant process along most of the coast? Cite evidence for your answer in terms of landforms and map symbols. Consider both on-land and offshore topography.

    c. Is this coast emergent or submergent? Cite evidence.

    d. Locate the following features: wave-cut platform, sea stack, headland, wave-cut cliff, estuary. Cite locations in terms of geographic names. Consider subaerial and submarine topography when searching for examples.

    e. What agents of erosion (for example, wind, streams, waves, glaciers, mass wasting) are operating in this area at the present time?

Scale 1:62 500
C.I. 20 feet

0   1/2   1   2 Miles

0   1   2   3 Kilometers

◄ FIGURE 11.11
    Portion of Boothbay, Maine, 15-minute quadrangle, for Problem 3.

Has the topography developed mainly as the result of the erosive agents you listed above, or were other agents active in the past? Explain your answer. Hint: Look at the map pattern and cross-profiles of valleys.

f. If climatic and sea-level conditions remain similar to those of the present, predict how this coastline might look in ten or twenty thousand years. Consider the effects of erosion and deposition by both marine and terrestrial processes. Sketch a possible new shoreline on the map, and clearly label those areas removed by erosion or built by deposition.

4. Go to *http://ceres.ca.gov/ceres/calweb/coastal/geography.html* (or link to it through the author's homepage—see *Preface*), and answer the following questions about marine terraces and beaches along the California coast.

a. **Marine terraces**

What are marine terraces and where along the California coast are they seen?

How are marine terraces formed?

In some places, up to 20 marine terraces are present. What combination of events occurred that allowed so many wave-cut platforms to form, then rise above sea level to form terraces?

b. **Beaches**

From where does the sand in California beaches come?

What is *littoral drift?*

How is sand permanently lost from the beach system?

How have humans altered the natural beach processes and what effect might this have on the system?

In Greater Depth

5. Figure 11.12 is a portion of the Redondo Beach, California, quadrangle from 1951. The portion of the map shown in Figure 11.12 is about 25 miles south of Los Angeles proper.

    a. Is the coast rocky or sandy in this area? Cite your evidence.

    b. Draw a topographic profile between A and A′ using the graph paper of Figure 11.13. Note that the contour intervals on land and offshore are different, and that on land, every *fourth* line is an index contour. Note that the profile is step-like, with areas of gentler slope separated by areas with steeper slopes.

    c. Refer to your topographic profile. How did the gently sloped area immediately seaward of the coastal cliff form, and what landform name is given to such a feature?

    d. How do you think the other gently sloped areas on your topographic profile formed?

    How many of these areas are there, including the one just off the coast? Number them for reference purposes; assign "1" to the one just off the coast, "2" to the first one inland, and so forth.

What is the approximate change in elevation between each of the flat areas? For example, between 1 and 2, 2 and 3, and so forth. Measure elevation differences between the top and bottom of each area with steep slope.

e. Look at the stereopair in Figure 11.14. These photos were taken in 1990. Note the amount of development since 1951, and where it has occurred.

   Is erosion or deposition likely to be the dominant process along the coast? Circle those areas that ought to pay the highest rates for natural hazards insurance, and explain your answer.

f. What does the topographic profile along A-A′ suggest about the manner in which emergence took place? That is, was it gradual, episodic, or can't you tell? Explain. (Be careful; this is tricky, because two variables must be considered.)

g. Assign relative ages to the flat areas on the profile. That is, which appears to be the youngest, which the oldest, and where in the sequence do the others fit? Hint: Which have been subjected to the least on-land erosion, which the most, and which are intermediate?

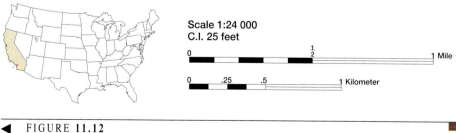

Scale 1:24 000
C.I. 25 feet

◄ FIGURE 11.12
Portion of the 1951 Redondo Beach, California, 7½-minute quadrangle, for Problem 5.

FIGURE 11.13
Graph paper for a topographic profile, Problem 5.

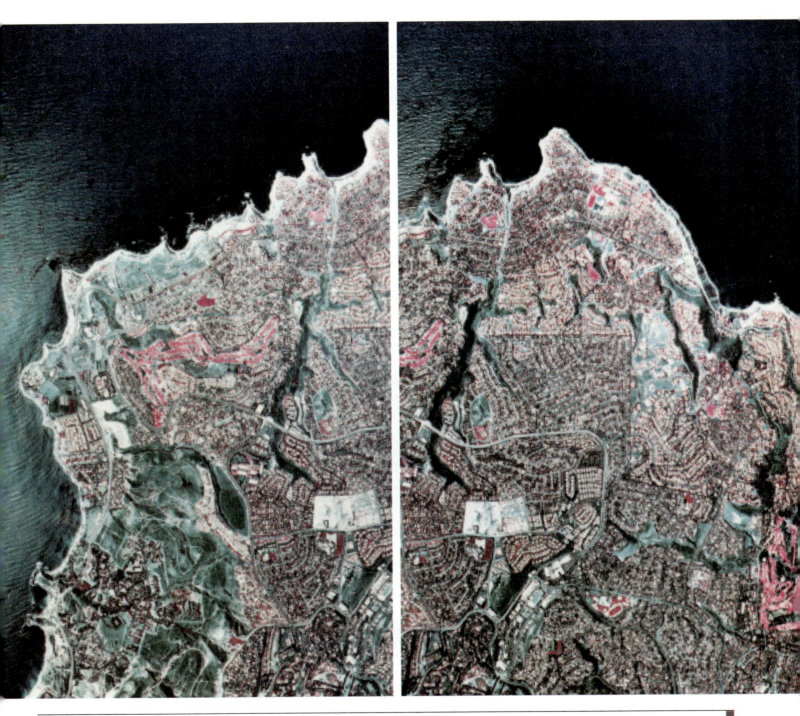

FIGURE **11.14**
Stereopair of Palo Verdes Point area, California, 1990. Scale is 1:40,000. North is to the right.

chapter

12

Arid-Climate Landscapes

Overview

The bold, commonly stark, landscapes of arid and semiarid lands contrast with the rounded, more subtle forms of humid regions. Although running water is the principal agent of erosion in arid regions, it commonly is present only after infrequent cloudbursts. Flash floods remove the products of mechanical weathering and mass wasting and leave steep banks and cliffs. Plateaus, mesas, and buttes develop from deep erosion of horizontal layers of resistant bedrock. As erosion wears down mountains, alluvial fans form at canyon mouths and, with time, gently sloping erosional and depositional surfaces extend outward from the mountain front. Where strong winds blow and sand is available, dunes of several types may form. In this lab, you will learn to recognize and analyze common arid-climate landforms on maps and aerial photographs.

Materials Needed

Pencil and eraser • Colored pencils • Metric ruler • Calculator • Stereoscope (provided by instructor)

INTRODUCTION

Warm, arid, and semiarid regions (precipitation <50 cm/year; see Fig. 8.1) are characterized by sparse vegetation, little water, thin soils, frequent strong winds, and sharp, angular landforms. Mechanical weathering predominates, and resulting particles tend to be coarser than in humid regions. Slopes typically are steeper, a requirement for removal of coarser particles. Even rock such as limestone, which is readily attacked by chemical weathering in humid climates, forms steep cliffs in dry climates.

In spite of the aridity, water is the principal agent of erosion in most deserts. Rain commonly comes as downpours that quickly fill streams and cause flash floods. These floods cause extensive erosion, and sediment is quickly transported down normally dry stream beds and out into valleys before it is deposited.

Wind is also an important agent of erosion and transportation in deserts, but does not move as much material as water does. In fact, large, wind-carved rock landforms are unusual. Sand dunes, however, are common, and demonstrate that wind is an active agent of sediment transport and deposition.

EROSIONAL LANDFORMS

Most landforms in arid and semiarid regions are shaped by running water, but streams in arid regions are different from those in humid climates. Most streams are **intermittent,** carrying water only part of the year. Many, whether intermittent or **perennial,** have **braided channels** (Fig. 12.1), steep,

FIGURE **12.1**

Dry streambed in Death Valley National Monument, California, with braided channels and easily eroded banks.

FIGURE **12.2**

Cliff-and-bench topography in Canyonlands National Park, Utah, along the Green River.

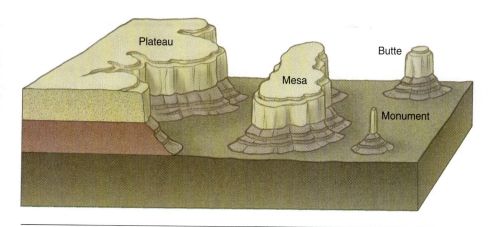

FIGURE **12.3**

Characteristic erosional landforms in arid or semiarid areas with horizontal rock layers. Retreat of a plateau leaves mesas, buttes, and monuments as remnants.

easily eroded banks, and a bed of sand and gravel. Deep canyons in dry climates may have near-vertical walls, and box-like cross profiles.

Where rivers traverse and cut into horizontal layers of sedimentary rocks or lava flows, *cliff-and-bench* topography, like that in Figure 12.2, develops. The more resistant layers of sandstone, limestone, or lava flows form steep cliffs where erosion has cut through them; flat benches develop on their upper surfaces. Less resistant layers of shale form gentler slopes between the benches. The cliff-and-bench topography is developed in a spec-

tacular way on the *Colorado Plateau* of the western United States. This region, in eastern Utah, western Colorado, north-western New Mexico, and northern Arizona, has 16 national parks or monuments featuring this scenery, including Grand Canyon and Canyonlands.

The comparatively flat upland areas away from the stream valleys, in regions with near-horizontal rock layers, are **plateaus.** Figure 12.3 illustrates several common features developed by the dissection of a plateau. As stream valleys enlarge by lateral and headward erosion, parts of the plateau may be cut off to form mesas. A

mesa (table, in Spanish) is a generally flat-topped area, bounded by cliffs, that is wider (commonly by quite a bit) than it is high. **Buttes** are smaller, flat-topped landforms isolated by erosion. They are more or less as wide as they are high. **Monuments** (or spires) are slender features that are much higher than they are wide.

Where the bedrock does not consist of horizontal layers of unequal resistance, different erosional landforms develop. In the *Basin and Range* area of the south-western United States, rugged mountain ranges are separated by wide valleys with flat floors. The mountains were uplifted along steep faults. Although some faults are still active and are readily discernible, others have been obscured by erosion and deposition. Figure 12.4 shows a profile from a mountain into the adjoining valley. A sharp break in slope at the edge of the mountain marks the *mountain front.* The gentle valleyward slope from the mountain front is the **piedmont.** It consists of two parts, the *pediment* and the *bajada.* A **pediment** is a gently sloping, erosional surface. It cuts across bedrock, and typically has a thin veneer of sediment on its surface. It is generally thought that a pediment develops as the steep slope at the mountain front retreats under the attack of erosion. The erosional debris is transported across the pediment to the bajada. Erosional remnants, or **inselbergs,** form isolated hills on the pediment.

DEPOSITIONAL LANDFORMS

Water Deposits

Most sediment in arid lands is transported by intermittent streams. In mountainous areas, these streams cut and occupy rock-walled channels, and sediment in transport is confined within the rock walls. As the canyons empty into a valley, the stream is no longer confined; it spreads out, losing its capacity to carry sediment. An **alluvial fan,** like those shown in Figure 12.5, forms as sediment is deposited. With time, alluvial fans from adjacent canyons coalesce to form a **bajada,** the gently sloping apron of sediment along the mountain front in Figure 12.6. If a pediment is present, the bajada lies basinward of the pediment, and together, they form the piedmont at the base of the mountain (see Fig. 12.4).

When an intense rainstorm occurs in arid areas, water runs out of the highlands, down canyons, and across the piedmont. If a pediment exists, much of the water washes across it to the bajada. As it crosses the bajada, a substantial amount of water may soak into the alluvium and become part of the groundwater. But, if it is a real downpour, much water flows out into the valley. If there is no through-going drainage, water may pond to form temporary lakes called **playas.** Evaporation of water from playas results in the deposition of evaporite minerals, such as gypsum or halite.

Wind Deposits

The most obvious wind deposits are sand dunes. They develop from the accumulation of sand-size grains that have been transported by the wind. Grains move by rolling and hopping, and rarely get more than a meter above the surface; many are set in motion by impact from other grains.

Initial deposition generally is caused by an obstacle or irregularity on the surface. Wind velocity decreases as it blows over an obstacle, and some of the sand is deposited on the downwind side; continued deposition produces a dune. Figure 12.7 is a cross section of a dune, showing how a steep, downwind face develops as sand rolls and hops up the windward side and tumbles and slips down the leeward side. As this process continues, the dune migrates downwind.

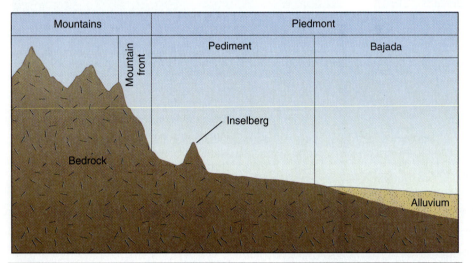

FIGURE 12.4
Profile from mountain into valley, showing the mountain front and the piedmont, with pediment, inselberg, and bajada.

FIGURE 12.5
Two large alluvial fans occur along the base of the Black Mountains in Death Valley National Monument, California.

FIGURE 12.6
Pediment and bajada are developed near the base of the Panamint Range (background) in Death Valley National Monument, California.

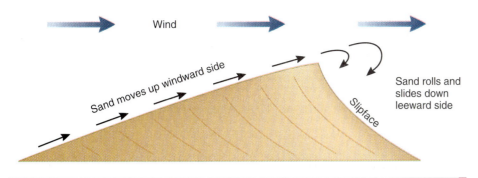

Wind

Sand moves up windward side

Sand rolls and slides down leeward side

Slipface

FIGURE 12.7
Cross section of a dune, showing sand moving up the gentle windward slope and falling off the steeper leeward slope.

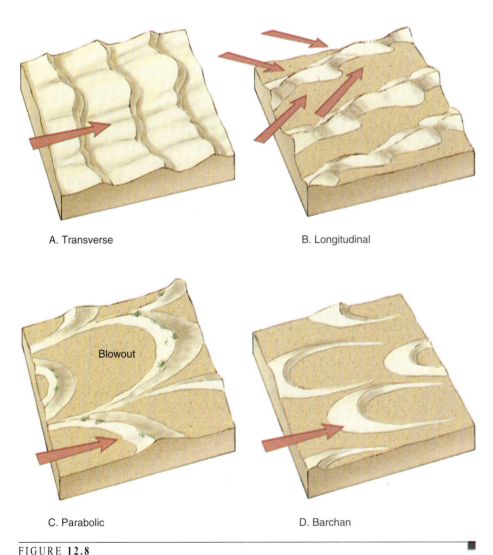

A. Transverse

B. Longitudinal

Blowout

C. Parabolic

D. Barchan

FIGURE 12.8
A. Transverse dunes; B. Longitudinal dunes; C. Parabolic dunes; D. Barchan dunes.
SOURCE: DATA FROM U.S. GEOLOGICAL SURVEY

Differently shaped dunes develop under different conditions. The principal controls are wind velocity, vegetation, and sand supply.

Transverse dunes (Fig. 12.8A) are wave-like ridges of sand perpendicular to the predominant wind direction. They form in areas of abundant sand supply, scarce vegetation, and moderate, unidirectional winds. They are found not only in deserts, but also commonly near shorelines and on barrier islands.

Longitudinal dunes (Fig. 12.8B) are long, narrow ridges of sand parallel to the prevailing wind direction. They form where sand supply and vegetation are meager and winds are strong. Although they parallel the general wind direction, the slip face varies from one side to the other along the axis of the dune, reflecting variations in wind direction.

Parabolic dunes (Fig. 12.8C) are crescent shaped, with the steep slip face on the convex side so that the horns point upwind. They typically develop where vegetation is available to anchor the horns. In many, the area in front of the dune, between the horns, is a **blowout,** a small depression excavated by the wind.

Barchan dunes (Fig. 12.8D) are crescent shaped, like parabolic dunes, but the steep slip face is on the concave side so that the horns of the crescent taper and point downwind. Barchans develop best on barren desert floors where prevailing wind direction is constant, vegetation is scarce, and sand supply is low. They may occur singly or in groups, where they may join to form more complex shapes.

Dune fields may become inactive as conditions change. Vegetation, in particular, acts to stabilize dunes. However, if patches of vegetation die or are killed by overgrazing or excessive human activity, blowouts develop and active dunes may reappear.

APPLICATIONS

DESERT ENVIRONMENTS ARE VERY FRAGILE, AND AS THE HUMAN POPULATION INCREASES, MORE AND MORE PEOPLE LIVE IN, OR INTERACT WITH, DESERTS. TO LIVE SUCCESSFULLY IN ARID REGIONS, THE GEOLOGIC PROCESSES OPERATING THERE MUST BE UNDERSTOOD. IN THIS LAB, YOU WILL LEARN HOW CAREFUL OBSERVATION AND MEASUREMENT OF DESERT LANDFORMS ELICITS A VARIETY OF QUESTIONS REGARDING THE PROCESSES THAT FORMED THEM. THE ANSWERS TO THE QUESTIONS ARE HYPOTHESES, SOME OF WHICH CAN BE TESTED USING AVAILABLE EVIDENCE, OTHERS OF WHICH REQUIRE FIELD EVIDENCE.

OBJECTIVES

If you complete all the problems, you should be able to:

1. Recognize the following desert landforms on a map or photograph and explain how they develop: plateau, mesa, butte, monument.

2. Distinguish pediment, inselberg, bajada, alluvial fan, and playa on a photograph or map, and explain how they form.

3. Recognize transverse, longitudinal, parabolic, and barchan dunes on a photograph or map, and describe the conditions favoring their development.

4. Recognize a braided stream, and use its presence to interpret the nature of the sediment in the streambed.

PROBLEMS

1. Figure 12.9 is a stereographic composite of aerial photographs of the Stovepipe Wells area of Death Valley National Monument, California. The buildings near the center of the photograph mark the small community of Stovepipe Wells. Its elevation is 0 feet—sea level—and the highest elevation in the photograph is about 2500 feet. North is to the left in the photographs.

    a. What are the prominent landform features extending from the mountain front, and how did they form?

    b. What happens to the single stream channel as it leaves the mountain canyon?

    Formulate a hypothesis to explain your observation, and suggest a way to test it.

    c. Rock or sediment exposed at the surface in Death Valley (and many other places) becomes coated with a very thin, dark-colored layer of clay, iron, and manganese minerals called rock varnish. The longer a rock surface is exposed without being eroded, the darker it becomes. The mountain-front features in Figure 12.9 have dark, light, and intermediate gray areas. Examine the photos with a stereoscope, noting the positions and relative elevations of the different colored areas. What are the relative ages of these areas?

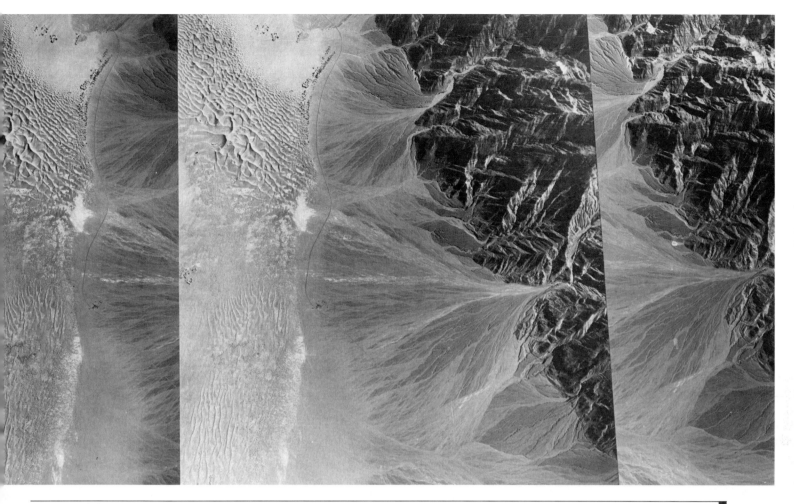

FIGURE **12.9**
Stereophotographs of the Stovepipe Wells area, Death Valley, California, for Problem 1. Scale 1:65,000; photos taken November 11, 1948.

What seems to be happening at present in each of the three differently colored areas? For example, are some areas active channels, and others abandoned channels? Is deposition occurring in some areas, and erosion in others?

    d.   Using a blue pencil, color the presently active channels.

2.   Figure 12.10 is a portion of the Bennetts Well, California, 1:62,500 quadrangle. This is in the heart of Death Valley and is about 30 miles south-southeast of Stovepipe Wells (see Fig. 12.9). The lowest elevation in the United States, 282 feet below sea level, occurs in this quadrangle. Because Death Valley is below sea level, all of the drainage in this area flows into the valley. No streams flow out. The Amargosa River flows north—when it flows. Rainfall averages about five centimeters (_____ inches?) per year, but it commonly comes in big storms after years of virtually no rain. Most erosion occurs during storms.

    a.   Most valleys form as erosion removes loose material and streams transport it out of the valley. But that cannot happen here. This valley is formed by faulting, and the faults are still active. Movement on a near-vertical fault on one side of the valley caused that side of the valley to drop. On which side of the valley is this fault? Cite at least two pieces of evidence to back your choice.

    b.   What name is given to the landform between the valley and the mountain front on the west side of the map?

What is the gradient of this feature, in feet/mile, between BM-242 (on the road) and the 19 in Section 19 to the west?

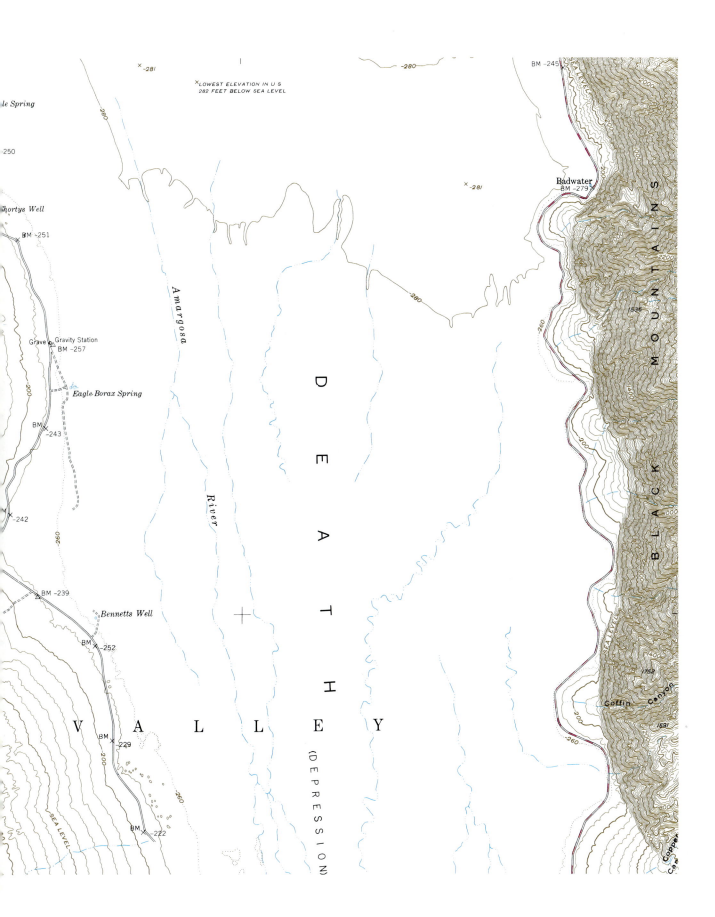

le Spring

-250

Shortys Well

BM -251

Grave Gravity Station
BM -257

Eagle Borax Spring

BM -243

-242

BM -239

Bennetts Well

BM -252

VALLEY

BM -229

SEA LEVEL

BM -222

×-281

×LOWEST ELEVATION IN U S
282 FEET BELOW SEA LEVEL

-280

×-281

DEATH

(DEPRESSION)

Amargosa

River

BM -245

Badwater
BM -279

BLACK MOUNTAINS

1536

Coffin   Canyon

1752

1591

Copper

Ca

Scale 1:62 500
C.I. 40 feet

0 1/2 1 2 Miles

0 1 2 3 Kilometers

◄ FIGURE 12.10
Portion of Bennetts Well, California,
15-minute quadrangle, for Problem 2
(*previous pages*).

3. Figure 12.11 is a portion of the Antelope Peak, Arizona, quadrangle. U.S. Interstate 8 crosses the area on its way from San Diego, California, to Tucson, Arizona. The area is within the *Basin and Range* province, but the basin-bounding faults have long since been covered by alluvium. The piedmont is extensive and occupies a large part of the map area. The rocks exposed in the mountains in this quadrangle are mostly metamorphic and igneous rocks of Precambrian age.

   a. How does the map-view shape of the mountain front differ from the mountain front in Figure 12.10?

   b. Determine the gradient in feet/mile along a line extending from the southwest corner of Sec.26 (near center of map) to the southwest corner of Sec.6 (northeast corner of map). How does the slope of the piedmont surface compare with the slope of the piedmont in Figure 12.10?

   How does the general contour pattern on the piedmont differ from the contour pattern in Figure 12.10?

   c. Note that the piedmont area north of the interstate is punctuated with a few hills to the northeast of the mountain front. How did these hills form?

   What part of the piedmont (see Fig. 12.4) lies between these hills and the mountain front? Explain.

   d. Based on the differences you have observed between the Antelope Peak and Death Valley areas, which area appears to have been subjected to erosion for the longest period without rejuvenation (for example, by tectonic activity or change in base level)?

   e. Note the water wells in the northeast part of the map. In what part of the piedmont are they likely to be located? Explain.

   What is the probable source for the groundwater tapped by the wells? Explain.

   f. Locate the canyon just south of Indian Butte. What landform has developed at its mouth?

Scale 1:62 500
C.I. 25 feet

0 1/2 1 2 Miles

0 1 2 3 Kilometers

► FIGURE 12.11
Portion of Antelope Peak, Arizona, 15-minute
quadrangle, for Problem 3.

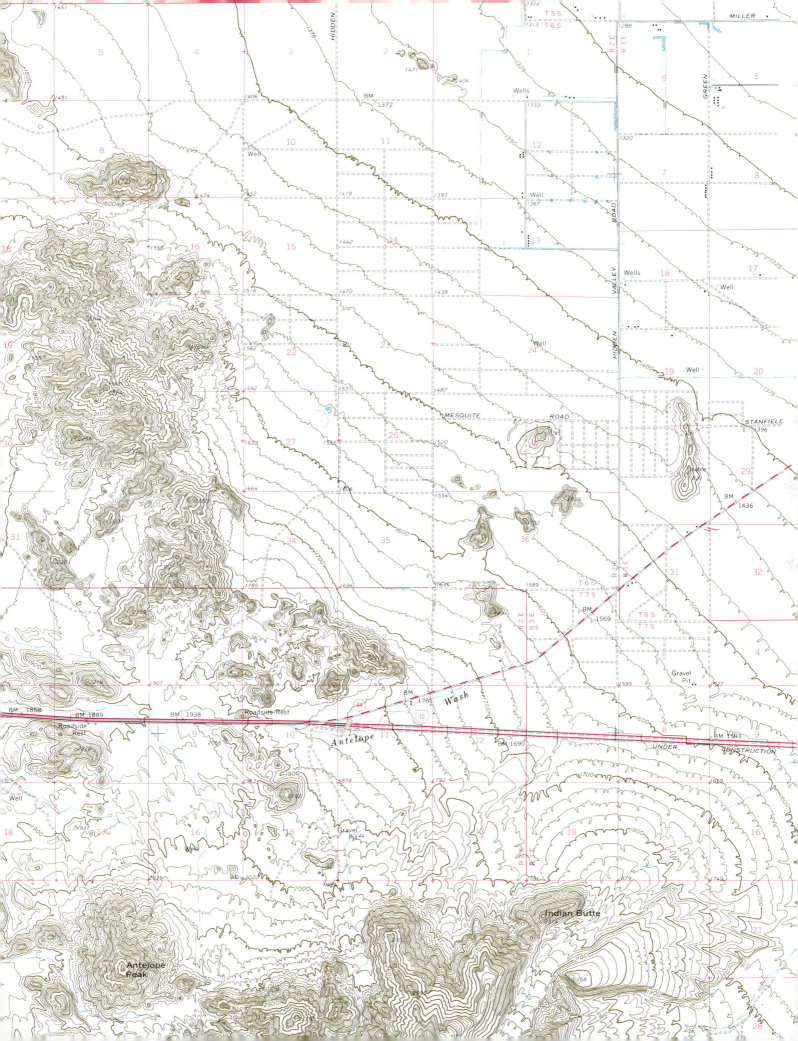

Directly across the canyon from Indian Butte, a wedge-shaped area slopes to the east from a hilltop with elevation 2258 feet. How do you think this feature formed?

The canyon has a box-like cross profile, with steep walls and a rather flat floor. One intermittent stream is shown on the canyon floor, yet the upstream bends in the contours indicate several potential channels. With a blue pencil, sketch as many possible channels as you can.

What pattern is shown by these channels, and what does that suggest about the nature of the sediment on the floor of the canyon?

4. Figure 12.12 is a portion of the Upheaval Dome, Utah, 1:62,500 quadrangle. It is within Canyonlands National Park, an area with beautiful canyons cut by the Colorado and Green Rivers (see Fig. 12.2). Reddish Paleozoic and Mesozoic sedimentary rocks, with nearly flat-lying layers, are carved into a series of cliffs and benches that gradually descend to the two rivers.

   a. Locate *The Neck* on the road just northeast of Grays Pasture. The Neck connects the plateau to the north to an area known as *Island in the Sky.* Island in the Sky is the Y-shaped, relatively flat area that includes Grays Pasture, Grand View Point, and the unnamed area to the west.

   What is the approximate width of The Neck?

   With the head of Taylor Canyon to the west, and the South Fork of Shafer Canyon to the east, what does the future hold in store for The Neck?

   Should your prediction come true, what kind of landform will the Island-in-the-Sky area become?

   b. Extending to the east of Grays Pasture are three, small, rather flat areas bounded by steep cliffs and connected by "necks;" the high point on the easternmost area is marked with its elevation of 5932 feet. What landform name would be given to these areas if the necks were to be severed by erosion? Assume that their base is at 4500 feet.

   c. Washer Woman is just southeast of Grays Pasture. What type of landform is just south of the word "Woman"?

   d. What is the relief in the map area? (The high point is on Willow Flat, near where the branches of the "Y" join.)

   Using the area around Grand View Point as a reference, speculate on the topographic level at which the next set of points and mesas would develop, once Grand View Point and the rest of the Island-in-the-Sky area have been eroded away? Use current elevations to express your answer.

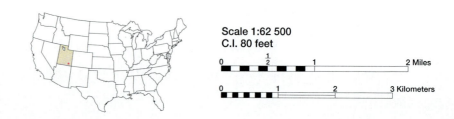

Scale 1:62 500
C.I. 80 feet

▶ FIGURE **12.12**
Portion of Upheaval Dome, Utah, 15-minute quadrangle, for Problem 4.

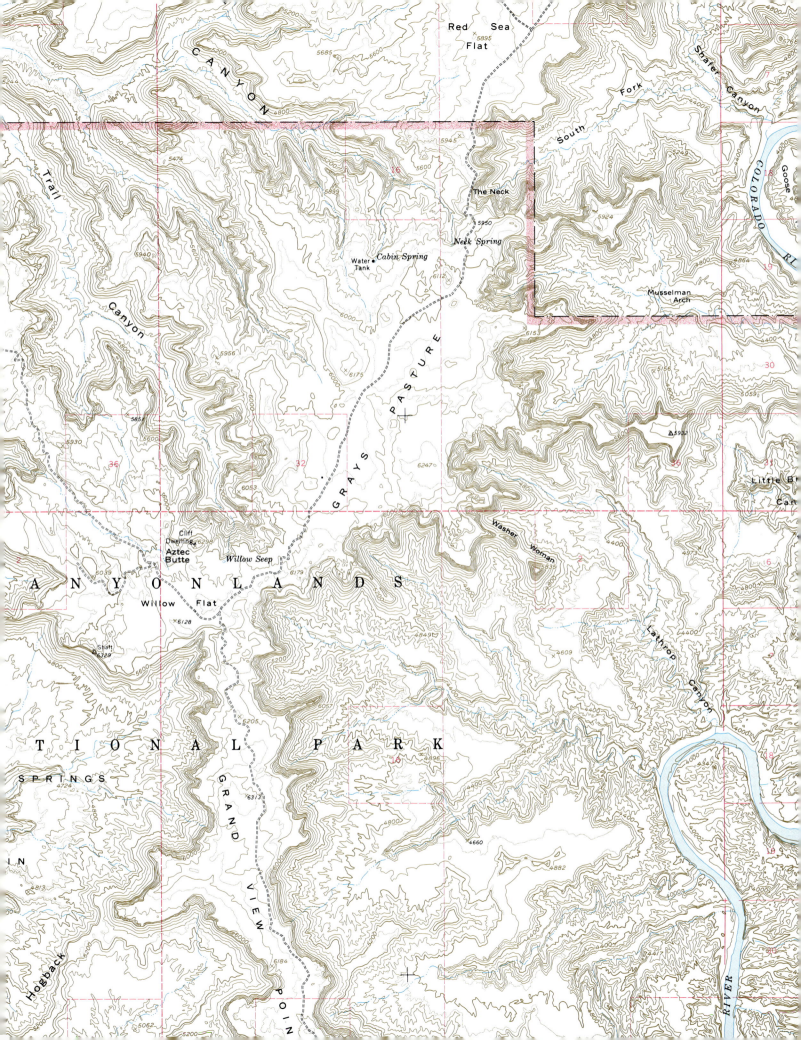

5. Figure 12.13 is a stereopair showing sand dunes on the southwest side of Salton Sea, south of Salton City, California, in the Imperial Valley.

    a.   What type of dunes are most prominent?

    b.   What is the prevailing wind direction?

    c.   These are active, migrating dunes. A comparison of the positions of selected, well-defined dunes in the years 1956 and 1963 shows just how active these dunes are. The distance each dune moved during this 7-year period is given in Table 12.1. The table also shows the height of the slip face of the dune, which is a measure of dune size. Prepare a graph on Figure 12.14 showing distance moved during the 7-year period on the x-axis, and the height of the slip face on the y-axis. Is there a relation between distance moved and size of dune, and if so, what is it?

        Give a possible explanation for this relation.

    d.   Dune number 9 in Figure 12.13 moved in the direction N75°E during the 7-year period. Using the data in Table 12.1, determine the rate, in meters per year, at which it moved. Show your work.

        What is the distance in centimeters in a N75°E direction from the top of the slip face in the center of dune 9 to the road?

        Convert this photo distance to actual ground distance in meters (scale of photo is 1:20,000). Show your work.

        Assuming the dune has continued to move at the same rate, how far has it moved since 1959, the date the photographs were taken? Show your work.

        In what year did (or will) the dune cross the road?

Table 12.1

DATA FOR PROBLEM 5
DATA FROM J. T. LONG AND R. P. SHARP, *GSA BULLETIN*, 1964.

DUNE NUMBER ON MAP	DISTANCE MOVED IN 7 YEARS (METERS)	HEIGHT OF SLIP FACE (METERS)
1	137	4.3
2	107	12.2
3	183	6.4
4	183	6.1
5	236	4.0
6	168	6.4
7	160	7.3
8	229	3.4
9	152	8.2
10	244	4.6
11	137	8.2
12	274	4.3
13	160	7.6
14	274	3.0

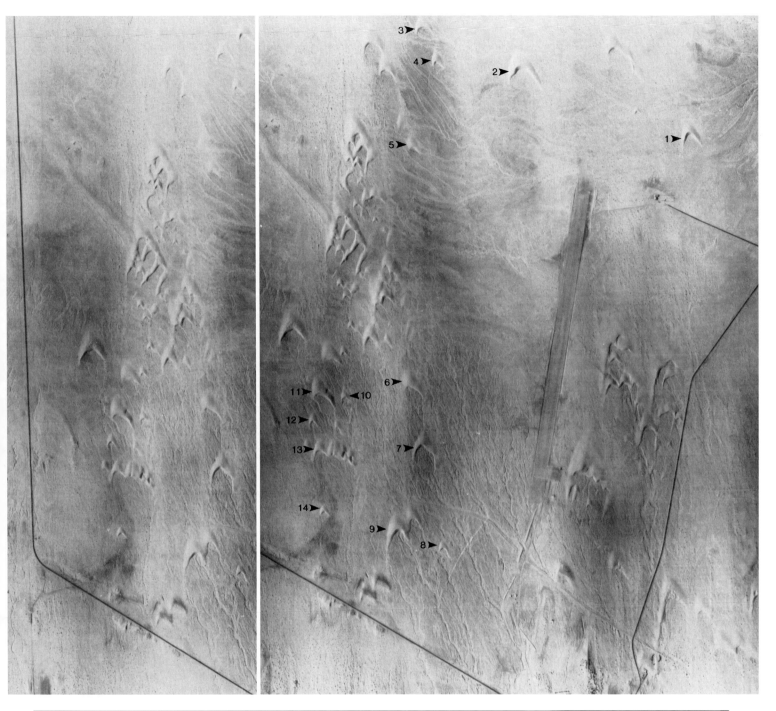

FIGURE **12.13**
Stereopair of area southwest of Salton Sea, California, for Problem 5. Scale 1:20,000; photos taken November 10, 1959. North is to the right.

6. Search on *"White Sands National Monument"* or go to *http://www.blarg.net/~lkj/gallery/white_sands.html* (or link to it through the author's homepage—see *Preface*), and from information contained there or in *links* from there, answer the following questions about the dunes at White Sands National Monument.

    a. The dunes at White Sands National Monument are unusual because of the mineral of which they are composed; it is the largest dune field in the world made of that mineral. What is that mineral and what is its hardness?

        From what you know about that mineral, or can learn from its description in Chapter 2, would you expect that sand grains made of it could have been transported very far from their source?

        According to Chapter 4 in the lab manual (for example, Table 4.5), how do most sedimentary deposits made of this mineral form?

    b. What sequence of events led to the formation of the dunes?

        How long ago did the dunes begin to form?

    c. In what direction does the prevailing wind blow?

        How far do some dunes migrate during a year?

        What types of dunes are present, and how large do some get?

IN GREATER DEPTH

7. For the purposes of this problem, assume that Figure 12.9 represents the west side of Death Valley on Figure 12.10 and that east is to the left in the photographs.

    a. Are the older deposits generally found near the upper or lower part of the fans (refer to your answer to Problem 1c for age relations)?

    b. Where on the fans is erosion most obvious, the upper or lower part?

    c. Can you explain this in terms of your answer to Problem 2a? Hint: How is the gradient of the west side of the valley in Figure 12.10 changed by movement along the fault, and how would this affect the balance between erosion and deposition?

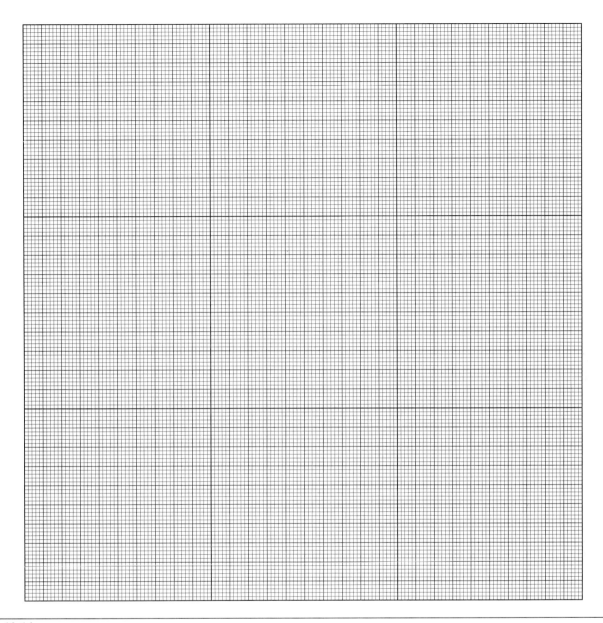

FIGURE **12.14**
Graph paper for Problem 5c.

Part 5

GEOLOGIC TIME AND SEQUENCES

chapter

13
Geologic Age

Overview

Rocks exposed at the surface of the Earth have had long and sometimes very complex histories. To understand what happened in the geologic past, and to understand the rocks themselves, geologists try to unravel that history. They want to know not only what happened, but when it happened. There are two ways to answer the question, "When did it happen?" One is to give a relative age by comparison with other events. For example, "The granite was intruded before deposition of sandstone A, but after limestone B." Another is to give the numerical age in number of years before the present. For example, "The granite was intruded 440 million years ago." The problems in this chapter illustrate how both relative and numerical ages are determined.

Materials Needed

Pencil and eraser • Metric ruler • Calculator • Slice of tree trunk showing growth rings (supplied by instructor)

INTRODUCTION

The "father of modern geology," James Hutton (1726–1797), realized that the events he saw recorded in rocks must have taken place over a very long period of time. He and his contemporaries were able to use scientific methods to recognize many different kinds of events in the rocks they examined, such as deposition of sediment, igneous activity, erosion, and deformation. They also were able to arrange these events in a *relative* sequence from oldest to youngest. However, they were unable to assign exact dates to anything that occurred before the beginning of recorded history. It was like knowing only

that you are younger than your mother, and that she is younger than your grandfather, but not knowing anyone's actual age.

Geologists now are able to determine quite precisely the dates, or **numerical geologic ages** (also known as *absolute geologic ages*), of many types of geologic events. Determining a numerical geologic age is complex and expensive, so numerical ages are not always readily available. However, enough numerical ages have been determined so that if the *relative geologic age* is known, an estimate of the numerical age can be made. Both types of ages, relative and numerical, are important pieces of information for interpreting and understanding the geology of an area, and

for unraveling the complexities of past tectonic-plate movements and interactions.

RELATIVE GEOLOGIC TIME

General Concept

Although it is important to know the numerical age of a particular geologic event, it is equally important to be able to indicate whether that event occurred before or after another event. This is the basis for a **relative geologic age**—the age of one event relative to another.

A description of the geology of an area includes a list of the geologic events

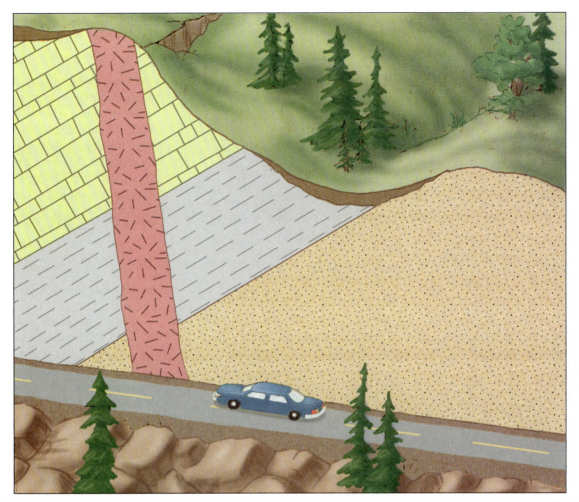

FIGURE **13.1**

View of rocks exposed in a vertical roadcut in a mountainous area. Sedimentary rocks are limestone (brick pattern), shale (parallel dashes), and sandstone (dots); granite dike shown with random dashes. What sequence of events led to what we see?

that took place, in the sequence in which they occurred, from oldest to youngest. Geologic events that are commonly described include deposition of sedimentary units, extrusion or intrusion of igneous rocks, metamorphism, folding, faulting, uplift, and erosion.

To understand how a sequence of events is determined, consider Figure 13.1, a sketch of a roadcut in a mountainous area. The sketch shows three inclined layers of sedimentary rock—sandstone, shale, and limestone—intruded by a granite dike. What geologic events must have occurred to produce what is seen in the roadcut, and in what order did they occur?

Start with the simplest, the dike. Because it cuts (intrudes) all three sedimentary layers, it must be younger than all of them.

Next, what are the relative ages of the sedimentary layers? The one on the bottom must have been deposited first (assuming the layers are not upside down, a possibility that is considered later), and the one on top last.

So far we have the following sequence:

granite dike (youngest)

limestone

shale

sandstone (oldest)

But there is more. Was the original sediment of the sedimentary rocks deposited as inclined layers? Not likely. The layers must have been tilted or folded after deposition.

And the dike—was it intruded before or after the beds were tilted? Can you tell? Not without more information, of a kind that is too detailed to discuss here. So we

must be satisfied with two possibilities for now; the dike could be older or younger than the tilting event.

What about the present-day surface, at the top of the roadcut? It is still forming, by erosion. Because the original sediments must have accumulated in a low place, and because they must have been buried under younger sediment, the erosion must have been accompanied by uplift of the whole area.

Now we can list the entire sequence in order of relative age.

Youngest: uplift and erosion

granite dike (or folding)

folding (or granite dike)

limestone

shale

Oldest: sandstone

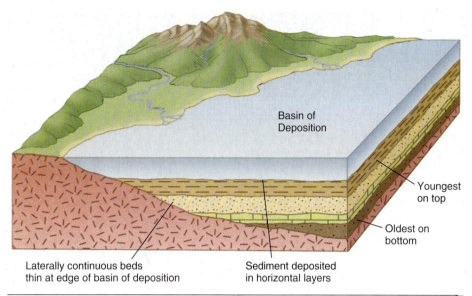

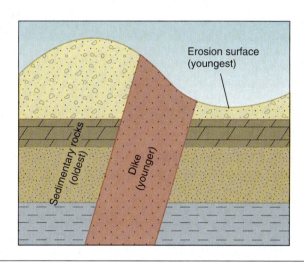

Basin of
Deposition

Youngest
on top

Oldest on
bottom

Laterally continuous beds
thin at edge of basin of deposition

Sediment deposited
in horizontal layers

FIGURE **13.2**

This three-dimensional view illustrates the principles of *Original Horizontality* (sediment is deposited in horizontal layers), *Superposition* (younger beds are deposited on older beds), and *Original Continuity* (sedimentary layers continue to the edge of the basin of deposition).

Erosion surface
(youngest)

Sedimentary rocks
(oldest)

Dike
(younger)

FIGURE **13.3**

Principle of Cross-Cutting Relations. This cross section shows a dike cutting preexisting layers of sedimentary rock; the dike is younger than the rocks it cuts. The erosion surface cuts both the sedimentary rocks and the dike, so is the youngest.

As you can see, there is nothing magical or mystical about the way in which this sequence was determined. Logical thought processes made use of a few basic principles.

Basic Principles

From the preceding example, you can see that several fundamental principles exist to help interpret the relative time relations among rocks. Among the more useful are the following, the first three of which were formulated by Nicholas Steno in the late 17th century.

Steno's Principle of Original Horizontality (Fig. 13.2): Sediments are deposited in layers that are horizontal or nearly horizontal. Therefore, when sedimentary rock layers are not horizontal, it can be assumed that they have been folded or tilted from their original position.

Steno's Principle of Superposition (Fig. 13.2): In any succession of sedimentary rock layers lying in their original horizontal position, the rocks at the bottom of the sequence are older than those lying above.

Steno's Principle of Original Lateral Continuity (Fig. 13.2): Sediments are deposited in layers that continue laterally in all directions until they thin out as a result of non-deposition, or until they reach the edge of the basin in which they are deposited. A layer that ends abruptly at some point other than the edge of the original basin was offset by a fault or intruded by an igneous rock, or has been partially removed by erosion.

Principle of Cross-Cutting Relations (Fig. 13.3): Any geologic feature (intrusive igneous rock, fault, fracture, erosion surface, rock layer) is younger than any feature that it cuts.

Principle of Inclusions (Fig. 13.4): An inclusion in a rock is older than the rock containing it. Examples of inclusions are pebbles, cobbles, or boulders in a conglomerate, or *xenoliths* (pieces of other rocks) in igneous rocks.

These principles help establish relative ages among rocks that occur together, but how about rocks that are widely separated? For example, although it was impossible to determine numerical ages at that time, 19th-century geologists knew that certain sedimentary rocks in England and the United States were the same age. How did they know? They compared **fossils** (naturally preserved remains or traces of preexisting plants or animals) and found them to be essentially identical. Two fundamental principles are based on fossils.

Principle of Fossil Succession: Fossil organisms succeed one another in time in a definite and recognizable order. Thus, the fossil content of a sedimentary unit will be different from older and younger units, and can be used to determine the unit's relative age.

Principle of Fossil Assemblages: Similar assemblages of fossil organisms indicate similar geologic ages for the rocks that contain them.

Testing Hypotheses

In the example in which the sequence of events illustrated in Figure 13.1 was hypothesized, some assumptions were made that should not have been made without further observations.

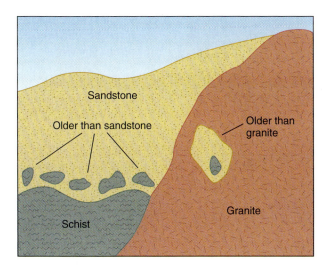

FIGURE 13.4

Principle of Inclusions. This cross section shows boulders of schist at the base of a sandstone; the boulders must be older than the sandstone containing them. Similarly, the xenolith of sandstone in the granite must be older than the granite.

The first assumption, a very reasonable one, was that the dike cut the layers of sedimentary rock. There is a *very remote* possibility that the dike was there first, standing as an inclined rock wall, while sediments were deposited on both sides of it. How would you tell? One way is to carefully examine the sedimentary rocks adjacent to the contact with the dike. For example, were they metamorphosed by the dike? If so, the dike must be younger. Do they contain any inclusions of the dike, such as pebbles, that might have been eroded from it? If so, the dike is older. Are there any small fingers of granite extending from the dike into the sedimentary rock that might have squeezed into weak places during intrusion of magma? If so, the dike is younger. Notice that the sedimentary layers cannot be projected straight across the dike but appear to be offset. Why? Because when the magma was intruded, it forced the rocks apart, at 90° to the dike margins.

A second assumption was that the layers, though not horizontal, have not been completely overturned. If they have, then the limestone is the oldest, and the sandstone the youngest. How can you tell? One way is to look for sedimentary structures that allow you to tell top from bottom, up from down. Some useful sedimentary structures, shown in Chapter 4, are:

Cross-stratification (see Fig. 4.4)— strata are commonly cut off on the top of the bed and become parallel to adjacent layers on the bottom.

Oscillation ripple marks (see Fig. 4.5B)— symmetrical, wave-like features whose crests point to the top of the bed.

Graded beds (see Fig. 4.6)—grain size commonly becomes progressively finer upward.

Mud cracks (see Fig. 4.7)—wider at the top than at the bottom.

Unconformities

The last event in the example shown in Figure 13.1 is erosion. What if sea level rose, or the land surface fell, and that erosion surface eventually was buried under younger sedimentary rock? The erosion surface would then be an **unconformity,** a surface that represents a substantial gap in the geologic record. It may be an ancient erosion surface, or it may be a surface on which neither erosion nor deposition occurred for a long period of time. If it is an erosion surface, it is recognizable because it cross-cuts older rocks, and its relative age can be determined by the Principle of Cross-Cutting Relations. If neither erosion nor deposition occurred, the unconformity may be difficult to recognize without studying fossils and applying the Principle of Fossil Assemblages. At any rate, there is no rock record of the time interval between the underlying rocks and those deposited on the unconformity. Unconformities record significant geologic events.

Unconformities are of three principal types (Fig. 13.5). Figure 13.5A shows an **angular unconformity,** an erosion surface separating rocks whose layers are not parallel. Layers above and below meet at an angle. A **disconformity** is either an erosion surface or a surface of nondeposition separating rocks whose layers are parallel. An erosion surface is uneven and cuts layers of underlying rocks (Fig. 13.5B); a surface of nondeposition parallels underlying rock layers (Fig. 13.5C). A **nonconformity,** Figure 13.5D, is an erosion surface separating sedimentary rocks from older plutonic or massive metamorphic rocks (that is, crystalline rocks that are not layered).

NUMERICAL GEOLOGIC TIME

Numerical ages or dates can be determined in several ways. For example, you can tell how old a tree is by counting the number of growth rings. Or you can determine how long some glacial meltwater lakes existed by counting the number of varves—annually deposited sets of layers—present in the lake sediment. For obvious reasons, these and similar methods have limited applicability. Radioactivity, however, provides a much more widely applicable method for determining numerical dates.

Radioactivity

Some isotopes are **radioactive;** that is, their nuclei spontaneously break down or decay. In these reactions, a radioactive isotope, or **parent,** decays to form a different isotope, the **daughter.** The decay process is described in the box titled "Radioactive Decay."

Isotopes are alternative forms of an element with different numbers of neutrons in their nuclei. For example, the element uranium has two common radioactive isotopes, ^{235}U and ^{238}U. The superscripts 235 and 238 are the atomic masses or **mass numbers** (number of protons plus neutrons in the nucleus) of the isotopes. Each isotope has 92 protons in its nucleus—92 is

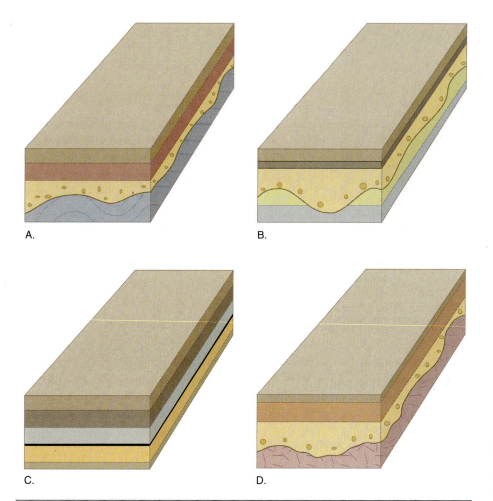

FIGURE 13.5

Three-dimensional diagrams of: A. angular unconformity; B. erosional disconformity; C. non-erosional disconformity; D. nonconformity. The unconformity is an ancient erosion surface in A, B, and D, but in C, it is a surface on which no erosion or deposition occurred for a long period of time.

er the mineral. Ages can be determined graphically or mathematically. The latter utilizes the equation

$$N_d = N_p (e^{\lambda t} - 1)$$
$$\text{or}$$
$$t = \frac{ln \left(\dfrac{N_d}{N_p} + 1 \right)}{\lambda}$$

in which N_d and N_p represent the number of atoms of daughter and parent, respectively; e (≈ 2.72) is the base of the natural logarithm (abbreviated ln); λ (the Greek letter lambda) is the *decay constant,* with units of 1/time; and t is time, in years. For isotopes with long half-lives and small decay constants, the previous equations can be approximated as

$$N_d \approx N_p \lambda t$$
$$\text{or}$$
$$t \approx (N_d/N_p)/\lambda.$$

The final equation can be used for the problems in this chapter.

GEOLOGIC TIME SCALE

The **geologic time scale** (Fig. 13.6) subdivides Earth's history into unequal intervals of time based on distinctive fossil assemblages or major geologic events. The scale is based on relative geologic time, but numerical ages are now known for all parts of the time scale. The largest subdivisions of geologic time are **eons;** eons are divided into **eras,** eras into **periods,** and periods into **epochs.** Subdivisions of the Phanerozoic eon are generally agreed upon worldwide, but this is not true with the time preceding the Phanerozoic. The subdivisions shown in Figure 13.6 for that time are more or less accepted in the United States. Because there has been so much disagreement about how to subdivide this time period, however, it is commonly just called the **Precambrian,** or less commonly, the Prephanerozoic.

the **atomic number** of uranium—but ^{235}U has 143 neutrons (235 − 92 = 143), whereas ^{238}U has 146 neutrons.

Ages from Isotopes

Laboratory measurements show that radioactive isotopes decay at a constant rate, and that characteristic makes them useful for determining numerical ages. The decay rate is measured by the **half-life**—the time required for half of the atoms of a radioactive isotope to decay. After one half-life,

half of the parent atoms are gone, having produced an equal number of daughter atoms. After two half-lives, only half of one-half, or one quarter, of the original parent remains; three-quarters of the parent has converted to daughter, so there are three times more atoms of daughter than parent.

The age of a mineral is determined from the relative amounts of parent and daughter isotopes it contains; the greater the percentage of daughter isotopes, and the lower the percentage of parent, the old-

There are three common ways by which a radioactive parent isotope of one element decays to a daughter isotope of another element: alpha decay, beta decay, and electron capture.

(1) In *alpha decay,* an *alpha particle,* which consists of two protons and two neutrons, is released spontaneously from the nucleus. The result is a daughter with two fewer protons and two fewer neutrons in its nucleus. For example, the first step in the decay of the ^{238}U isotope is

$$^{238}_{92}U \rightarrow {}^{234}_{90}Th + alpha$$
$$(_{92}U = uranium; {}_{90}Th = thorium).$$

Note that the loss of two protons and two neutrons reduces the atomic number of the daughter by two, and the mass number by four; when the atomic number (shown as a subscript) changes, a new element is formed.

(2) *Beta decay* requires the release of an electron or *beta particle* from the nucleus. This is accomplished by the breakdown of a neutron into a proton and an electron; the electron escapes the nucleus, but the proton remains. The result is a daughter with one more proton and one less neutron in its nucleus. For example, the decay of ^{87}Rb is

$$^{87}_{37}Rb \rightarrow {}^{87}_{38}Sr + beta$$
$$(_{37}Rb = rubidium; {}_{38}Sr = strontium).$$

Note that gaining one proton increases the atomic number of the daughter by one, but the simultaneous loss of one neutron causes the mass number to remain the same.

(3) In *electron capture,* an electron is captured by a proton in the nucleus. In the process, the proton is converted to a neutron. The result is a daughter with one less proton in the nucleus. For example, the decay of ^{40}K is

$$electron + {}^{40}_{19}K \rightarrow {}^{40}_{18}Ar$$
$$(_{19}K = potassium; {}_{18}Ar = argon).$$

Note that the loss of a proton in the nucleus reduces the atomic number by one in the daughter, but since the proton is converted to a neutron, the mass number does not change.

Q u e s t i o n s : What kinds of decay take place in the following reactions?

1. $^{14}_{6}C \rightarrow {}^{14}_{7}N$

2. $^{147}_{62}Sm \rightarrow {}^{143}_{60}Nd$

Geologic Time Scale

Eon	Era	Period		Epoch	Age (millions of years ago)
Phanerozoic	Cenozoic (Cz)	Quaternary (Q)		Recent or Holocene	
					—0.01—
				Pleistocene	
					—1.6—
		Tertiary (T)	Neogene (N)	Pliocene	
				Miocene	
					—23.7—
			Paleogene (Pε)	Oligocene	
				Eocene	
				Paleocene	
					—65—
	Mesozoic (Mz)	Cretaceous (K)			
					—146—
		Jurassic (J)			
					—208—
		Triassic (Ŧ)			
					—245—
	Paleozoic (Pz)	Permian (P)			
					—290—
		Carboniferous (C)	Pennsylvanian (℗)		
					—325—
			Mississippian (M)		
					—360—
		Devonian (D)			
					—410—
		Silurian (S)			
					—440—
		Ordovician (O)			
					—505—
		Cambrian (Є)			
					—544—
Precambrian	Proterozoic				
					—2500—
	Archean				
					—3800—
	Hadean				
					—4600—

FIGURE 13.6

The geologic time scale, with symbol abbreviations in parentheses. Adapted from The Museum of Paleontology of the University of California at Berkeley Web page.

ROCKS ARE THE PAGES IN EARTH'S HISTORY BOOK. UNFORTUNATELY, THERE IS NO PLACE ON EARTH WHERE THE WHOLE BOOK IS INTACT. PAGES, OR EVEN WHOLE CHAPTERS, HAVE BEEN REMOVED BY EROSION, AND MANY PAGES HAVE BEEN BENT OR TORN APART BY DEFORMATION. THE GEOLOGIST'S JOB IS TO READ THE BOOK THE WAY IT IS, AND THAT REQUIRES SORTING OUT THE TIME RELATIONS USING SOUND SCIENTIFIC METHODS. IN THIS LAB, YOU FIRST WILL LEARN HOW BASIC PRINCIPLES ARE USED TO ESTABLISH THE ORDER IN WHICH ROCKS FORMED OR GEOLOGIC EVENTS OCCURRED. ADDITIONAL PROBLEMS ILLUSTRATE HOW NUMERICAL GEOLOGIC AGES ARE DETERMINED.

OBJECTIVES

If you complete all the problems, you should be able to:

1. Determine relative ages and the sequence of geologic events as illustrated in a geologic cross section.

2. Use daughter-parent ratios, half-life, and the decay constant to determine the numerical age of a rock or mineral graphically and mathematically.

PROBLEMS

1. Figures 13.7 and 13.8 are geologic cross sections. Determine the sequence of events that led to the present situation, and list them in order, from oldest to youngest, in the blanks provided. Use the letters on the illustrations to specify rock units or events. Identify the rock types represented by each letter from the symbols given in Figure 13.9; identify other events by name (for example, folding or erosion). The number of blanks equals the number of events required.

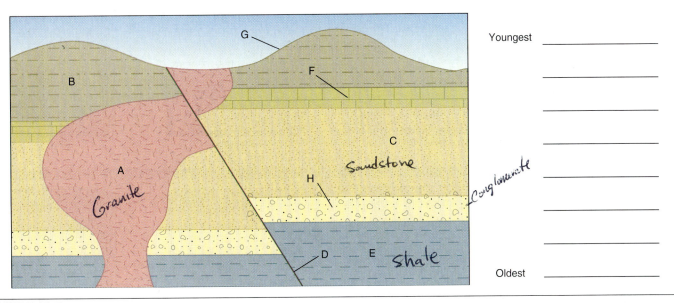

Youngest _____

_____

_____

_____

_____

_____

_____

Oldest _____

FIGURE **13.7**
For Problem 1, determine the relative ages of the lettered geologic features illustrated in this cross section, and list them chronologically in the spaces provided. D (dark line) is a fault, and G is the present surface.

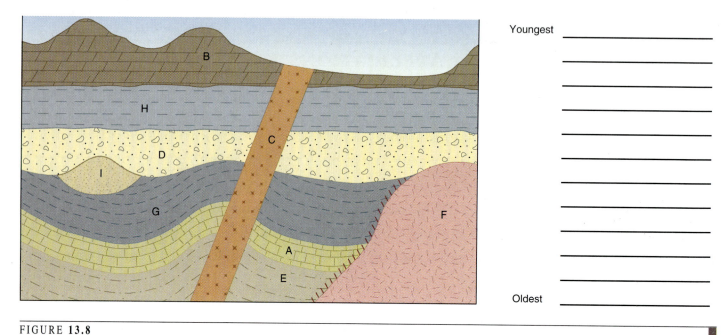

Youngest _____

_____

_____

_____

_____

_____

_____

_____

_____

_____

Oldest _____

FIGURE 13.8

For Problem 1, determine the relative ages of the geologic features illustrated in this cross section, and list them chronologically in the spaces provided. Letters indicate rock units; other events are not labeled. Also see Problem 7.

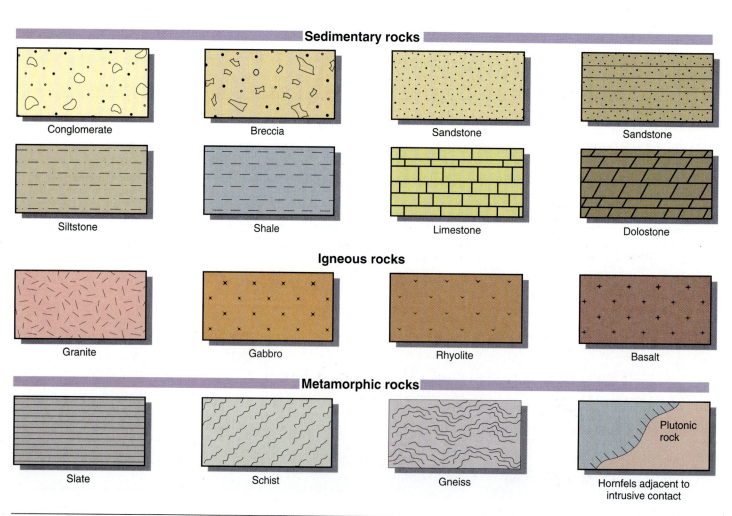

FIGURE 13.9

Symbols commonly (but not universally) used to show different kinds of rocks.

216

FIGURE **13.10**
For Problem 2, tape a piece of paper to the tree slice, mark the rings on the paper, and label the years.

2.   Trunks and branches of trees grow by adding new layers each year just under the bark. When the tree is cut, these concentric tree rings are visible and can be used to determine the age of the tree. The width of a tree ring depends on the growth conditions that year, principally the availability of water. The University of Arizona Laboratory of Tree-Ring Research has used tree rings extensively to learn about North American climates during the past 400 years. Tree-ring analysis also has been used to date volcanic eruptions, earthquakes, major glacier expansions, floods, and landslides that occurred as much as 8,600 years ago. In this problem, a slice of a tree is used to illustrate one method of determining a numerical age, and to illustrate the correlation between precipitation and tree-ring width.

   Tape a strip of paper on the tree slice so its edge goes from the middle of the innermost ring to the outer edge of the slice (Fig. 13.10). If the widths of individual rings vary considerably, choose a position for your strip of paper that appears representative, or "average." After checking with your instructor to make sure you know what an annual tree ring looks like, mark the position of each ring on the paper; do this in such a way that the distance between your marks is equal to the width of the ring. Do not mark on the wood—it has to be used by others. Take the paper off, and label the band representing the outermost ring with the year in which the tree was cut down. Then label those bands on the paper representing 1990, 1980, 1970, and so forth. Use your paper record to answer the following questions.

   a.   Determine the year in which the tree started growing.

   b.   Measure the radius of the tree in millimeters.

   c.   Determine the *average rate* at which the radius of the tree increased by dividing the radius by the age of the tree.

   d.   Find the range of radius-growth rates during the tree's lifetime by measuring the width of the smallest and the widest bands.

   e.   Fill in the "Year" column of Table 13.1, starting at the top with the year the tree was cut. Obtain annual precipitation data for these years from your instructor, and enter in the "Precipitation" column. Measure the width of each annual ring, in millimeters, and write its value in the appropriate row in the "Ring Width" column of Table 13.1. Make a graph of these data on Figure 13.11. Plot precipitation on the abscissa (X axis) and ring width for that year on the ordinate (Y axis). Does there appear to be a direct relation between precipitation and tree-ring width? Explain the relation. What other variables might account for differences in tree-ring width?

Table 13.1

YEAR	PRECIPITATION	RING WIDTH (MM)	YEAR	PRECIPITATION	RING WIDTH (MM)

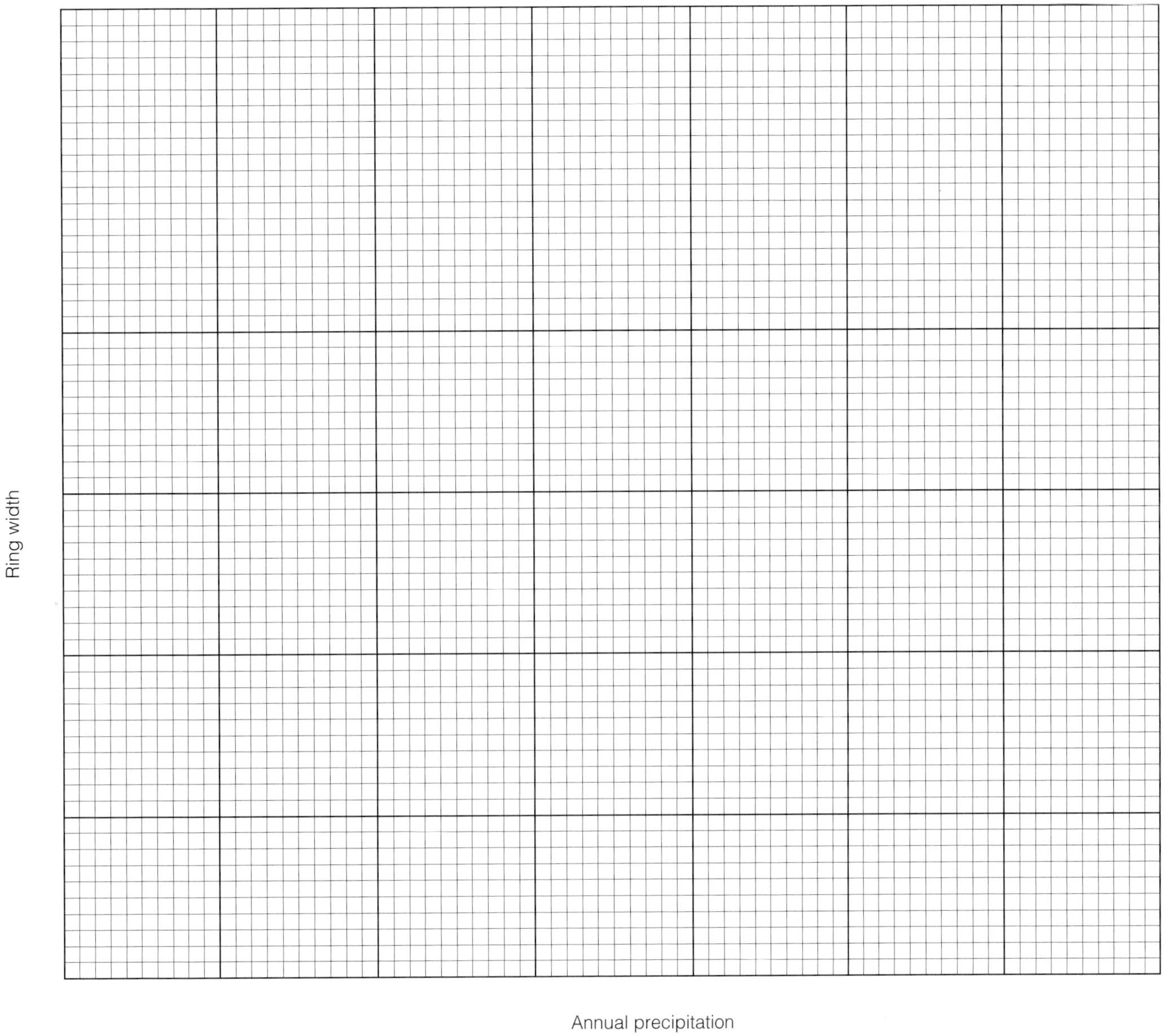

FIGURE **13.11**
Graph for Problem 2e.

3. The purpose of this problem is to illustrate how numerical ages based on radioactivity can be determined graphically.

   a. Complete columns A and B in the table on the following page. For example, after one half-life, the parent fraction is 0.5 and the daughter fraction is 0.5; after 2 half-lives, the parent fraction is 0.25 and the daughter fraction is 0.75. Remember, the sum of the two fractions must equal 1.0.

   b. Next, complete column C in the table by dividing values in Column B by corresponding values in Column A. For example, for 1 half-life elapsed, (Col. B)/(Col. A) = 0.5/0.5 = 1.

HALF-LIVES ELAPSED	A. PARENT FRACTION	B. DAUGHTER FRACTION	C. DAUGHTER/ PARENT RATIO
0			
1	0.5	0.5	1
2	0.25	0.75	
3			
4			
5			

c. Using Figure 13.12 and the data in the preceding table, construct a graph in which the ordinate is "Daughter-Parent Ratio," and the abscissa is "Half-Lives Elapsed." The point representing one half-life is already plotted. Plot the rest, and draw a *smooth* curve connecting the data points; that is, do not connect the points with straight-line segments, but estimate the curvature between points as best as you can so that the entire curve bends smoothly.

d. For samples 1–3 in the following table, first calculate and record N_D/N_P. Then, using your graph in Figure 13.12, determine the number of half-lives that have elapsed for each sample and write your answer in the "Half-Lives Elapsed" column.

SAMPLE NUMBER	ATOMS OF PARENT N_P	ATOMS OF DAUGHTER N_D	N_D/N_P	HALF-LIVES ELAPSED	AGE IN YEARS
1	2135	3203			
2	4326	10,815			
3	731	14,620			

e. If the half-life is 8,200 years, calculate the age in years of the samples in the preceding table and write your answer in the "Age in Years" column. Show your work.

4. The purpose of this problem is to illustrate how numerical ages based on radioactivity can be determined mathematically. Show all your calculations.

a. If the decay constant for radioactive decay of $^{40}_{19}K$ is $\lambda = 5.543 \times 10^{-10}$/year, use the final equation on p. 212 to calculate the ages of the samples in the following table, and write them in the "Age in Years" column.

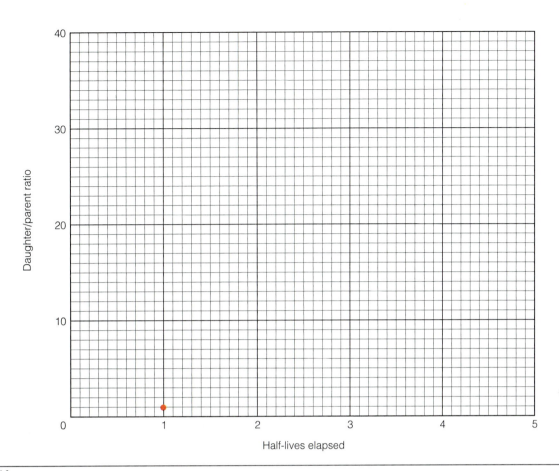

FIGURE **13.12**
Graph for Problem 3c.

SAMPLE NUMBER	ATOMS OF PARENT N_P	ATOMS OF DAUGHTER N_D	N_D/N_P	HALF-LIVES ELAPSED	AGE IN YEARS
1	6439	2303			
2	4395	1303			
3	8763	1893			

b. The half-life, $t_{1/2}$, of $^{40}_{19}K$ can be determined from the decay constant using the following relation:

$$t_{1/2} = 0.693/\lambda. \quad \textit{(see footnote 1)}$$

What is the half-life of $^{40}_{19}K$? _____

c. Using the half-life calculated in Problem 4b, complete the column labeled "Half-Lives Elapsed" in the previous table.

---

[1] *The equation $t_{1/2} = 0.693/\lambda$ is derived from the equation $t = 1n(N_d/N_p + 1)/\lambda$ as follows: When $t = t_{1/2}, N_d/N_p = 1$, and $1n(N_d/N_p + 1) = \ln 2 = 0.693$, then $1n(N_d/N_p + 1)/\lambda = 0.693/\lambda = t_{1/2}$.*

5. Go to *http://www.ucmp.berkeley.edu/help/timeform.html* (or link to it through the author's homepage—see *Preface*), and use the *links* to answer the questions below. You will notice in the geologic time scale that the subdivisions for the Precambrian are different from those in Figure 13.6, and that some of the dates for periods in the Phanerozoic are different as well. These differences illustrate that the science is still evolving and that there is still disagreement about some things.

a. What is the significance of the suffix "zoic" in the names for the eons and eras?

b. How old are the oldest known fossils?

When did oxygen first become a significant component of the Earth's atmosphere, and where did it come from?

What are stromatolites, when did they first appear, and are stromatolites still alive on the Earth today?

When did the first animals appear?

c. How did life change at the beginning of the Paleozoic?

d. When were dinosaurs alive on Earth?

When did each of the following first appear on Earth:
Birds?
Reptiles?
Mammals?
Insects?
Fish?
Flowering plants?
Humans?

e. What percent of geologic time are each of the following: Precambrian, Paleozoic, Mesozoic, and Cenozoic?

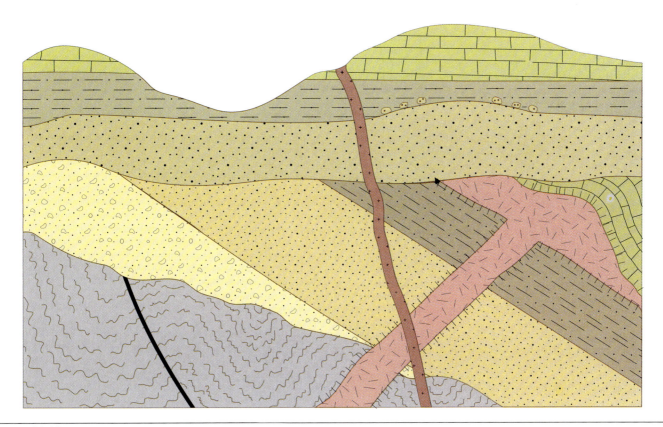

FIGURE **13.13**

For Problem 6. Determine the sequence of events illustrated in this cross section. Identify rock units from their symbols, and list all the events that led to the present situation. Wide line is a fault.

IN GREATER DEPTH

6. Using Figure 13.13, identify rock units from their symbols, and identify and list in sequence *all* the events that led to the present situation. On the cross section, write number 1 on the oldest rock or event, number 17 on the youngest, and intermediate numbers on rocks or events in between. If you cannot determine which of two or more rocks/events is the older, explain why not.

7. Refer to Figure 13.8.

    a. The daughter-parent isotope ratios (N_D/N_P) for rocks F and C are 16.0 and 0.92, respectively. Assume that the half-life of the parent is 6.5×10^7 years, and use the curve you derived for Problem 3 to determine the numerical ages of each of these rocks.

    b. During what geologic periods were these rocks intruded (see Figure 13.6)?

    c. Assume, for illustrative purposes, the highly unlikely scenario in which each of the lettered rock units A through I represents a geologic period of the Phanerozoic. What period (or periods, if more than one answer is possible) does each represent?

    d. During what geologic period or periods did folding and erosion occur?

14

Structural Geology

Overview

Stresses within the Earth bend and break rocks, producing folds, faults, and other geologic structures. These structures are three-dimensional, and commonly too large or too poorly exposed to be easily recognized from a single vantage point on the ground. For these reasons, basic geologic data must be collected in the field and plotted on a map to understand the structure fully.

This chapter will help you visualize geologic structures in three dimensions so that you can understand them when seen in two dimensions on a geologic map, aerial photograph, or on the Earth's surface. You will learn how the orientations of planar features, such as sedimentary bedding, are described and used to portray the geologic structure on a geologic map. Vertical cross sections of geologic maps illustrate what you expect to see below the surface, and block diagrams combine maps and cross sections to provide a three-dimensional view.

Materials Needed

Pencil and eraser • Protractor • Metric ruler • Silly Putty® (provided by instructor)
• Scissors • Removable tape (optional)

INTRODUCTION

Structural geology is the study of the form, arrangement, and internal structure of rocks. It is concerned especially with the way in which rocks are deformed. Evidence for deformation is abundant. We see rocks broken during earthquakes, and evidence for deformation in the past—such as bent and broken layers of sedimentary rocks—is common.

Deformation is caused by **stress,** a force acting on an area of a body that tends to change its size or shape. Stress is a force per unit area, so the magnitude of stress depends not only on the magnitude of the force, but also on the area over which the force is applied. Stress will be greater if a given force is applied to a small, rather than large, area; a person wearing high-heeled shoes is more likely to punch a hole in a rotten floorboard than a person wearing tennis shoes. *Compressive stress,* or **compression,** causes shortening (Fig. 14.1A), and is an important type of stress at convergent tectonic plate boundaries. *Tensional stress,* or **tension,** the principal stress at divergent boundaries, causes stretching (Fig. 14.1B). **Shear stress** causes one side of a body to slide past the other (Fig. 14.1C), and is the principal type of stress along transform-fault plate boundaries.

Strain, the deformation that occurs in response to stress, is a change in size or shape. Strain may be temporary or permanent. **Elastic strain** is temporary; if the stress is removed, the original size or shape will be restored. A stretched rubber band and a bouncing ball both undergo

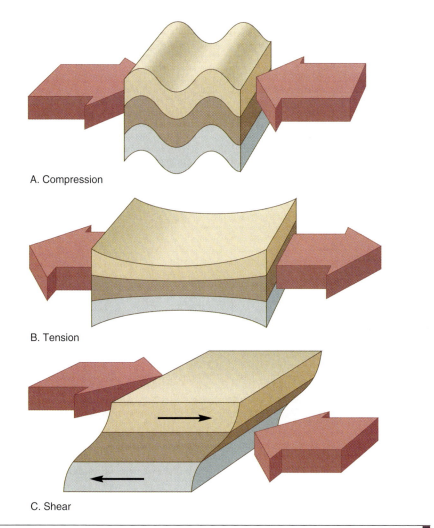

A. Compression

B. Tension

C. Shear

FIGURE **14.1**
The three principal types of stress: A. compression; B. tension; C. shear.

Table 14.1

STRESS/STRAIN CONDITIONS OF COMMON GEOLOGIC STRUCTURES

STRAIN	STRESS		
	Compression	*Tension*	*Shear*
Ductile or plastic strain	Folds Anticline Syncline		
Brittle strain or rupture	Faults Reverse Thrust	Fault Normal	Fault Strike-slip

elastic strain. **Plastic strain** is permanent; if the stress is removed, the original size or shape will not be restored. If you bend a piece of soft wire, the wire undergoes plastic strain; the bend is permanent. If it is bent back and forth at the same place, it eventually breaks or **ruptures.**

A rock is **brittle** if it breaks while undergoing elastic strain, but little or no plastic strain. A dry stick of wood behaves this way—you can bend it so far before it breaks. If you let go before it breaks, it springs back (elastically) to its original position. A rock that undergoes much plastic strain without rupturing is **ductile.** The aforementioned bent wire is ductile.

A rock may be brittle under some conditions and ductile under others. Ductile behavior is favored over brittle by high confining pressure, high temperature, and slowly applied stress. High confining pressure and high temperature are found at depth. The folded rocks visible in many mountain ranges were deformed well below the surface by forces acting over long periods of time (millions of years). The combination of depth and time enables deep rocks that would be brittle at the surface to behave plastically and gradually bend into folds. Eventually, erosion of overlying rock material and gradual uplift bring the deformed rocks to the surface where we can view them.

Table 14.1 summarizes the principal kinds of geologic structures in terms of the type of stress that produced them and the nature of the strain.

RECOGNIZING AND DESCRIBING DEFORMED ROCKS

A geologist must be able to visualize the architecture or structure of rocks in three dimensions. In some parts of the world, rocks are so well exposed at the surface that it is easy to see and understand their structure. In most areas, however, the rocks are buried beneath soil or young, unconsolidated sediment, and are exposed only in a few **outcrops.** In such places, it is difficult to understand the three-dimensional aspects without carefully examining the rocks, measuring their orientation, and plotting these data on a map.

FIGURE 14.2
The steeply inclined surface in the center of the photograph is a fault contact. It cuts the sandstone and mudstone layers of the Pittsburg Bluff Formation (Eocene–Oligocene) near Vernonia, Oregon.

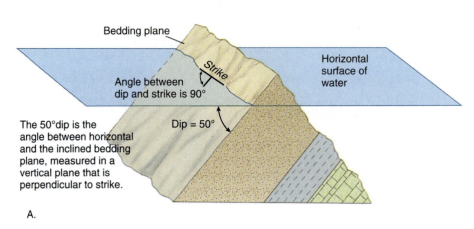

A.

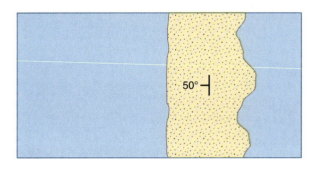

B.

FIGURE 14.3
A. A layer of sandstone with a smooth bedding plane is shown protruding above the water level in a lake. The water surface is horizontal, so the intersection with the bedding plane—the shoreline—is parallel to the strike of the bed. The bed dips into the water at an angle of 50° to the horizontal; the direction of dip is at 90° to the strike and is measured in a vertical plane. B. When viewed from above, as on a map, the strike and dip of the bed are shown with a T-like symbol. The top of the T parallels the strike, and the leg of the T parallels, and points in the direction of, the dip. The angle of dip is 50°.

What to Look for: Stratification, Contacts, and Fault and Joint Surfaces

How can you tell if sedimentary rocks have been deformed? It generally is easy—just look at the orientation of the strata. The original sediments were deposited as continuous horizontal, or nearly horizontal, strata or layers. If the strata are still horizontal and not broken, then they have not been deformed, although they may have undergone gentle uplift or subsidence. If strata are broken or not horizontal, then they have been deformed. Deformation takes many forms, from slight tilting of strata to complex folding, as described in a later section.

Contacts are surfaces separating adjacent bodies of rock of different types or ages. Understanding the nature of the contact is important, because the contact

records a change, an event. There are several kinds of contacts.

Depositional contacts separate older bodies of rock from younger sedimentary rocks deposited upon them. In a sequence of sedimentary rocks deposited continuously, or **conformably,** in a basin of deposition, the contacts between adjacent layers are planar **bedding planes** (see Fig. 4.3); they are parallel to—and, in fact, define—the bedding or stratification. Some depositional contacts are ancient erosion surfaces. These **unconformable** contacts can be uneven rather than planar, and can juxtapose very different rocks with very different orientations (see *Unconformities* in Chapter 13).

An *intrusive contact* separates intrusive igneous rock and the rock that it intruded; intrusive contacts are either *concordant* (parallel to layers in the intruded rock) or *discordant* (cut across layers in the intruded rock).

Rocks separated by a fault are in *fault contact* (Fig. 14.2). In the simplest case, the fault is exposed as a flat surface, and its orientation is apparent. However, faults commonly are expressed as zones along which rocks are brecciated (broken into angular fragments), or even pulverized, so that orientation may be more difficult to determine. Even more commonly, faults, like other types of contacts, are buried beneath soil or sediment. In such places, their orientation, and even their presence, must be inferred from data gathered from many outcrops and plotted on a map.

Describing Surface Orientation

To describe the orientation, or **attitude,** of a surface, whether a sedimentary bed, a contact, or a fault, a standard frame of reference must be used. The most convenient

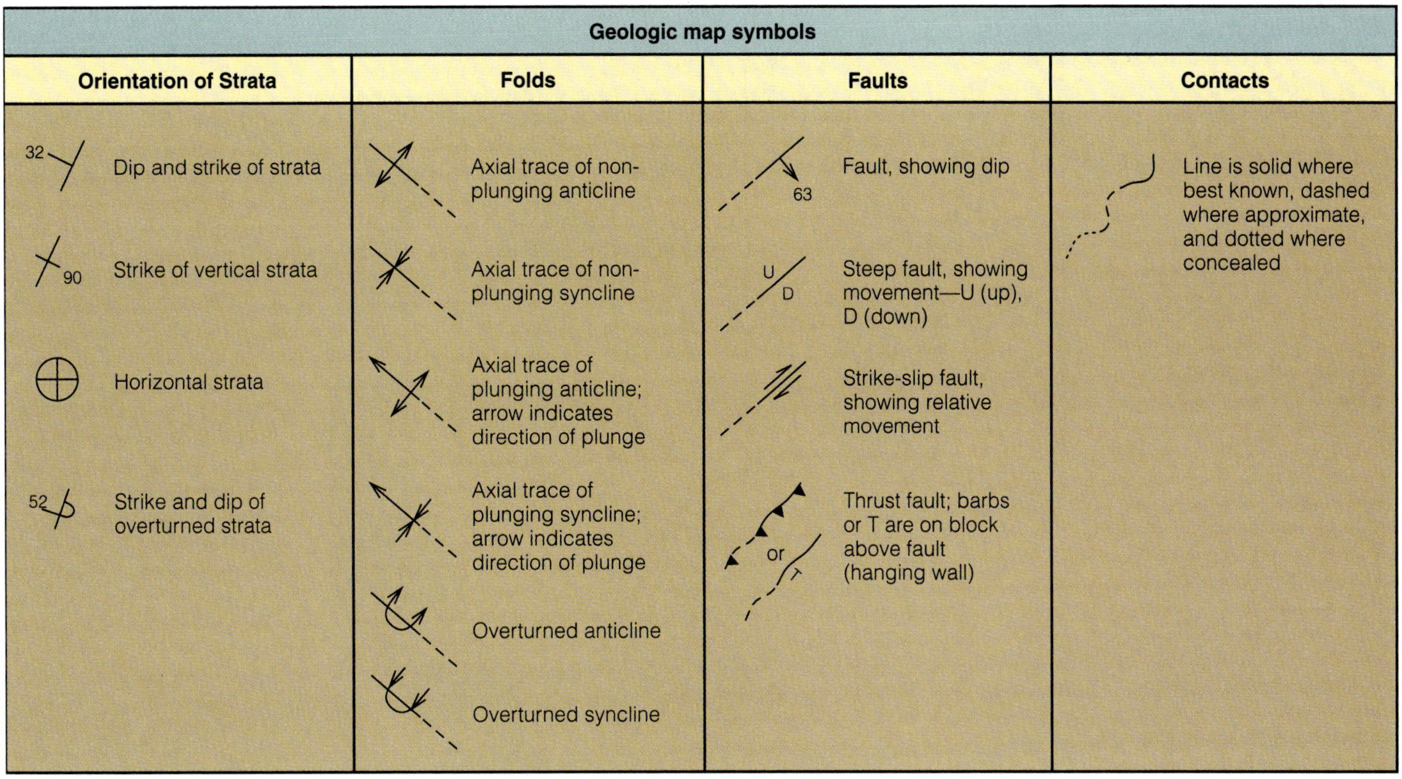

FIGURE 14.4
Geologic map symbols.
Source: Data from American Geological Institute and others.

standards are compass direction and a horizontal surface. Compass direction is easily determined with a compass, and *horizontal* is found with a level similar to that used by a carpenter. The orientation of a line on a surface is described by its compass direction; the tilt or inclination of a surface is described by its deviation from horizontal.

Geologists measure the orientation of a surface by determining the *strike and dip* of that surface. The **strike** of a surface is the compass direction of the line formed by the intersection of the surface and a horizontal plane (Fig. 14.3). In other words, strike is the compass direction of any horizontal line on the surface of interest. A common convention is to record the strike as the number of degrees east or west of north: N30°W means 30° to the west of north. With this convention, the number of degrees of strike is always less than 90; thus, N90°E is called E-W, and N91°E is called N89°W. **Dip** is the angle of inclination of the surface, as measured downward from a horizontal plane to the surface. Dip is always measured in a vertical plane that is perpendicular to the strike (Fig. 14.3). In addition to the angle of dip, it also is necessary to specify

the *direction of dip*. But because dip is perpendicular to strike, it is only necessary to indicate the compass quadrant toward which the surface dips (that is NW, NE, SE, or SW). For example, a dip of 52°NE indicates that a surface is inclined to the northeast at an angle of 52° from the horizontal. The complete orientation, or *attitude,* of a surface is then expressed with both strike and dip; for example, N30°W, 52°NE. Strike and dip are indicated on maps by various symbols, as shown in Figure 14.4.

Geologic Maps, Cross Sections, and Block Diagrams

Geologic information is compiled and displayed in three common ways, shown in Figure 14.5: geologic maps, geologic cross sections, and block diagrams.

Geologic maps show the distribution of rock units at or very near the surface of the Earth. They also portray the structure of the rock units by means of strike and dip symbols. An accompanying legend, or *explanation,* lists and briefly describes the rock units in order of age; major periods of

erosion are also listed. Topography may be shown with an overlay of contour lines.

Geologic cross sections are vertical slices that show how the rocks would appear if they could be viewed on a cliff face. They are topographic profiles; the area below the surface is filled in with representations of rocks. Because such cutaway views do not usually present themselves in nature, they are interpretations based on information obtained on the surface, or from wells or geophysical data.

Block diagrams combine maps and cross sections to show the three-dimensional aspects. The three-dimensional blocks are illustrated with two-dimensional, perspective drawings, with the map view shown on the top, and cross sections on the ends and sides. Most block diagrams have a flat upper surface and do not attempt to show topography, but some, drawn by clever people, manage to illustrate topography as well as geology.

FOLDS

A fold is a bend in a layer of rock. Upward bends, or upfolds, with older rocks in the

229

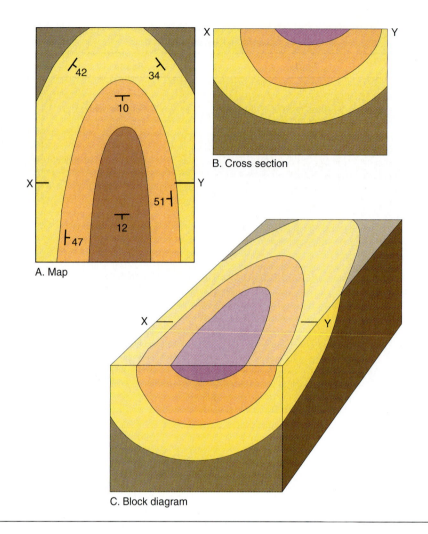

A. Map

B. Cross section

C. Block diagram

FIGURE 14.5

Three ways in which geologic information is illustrated: A. A geologic map shows rock units and geologic structure as they appear on the surface.
B. A geologic cross section of the map from X to Y is like a vertical slice into the Earth. C. Block diagrams combine map and cross-section views in a three-dimensional perspective drawing.

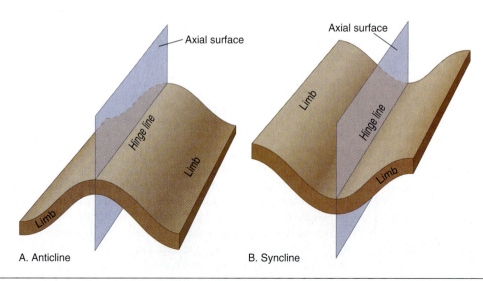

A. Anticline

B. Syncline

FIGURE 14.6

Folds are described in terms of limbs, hinge line, and axial surface, shown here for an anticline (A) and a syncline (B).

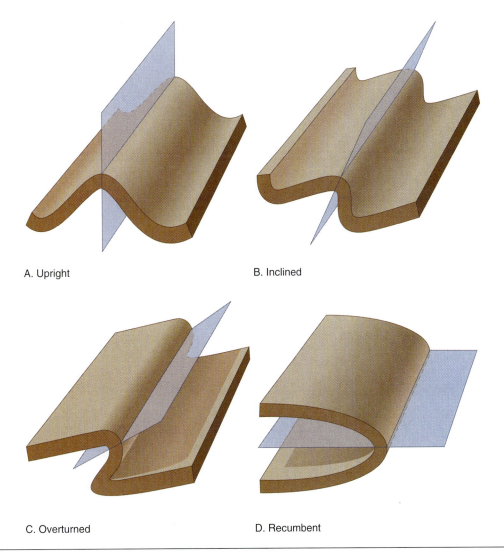

A. Upright

B. Inclined

C. Overturned

D. Recumbent

FIGURE 14.7
The axial surface of a fold can be: A. Vertical in **upright folds;** B. inclined in **inclined folds;** C. inclined so much that opposite limbs dip in the same direction in **overturned folds;** D. horizontal in **recumbent folds.**

center are called **anticlines;** downward bends, downfolds, with younger rocks in the center are called **synclines.** The sides or flanks of a fold are the **limbs** (Fig. 14.6). The limbs join at the **hinge line** of the fold, which is the line of maximum curvature (Fig. 14.6). The **axial surface** (or *axial plane* if it's not curved) of a stack of folded layers passes through the hinge lines and most nearly divides the fold into two equal parts (Fig. 14.6). An **upright fold** has a vertical or nearly vertical axial surface; beds on opposite limbs have similar dips, though in opposite directions (Fig. 14.7A). An **inclined fold** has an axial surface that is neither vertical nor horizontal (Fig. 14.7B). An inclined fold is **overturned** if beds on opposite limbs dip in the same direction; beds on the overturned limb are upside-down, as they have been

overturned more than 90° (Fig. 14.7C). A **recumbent fold** has an approximately horizontal axial surface (Fig. 14.7D).

A **non-plunging fold** has a horizontal or nearly horizontal hinge line (Figs. 14.8A, B). A **plunging fold** has an inclined hinge line (Figs. 14.8C, D).

Figure 14.9 illustrates folds using block diagrams in which the top surface is horizontal and flat. Note that when non-plunging folds intersect a horizontal surface, such as that approximated by the surface of the Earth, the contacts between adjacent beds are straight lines (Figs. 14.9A, B). When plunging folds intersect a horizontal surface, the contacts between adjacent beds are curved lines. In a plunging anticline, the contacts bend so that they "point" in the direction of plunge,

whereas in a plunging syncline, they "point" in the direction opposite the plunge (Figs. 14.9C, D).

Also note that in an eroded syncline, the *youngest* beds appearing on the surface are in the center of the syncline (Figs. 14.9B, D); in an eroded anticline, the *oldest* beds are in the center (Figs. 14.9A, C).

Domes and **basins** have cross sections like anticlines and synclines, respectively, but are approximately circular in map view (Fig. 14.10). Beds in domes dip outward in all directions, and like anticlines, the oldest rocks are found in the center. Similarly, the youngest rocks in basins are in the center, just as they are in synclines, and beds dip inward. Domes and basins may be quite large, with diameters of 100 km or more.

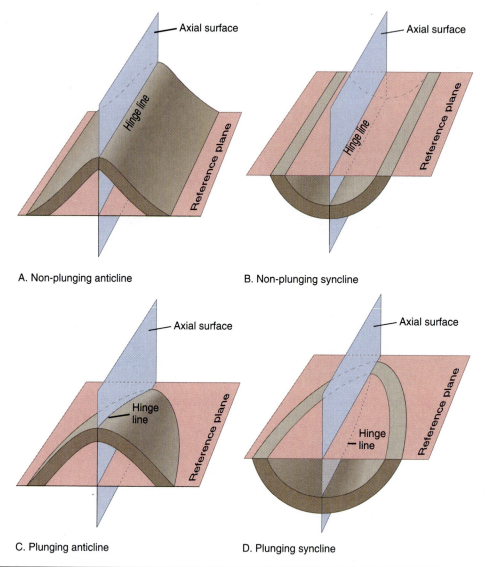

A. Non-plunging anticline

B. Non-plunging syncline

C. Plunging anticline

D. Plunging syncline

FIGURE 14.8

Non-plunging folds have horizontal hinge lines, as in a non-plunging anticline (A) and a non-plunging syncline (B). Plunging folds have inclined hinge lines, as in a plunging anticline (C) and a plunging syncline (D). Axial surfaces are vertical, and reference plane is horizontal.

FAULTS

Faults are breaks or fractures in the Earth's crust along which movement has occurred. The rocks on one side of a fault have moved relative to those on the other side. Relative movement may be up and down, sideways, or a combination of the two. The amount of movement ranges from centimeters to kilometers.

Faults are classified on the basis of relative movement. In a **normal fault,** rocks above the fault plane (referred to as the **hanging wall** because it would be hanging above your head if you could walk on the fault) have moved downward relative to the rocks below the fault plane (referred to as the **footwall** because it would be under your feet if you walked on the fault plane), as shown by displacement of the marker bed in Figure 14.11A. In a **reverse fault** (Fig. 14.11B), the hanging wall moved upward relative to the footwall, and the dip of the fault plane is greater than 45°. **Thrust faults** (Fig. 14.11C) are like reverse faults, but they dip at less than 45°. **Lateral,** or strike-slip, faults (Fig. 14.11D) are those in which the relative movement was sideways or essentially horizontal, parallel to the strike of the fault. *Right-lateral strike-slip faults* are those in which the block on the opposite side of the fault has moved relatively to the right. Thus, if you stood on one side of a right-lateral strike-slip fault, and looked across the fault, the rocks on the other side would be displaced to the right, as in Figure 14.11D. In *left-lateral strike-slip faults,* the block on the opposite side of the fault has moved relatively to the left.

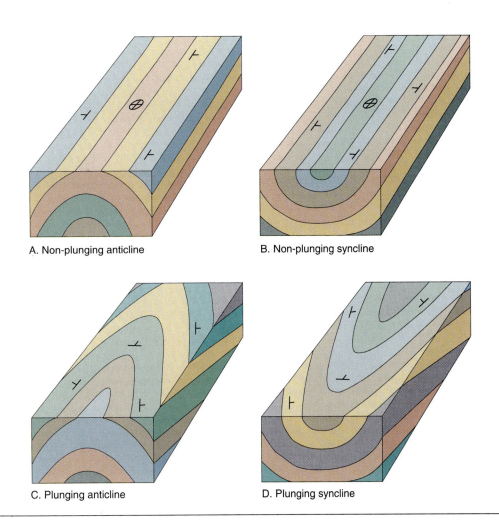

A. Non-plunging anticline

B. Non-plunging syncline

C. Plunging anticline

D. Plunging syncline

FIGURE 14.9
The top surface of these block diagrams, which is analogous to the Earth's surface, shows that older beds are exposed in the centers of eroded anticlines and younger beds are exposed in the centers of eroded synclines. Contact lines on the top surface are straight if folds do not plunge, but bend if they do. In plunging anticlines, the contact lines bend and point in the direction of plunge; in plunging synclines, they bend and point away from the direction of plunge. In C and D, both folds plunge into the page.

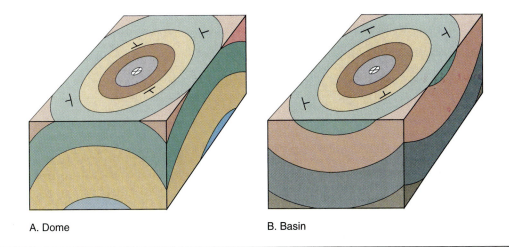

A. Dome

B. Basin

FIGURE 14.10
These block diagrams show that older beds appear in the center of an eroded dome, while younger beds appear in the center of an eroded basin.

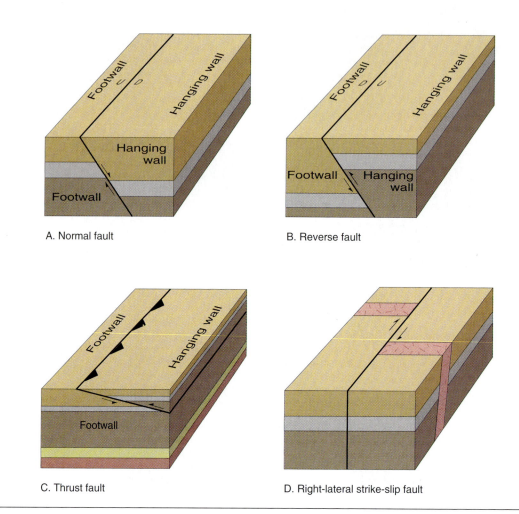

FIGURE **14.11**
Four types of faults are illustrated in these block diagrams. If the fault plane is inclined, the side above the fault is the hanging wall, and the side below is the footwall. Displaced marker beds show movement.

APPLICATIONS

IF GEOLOGISTS ARE TO UNDERSTAND A DYNAMIC EARTH, THEY MUST UNDERSTAND THE NATURE OF THE STRESSES WITHIN THE EARTH. ONE WAY TO LEARN ABOUT STRESS IS TO STUDY THE STRAIN, OR DEFORMATION, THAT IT HAS PRODUCED. THIS TYPE OF STUDY IS BASED ON DESCRIPTIVE AND INTERPRETIVE ANALYSIS OF FOLDS, FAULTS, AND OTHER DEFORMATIONAL FEATURES. USING DRAWINGS AND THREE-DIMENSIONAL MODELS, THIS LAB FOCUSES MAINLY ON THE GEOMETRIC ASPECTS OF DEFORMED ROCKS. YOU WILL LEARN HOW OBSERVATIONS AND MEASUREMENTS MADE AT THE SURFACE ARE USED TO PREDICT THE STRUCTURE OF ROCKS IN THE SUBSURFACE.

OBJECTIVES

If you complete all the problems, you should be able to:

1. Distinguish between elastic and plastic strain, and list the conditions that might favor one over the other.

2. Recognize structures formed by compression, tension, and shear.

3. Identify conformable and unconformable depositional contacts, intrusive contacts, and fault contacts on a geologic map, cross section, or block diagram.

4. Define dip and strike, plot them on a map, and determine their general orientation on a geologic map or top surface of a block diagram.

5. Define, sketch, and recognize—on a geologic map or a block diagram—a dome or basin and plunging and non-plunging anticlines and synclines.

6. Determine the direction of plunge of plunging anticlines and synclines on a geologic map or block diagram.

7. Define, sketch, and recognize— on a cross section or block diagram—normal, reverse, thrust, and strike-slip faults.

8. Distinguish the hanging wall and footwall of a normal, reverse, and thrust fault on a cross section, geologic map, or block diagram.

PROBLEMS

1. To demonstrate how the same material can respond to stress (or deform) differently under different conditions, try the following with Silly Putty®. First, roll it into a ball and see if it bounces. Next, roll it into a thick cigar shape and slowly stretch it out. Finally, roll it into a thick cigar shape again and pull it apart as fast as you can. In each case, the Silly Putty® behaves differently.

   a. Name the type of behavior exhibited in the three cases, using the terminology given in the introduction to this lab.

   b. How did the conditions under which deformation occurred differ in the three cases?

2. The following problems use cut-out block diagrams of blocks A, B, and C, which are found at the end of the lab manual. The blocks should be cut out before beginning the problem in which they are used. When constructing the box, be very careful to make the folds exactly along the lines, or the problems may not work out as they should. It will be easier to draw on the boxes if you do not tape the sides together until you have completed the problem. Even then, it is recommended that you use removable tape.

   a. Refer to Block A with Side 1 up. When Side 1 is up, Block A is similar to Figure 14.9A.
      (1) Fill in the blank faces on the sides of the block.
      (2) Measure the dip with a protractor at several places on the front surface (note that the dip changes below the surface and that the angle you should measure is that at the surface) and put strike and dip symbols at the places indicated with dots on the top (map) surface.
      (3) Illustrate the position of the axial surface of the fold by sketching its trace on the top and ends of the block (the **trace** is the imaginary line formed where the axial surface intersects another surface).
      (4) When Side 1 is up, what is the age of the beds in the center, or core, of the fold relative to the other beds?

      (5) What kind of fold is this (anticline or syncline; upright, inclined, overturned, or recumbent; plunging or non-plunging)?

   b. Refer to Block A with Side 2 up.
      (1) Fill in the top face.
      (2) Put strike and dip symbols at the places indicated with dots on the top (map) surface.

(3)   When Side 2 is up, what is the age of the beds in the center, or core, of the fold relative to the other beds?

(4)   What kind of fold is this?

(5)   Sketch the trace of the axial surface (axial trace) on the top of the block using one of the fold symbols in Figure 14.4.

c.   Refer to Block B with Side 1 up. When Side 1 is up, Block B is similar to Figure 14.9C.
   (1)   Fill in the blank faces on the side of the block.
   (2)   Put strike and dip symbols at the places indicated with dots on the top (map) surface.
   (3)   Illustrate the position of the axial surface of the fold by sketching its trace on the top and ends of the block.
   (4)   When Side 1 is up, number the beds from oldest (number 1) to youngest. Where are the oldest beds as seen on the top (map) surface?

(5)   What kind of fold is this?

d.   Refer to Block B with Side 2 up.
   (1)   Fill in the top face.
   (2)   Put strike and dip symbols at the places indicated with dots on the top (map) surface.
   (3)   When Side 2 is up, what is the age of the beds in the center, or core, of the fold relative to the other beds?

(4)   What kind of fold is this?

(5)   Sketch the trace of the axial surface (axial trace) on the top of the block using one of the fold symbols in Figure 14.4.
e.   Refer to Block C, with Side 1 up.
   (1)   Complete the sides of the block.
   (2)   Put strike and dip symbols of beds at the dots on the top (Side 1) surface.
   (3)   Put a dip symbol (Fig. 14.4) on the fault (dark line).
   (4)   Indicate the relative movement of the fault by placing arrows on opposite sides of the fault on the side faces.
   (5)   Label the hanging wall and footwall on the top and side faces.
   (6)   With Side 1 up, number the beds in order of age, with 1 the oldest. Are the beds at the surface of the upthrown block generally older or younger than those at the surface of the downthrown block? Why?

(7)   What kind of fault is this?

(8)   Fill in Side 2.

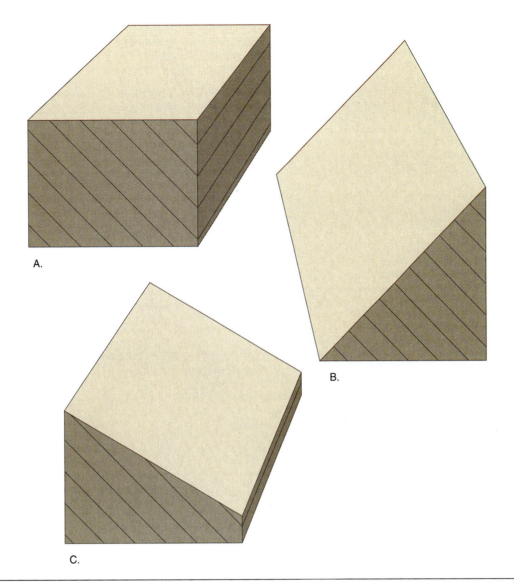

A.

B.

C.

FIGURE 14.12

These box diagrams for Problem 3 show beds dipping at 45°E, but from different perspectives. A. Upper surface is horizontal. B. Upper surface slopes 45°W, so it intersects the beds at 90°. C. Upper surface slopes 30°E, so it intersects the beds at 15°.

3.　Notice on all of the cut-out diagrams how the width of a given layer, as seen on one of the faces, changes depending on the angle at which it intersects that face. This problem is intended to show you why. The three box diagrams in Figure 14.12 all show beds dipping at 45°E, but from different perspectives. The diagrams also differ because in *A* the upper surface is horizontal, in *B* it slopes 45°W so that it intersects the beds at 90°, and in *C* it slopes 30°E so that it intersects the beds at 15°. Carefully complete the upper surface of the diagrams. The width of outcrop is the distance between contact lines, as seen on the upper surface. The width can be measured with a ruler at the top of the front face of the box, where the front face intersects the upper surface. Measure the following:

a.　width of outcrop on *A;*

b.　width of outcrop on *B;*

c.　width of outcrop on *C;*

d.　the true thickness of the beds.

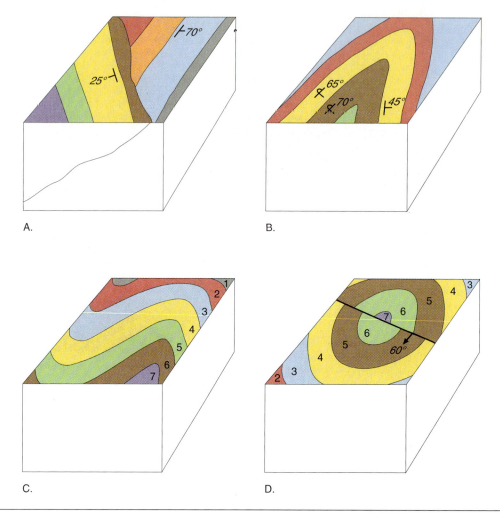

FIGURE **14.13**
Box diagrams to be completed for Problem 4.

4. Partially completed block diagrams are shown in Figure 14.13. Complete each of these by filling in the blank faces. Place general dip and strike symbols on the top surface if not already present. Lowest numbers correspond to oldest rocks.

   a. In Figure 14.13A, what name is given to the uneven feature that runs diagonally across the map surface?

   b. What structure is shown in Figure 14.13B?

   c. Describe the structure shown in Figure 14.13C.

   d. What kind of fault is shown in Figure 14.13D, and what kind of structure does the fault cut?

5. Fill in the blanks to formulate a series of handy-dandy rules to help in the interpretation of geologic maps and cross sections.

   a. The following rules refer to relative ages of sedimentary beds. Choose *"older"* or *"younger"* for your answers.

     • When anticlines and domes are eroded, the beds exposed in the center are _____ than the beds exposed away from the center.

     • When synclines and basins are eroded, the beds exposed in the center are _____ than the beds exposed away from the center.

- _____ beds dip under (or toward) _____ beds, unless they are overturned.

b. The following rules refer to beds and contacts between beds, as seen in map view.

- When _____ folds have been eroded, the contacts between beds, as seen on a horizontal surface, are straight and parallel. Choose *"plunging"* or *"non-plunging."*

- When plunging anticlines have been eroded, the contacts between beds, as seen in map view, bend so as to point _____ the direction of plunge. Choose *"toward"* or *"away from."*

- When plunging synclines have been eroded, the contacts between beds, as seen on the surface, bend so as to point _____ the direction of plunge. Choose *"toward"* or *"away from."*

- The outcrop width of a bed _____ as the angle between the surface and the dip of the bed decreases. Choose *"increases"* or *"decreases."*

6. Go to *http://ncgmp.usgs.gov/* (or link to it through the author's homepage—see *Preface*), and link to *State geological surveys* and select *Indiana*. (If this does not work, go to *http://ncgmp.usgs.gov/* and click on *State geological surveys* in the list of Geologic Mapping Projects). Find the map of bedrock geology. Because of the scale of the map and the low relief of the state of Indiana, the map is comparable to the top surface of a block diagram. The units shown on the map are listed in the explanation, with the youngest at the top and the oldest at the bottom.

  a. Where are the youngest units in the state? The oldest?

  b. The major structural feature in the state is a broad anticline or arch. Is the anticline a plunging or nonplunging anticline?

  If plunging, which way does it plunge?

  c. Which way do the rock units in the southwest part of the state dip?

  d. Two faults, the Royal Center and Fortville, can be found by examining some of the other links from the *Maps of Indiana* page. These faults have similar strikes and they dip in the same direction.

  In what general direction do they strike?

  In what direction do they dip?

IN GREATER DEPTH

7. The following problems refer to cut-out block diagram D, found at the end of the lab manual.

a. Complete the following:

  (1) Fill in the blank side panels.

  (2) Put strike and dip symbols at the places indicated with dots on the Side 1 surface.

  (3) Measure the dip with a protractor on:

    (a) the small, already-filled-in, front panel;

    (b) the two corner panels that you filled in;

(c)   the side surface. The small front panel is perpendicular to strike, so the dip measured there is the *true dip;* the other surfaces are not perpendicular to strike, so the dips observed and measured there are not true dips, but *apparent dips.*

(4)   Fill in Side 2.

(5)   Imagine that the top half of the box (either side up) was eroded away and you could see the new surface exposed. In what direction, in terms of strike or dip, would the outcrops have shifted from their pre-erosion positions?

b.   Fill in the blanks to formulate additional rules to help in the interpretation of geologic maps and cross sections.

- The apparent dip of beds seen on vertical surfaces _____ as the angle between the strike of the beds and the strike of the vertical surface decreases (choose *"increases"* or *"decreases"*); the true dip can be seen only on vertical surfaces that are _____ to the strike of the beds (choose *"parallel"* or *"perpendicular"*).

- As erosion continuously lowers the level of the land surface, a dipping bed or fault, as seen in map view, migrates _____ the direction of dip. Choose *"in"* or *"away from."*

8.   The three cross sections in Figure 14.14 show three outcrops of the same graywacke (sandstone) unit, all of which have beds that dip about 60° to the west. The surrounding rocks, probably shale, are easily eroded and are covered with soil.

a.   Formulate three hypotheses to explain the relations among the outcrops by completing the cross sections in three different ways. Use dashed lines to indicate contacts above the cross section that have been eroded away.

b.   What kind of information might the three outcrops contain that would enable you to distinguish among your multiple hypotheses? Hint: see *Sedimentary Structures* in Chapter 4.

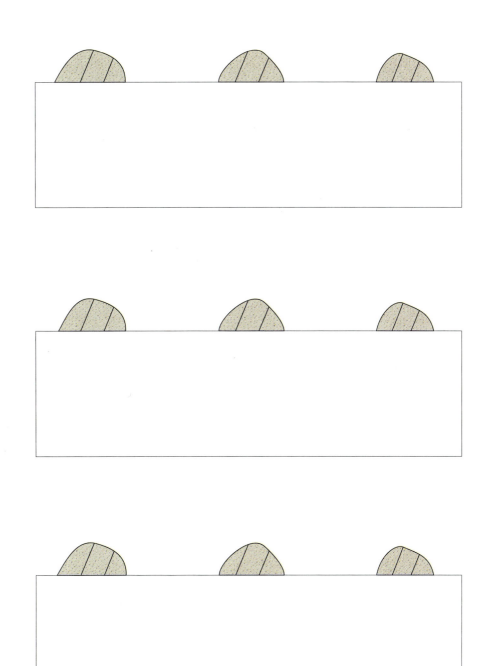

FIGURE **14.14**
Cross sections for Problem 8.

15
Geologic Maps

Overview

Geologic maps are a visual expression of the basic field data of geology. A good geologic map shows not only what kinds of rocks are present, but their ages and structure. Because geologic maps are two-dimensional views of three-dimensional rock sequences, they can appear quite complex. However, simple observations can help you understand what is shown on a geologic map. You will learn: how to recognize whether rock formations are horizontal, inclined, folded or faulted, even if dip and strike symbols are absent; how to determine the chronologic order of geologic events; how to visualize the third dimension by drawing a geologic cross section; and how geologic maps are made.

Materials Needed

Pencil and eraser • Ruler • Protractor • Calculator •
Stereoscope (provided by instructor)

INTRODUCTION

A geologic map shows the distribution and structure of rock units on or very near the Earth's surface. An accompanying *explanation* lists the rock units present and explains special symbols. Geologic maps commonly are drawn on contour maps so that geology and topography are easily compared.

The rock units shown on geologic maps are usually *formations,* although in some cases, other types of rock units or time-rock units (rocks formed during a specific time interval) are shown. A **formation** is a mappable rock unit. That is, it has specific characteristics that make it recognizable in the field, and upper and lower boundaries (contacts) that make it easy to tell from adjacent rock units. Formations have two-word names. The first word usually refers to the place where the formation originally was described. If the formation consists principally of one type of rock, the rock name is used as the second part; if not, the word "formation" is used. Examples: St. Peter Sandstone, Trenton Limestone, Morrison Formation.

MAKING A GEOLOGIC MAP

Making a geologic map requires:

1. collecting the basic data in the field;
2. correlating among outcrops;
3. making a geologic cross section of the area to show the three-dimensional relation of one rock unit to another; and
4. preparing an *explanation.*

Collecting the basic data is by far the most important and time-consuming task—and the most fun. Exposures, or **outcrops,** of bedrock are located by studying aerial photographs and by walking the countryside to locate where the rock actually can be seen. In the western United States, rock is exposed over large areas. In more humid areas, like the U.S. Midwest, outcrops are commonly small and widely scattered; natural outcrops commonly are found in stream valleys,

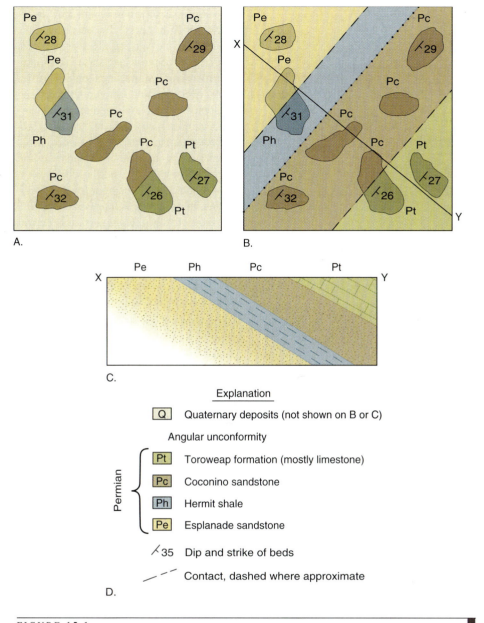

FIGURE **15.1**

Making a geologic map. See text for discussion. A. Outcrop map. B. Contacts drawn in. C. Cross section along line *X-Y*. D. Explanation, with rock units and angular unconformity listed chronologically.

lines to indicate they are not actually visible. In some places, no contacts are actually exposed, but their presence can be inferred. For example, say one outcrop is sandstone and an adjacent one is shale; the two must be in contact someplace, but the contact is buried. In such places, a "best-guess" dotted line is drawn to represent the contact (Fig. 15.1B).

One of the best ways to visualize what is shown on the map, and to double check an interpretation, is to draw one or more geologic cross sections. The cross section in Figure 15.1C was drawn along the line *X-Y* on the map. That particular cross section was chosen because it best illustrates the overall structure. Note that the dip and strike symbols tell the angle and direction in which the layers are inclined at the surface. Correlation of the units below the surface completes the structure.

Finally, rock units and unconformities are listed, with the oldest at the bottom and the youngest at the top, in the *Explanation* (Fig. 15.1D).

APPEARANCE OF GEOLOGIC FEATURES FROM THE AIR AND ON MAPS

You learned what common geologic structures look like in simple block diagrams in Chapter 14. Those diagrams illustrate ideal structures without the complexity of topography on the top surface. Even so, outcrop patterns of rock units seen on maps or from the air do resemble what you saw on the upper surface of the block diagrams. Hills and valleys complicate the smooth and regular appearance of formational contacts somewhat, but the basic pattern is there. The purpose of this section is to help you interpret what you see from an airplane, or on a geologic map.

Horizontal Strata

If horizontal strata are exposed on a horizontal surface, only one formation will be seen. However, if the surface is not flat and erosion has cut into two or more horizontal formations, the contact lines between adjacent formations will parallel the contour lines. Like contours, contacts bend so as to point upstream when cross-

but rock is also exposed in road cuts, mines, quarries, and construction sites. The outcrops are examined in the field and plotted on a topographic map or an aerial photograph. A description is recorded on the map or in a field notebook and includes such information as rock type, formation name, strike and dip of beds or other structural features, fossils present, and anything else deemed important by the geologist. Figure 15.1A is an example of a map showing outcrops; the area between outcrops is covered with soil.

The next step is correlation among outcrops; this is usually done in the field. Notice in Figure 15.1A that strikes and dips are similar in all outcrops and that the sandstone labelled *Pc,* in particular, occurs in a series of outcrops strung out in the same orientation as the strike. The simplest and most reasonable assumption to make is that the four outcrops of this sandstone belong to the same formation and should be correlated, as illustrated in Figure 15.1B. Identical contacts are joined from one outcrop to another, using dashed

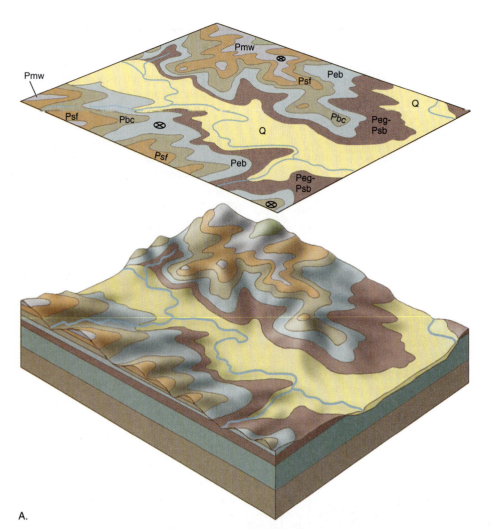

A.

ing valleys, as shown in Figure 15.2. In an area with little relief, few strata will be exposed at the surface. The greater the relief, the deeper the land surface cuts into the underlying strata, exposing more of them. Because of differential erosion, resistant beds have steep slopes (narrow expression in map view), and easily eroded beds are marked by gentle slopes (wide expression in map view). If beds are not quite horizontal—that is, the dip is very low—contacts cross contour lines at small angles. In other words, *the flatter the dip, the more closely geologic contacts parallel contour lines.* Similarly, *the steeper the dip, the greater the divergence between contacts and contours.*

Inclined Strata

Strata inclined in one direction appear as roughly parallel bands on a map or aerial photograph, as shown in Figure 15.3. If the relief is moderate, the bands will be essentially parallel to the strike of the beds. In general, with increasing steepness of dip, the outcrop bands will be straighter and the contacts more regular, and the contacts will more closely parallel strike. Differential erosion may produce ridges and valleys that parallel the units. The direction of dip

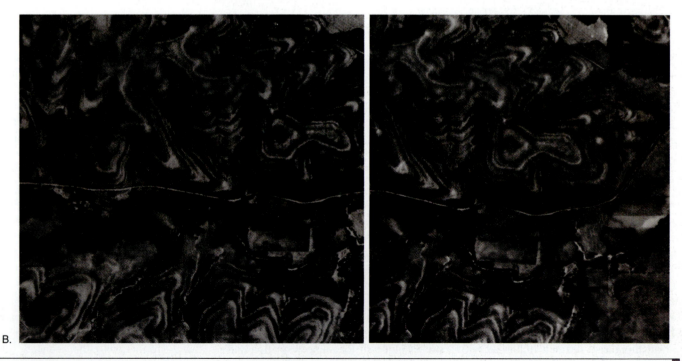

B.

FIGURE **15.2**

Horizontal Permian strata in the Flint Hills of northeastern Kansas. The geologic map and block diagram in A. are based on the aerial photographs (scale 1:20,000) in B. *Q* on the map represents Quaternary deposits, and *Pmw, Psf, Pbc, Peb, Psb,* and *Peg* are symbols for Permian formations. The white bands in the photographs (B) are limestone; the gray bands are shale.

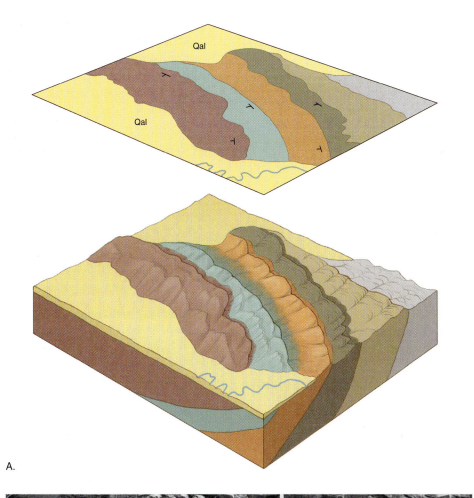

A.

can be determined by the Rule of Vs, as described later, or in some places, by differences in hill slope on opposite sides of ridges. Hills sloping in the direction of dip generally are more gentle than those sloping in the opposite direction.

Once the direction of dip is known, relative ages can be determined, because *older beds dip under (or toward) younger beds (assuming beds are not overturned).*

Folded Strata

Eroded, folded strata appear as winding or curved bands on geologic maps or aerial photographs if the folds are plunging, as shown on Figure 15.4, and as roughly parallel bands, like those in Figure 15.3, if they are not. Differential erosion produces a topography that commonly mimics the structures. However, a map view of the rock units looks much like the top surface of a block diagram. This is because, from the perspective of a map or aerial photograph, elevation differences are small compared to horizontal distances, and topography does not have much of an effect on the overall picture.

As you learned from Problem 5 in Chapter 14, several things are worth remembering when observing eroded folds *in map view.*

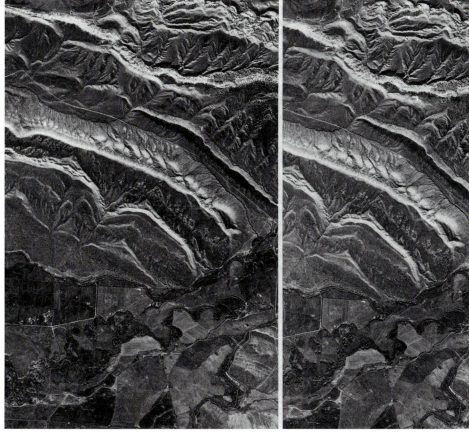

B.

FIGURE 15.3
Inclined Cretaceous strata along Owl Creek, Hot Springs County, Wyoming. Ridges are sandstone, and valleys are shale. The geologic map and block diagram in A are based on the aerial photographs (scale 1:48,500) in B.

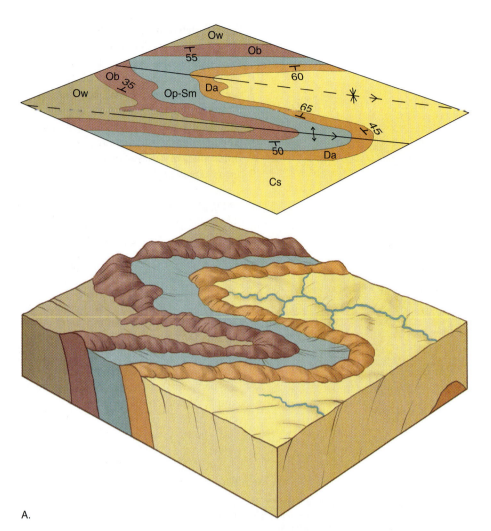

A.

1. *When anticlines or domes are eroded, older rocks appear in the center and are surrounded by younger rocks.*

2. *When synclines or basins are eroded, younger rocks appear in the center and are surrounded by older rocks.*

3. *The contacts between beds in a plunging anticline, as seen in map view, bend so as to "point" in the direction of plunge.*

4. *The contacts between beds in a plunging syncline, as seen in map view, bend so as to "point" in the opposite direction from which it plunges.*

Faulted Strata

Faults are shown on geologic maps with thick, black lines and symbols indicating the nature of the fault, if that is known (see Fig. 14.4 in the previous chapter). Faults cause beds and contacts to be offset, and commonly juxtapose quite different kinds of rocks. Differential erosion may produce a cliff on one side of the fault or a narrow valley marking the fault itself. Fault valleys typically appear as straight, dark lines on aerial photographs, because the vegetation along them is commonly more lush than that surrounding them. On a geologic map, contacts, formations, and structures may end or be offset at faults.

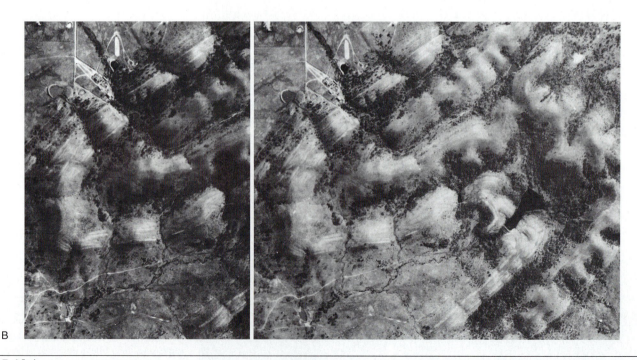

B

FIGURE 15.4

Folded strata in the Ouachita Mountains, Atoka County, Oklahoma. Ridges are Ordovician (*Ob*) and Devonian (*Da*) chert; low areas are Ordovician, Silurian, and Carboniferous shale (*Ow, Op, Sm, Cs*). The geologic map and block diagram in A are based on the aerial photographs (scale 1:17,000) in B.

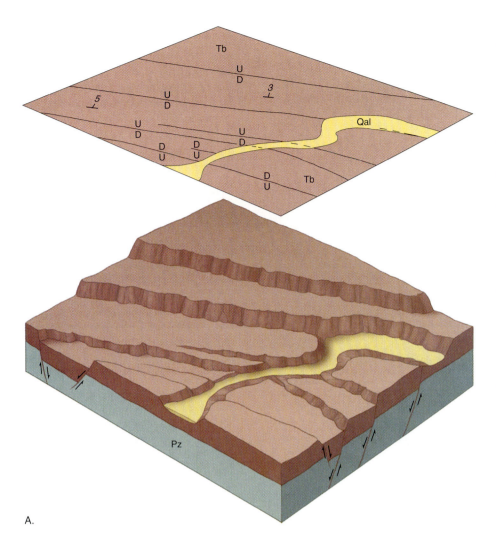

A.

Steeply dipping faults have straight traces (*trace* is the intersection between the fault plane and the surface) on maps or aerial photographs, whereas gently dipping faults may have irregular traces, depending on the topography. Faults dip under their hanging walls; therefore, on a geologic map, the direction of dip is toward the hanging wall. In map view, *normal faults* (Fig. 15.5) have older rocks on the footwall side. In contrast, *reverse faults* have younger rocks on the footwall side. This is because *erosion of the topographically higher side (the upthrown side) eventually exposes deeper and (usually) older rocks.* The traces of *strike-slip faults* are straight because of their common near-vertical dips. They are more likely than the other types of faults to be marked by straight, narrow valleys.

Thrust faults may have very irregular traces because of their low dips, often only a few degrees. They are recognizable on a geologic map when structures above and below the fault do not correspond (for example, formations are not the same, or strikes and dips differ), and when the rocks in the hanging wall are older than those in the footwall.

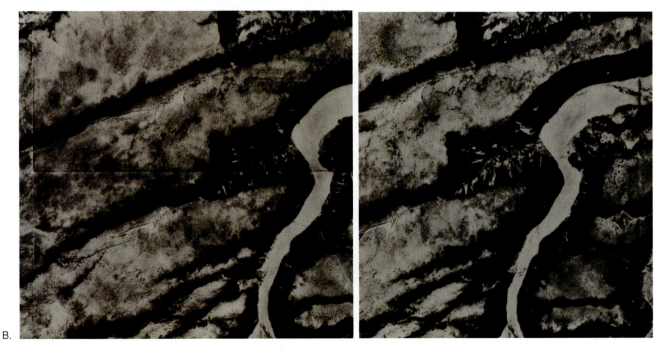

B.

FIGURE **15.5**

Basalt flows of Tertiary age (*Tb*) cut by normal faults, Sandoval County, New Mexico. Displacement ranges from near zero to about 50 m. Fault blocks are tilted eastward. The geologic map and block diagram in A are based on the aerial photographs (scale 1:31,680) in B. *Qal* indicates Quaternary alluvium, Pz (in block diagram) indicates rocks of Paleozoic age.

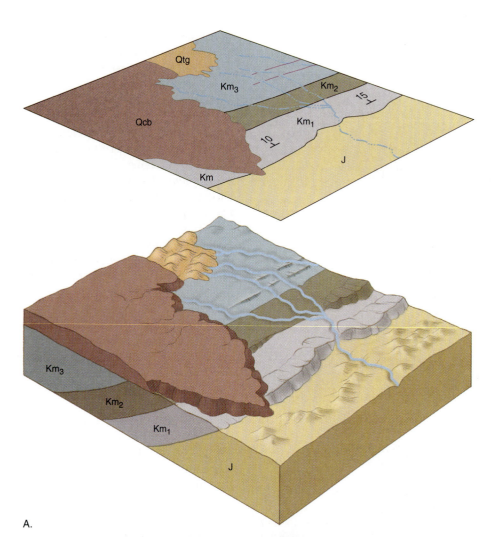

A.

Unconformities

Angular unconformities are recognized most easily on aerial photographs and geologic maps by non-parallel layers in rocks above and below the unconformity, as shown in Figure 15.6. Angular unconformities with low dips appear similar to low-angle thrust faults, except that rocks above the unconformity are younger than those below it.

Nonconformities separate non-bedded, crystalline rocks from younger sedimentary rocks. They can be confused with intrusive igneous contacts. In nonconformities, the sedimentary rock is usually of the same formation throughout the map area, whereas intrusive contacts may cut more than one formation. Relative ages are most important; igneous rocks are younger than the rocks they intrude, but older than rocks deposited on their eroded, *nonconformable* surfaces.

Disconformities are much more difficult to recognize on a geologic map. Variable thickness of a particular formation, or inexplicable contacts between otherwise conformable (parallel) layers, are clues.

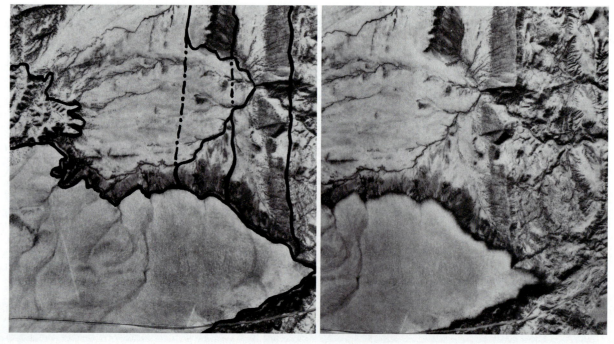

B.

FIGURE 15.6

Angular unconformity between nearly flat-lying Quaternary basalt lava flows (*Qcb*) and gently dipping Jurassic (*J*) and Cretaceous (*Km1, Km2,* and *Km3*) sedimentary rocks, Santa Fe and Sandoval Counties, New Mexico. The red lines on the map are dikes; *Qtg* is Quaternary gravel. The geologic map and block diagram in A are based on the aerial photographs (scale 1:54,000) in B.

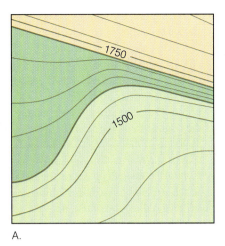

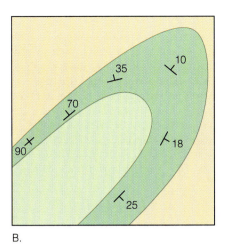

A.   B.

FIGURE 15.7

A. Map of a horizontal layer, 150 feet thick, that forms a narrow outcrop where the slope is steep and a wide outcrop where the slope is more gentle. B. The outcrop width of a folded layer on a horizontal surface equals true thickness where the dip is 90° but is wider where the dip is less, as shown on this map.

Igneous Intrusions

Irregular-shaped **plutons** (bodies of intrusive igneous rock) show up on the surface as **discordant** masses of igneous rock, cutting across the layering in surrounding rocks. Plutons may be somewhat circular or elliptical in map view if the map scale is small enough to show the whole igneous body.

Dikes are thin, tabular intrusions that cut across bedding (Fig. 15.6). They appear on maps as narrow, discordant lines, and commonly occur in sets (*swarms*) with parallel or near-parallel alignments. **Sills,** the other tabular intrusions, are **concordant** (parallel to bedding), so they cannot be recognized on maps unless it is known that the rock is intrusive.

Volcanoes and Volcanic Rocks

Young volcanoes with central **craters** (small depression at the top) or **calderas** (large depressions at the top), and with **lava flows** or **pyroclastic flows** (gas-charged flows that produce tuff) radiating outward, are easily recognized. Older, more deeply eroded volcanoes become increasingly more difficult to recognize with age.

Extensive sheets of lava or tuff may blanket large areas and have a map appearance similar to sedimentary rocks, whether folded or horizontal. Like sedimentary rocks, they are given formation names (for example, San Juan Formation).

GEOLOGIC CROSS SECTIONS

Geologic cross sections are like topographic profiles with the subsurface geology shown schematically. Geologic maps commonly are accompanied by one or more geologic cross sections, because they make it much easier to visualize the third dimension. Geologic cross sections are interpretive in nature, because geologic maps are based on a limited number of outcrops and most of what is portrayed is out of sight, below the surface. Cross sections help to clarify the author's interpretation.

Geologic cross sections are made so that they will provide the most information possible. *A line of section* must be chosen. This is the line, or set of connected line segments, along which the cross section will be made. For example, if the map shows an anticline, a logical line of section would be one perpendicular to the *trace of the axial surface* of the fold on the Earth's surface, because that section would best portray the anticline. Most cross sections are drawn perpendicular to the general strike. Not only do such sections convey the most information, they are a lot easier to draw because the actual dip can be shown.

To make a geologic cross section, a topographic profile is first drawn along the line of section. The thickness (or depth below the surface) represented on a cross section depends mostly on how much is

known or can be inferred about what is below the surface. Some cross sections show thousands of meters, others only a hundred or so. It is generally less confusing if no vertical exaggeration is used on a cross section. Vertical exaggeration makes it more difficult to draw the geology below the surface, because all the angular relations also must be exaggerated (for example, with a vertical exaggeration of 10×, a 10° dip would appear as a 60.4° dip on the cross section). If all geologic contacts are horizontal, vertical exaggeration can be used with impunity.

The geology is next added along the line of section. Whenever a geologic contact on the map is crossed by the line of section, the contact is marked on the section, and a notation is made to indicate what rock units are present. The direction and approximate angle of dip of the contacts can be determined from the dip and strike symbols closest to the line of section, or from the Rule of Vs (explained later). Contacts are then extended below the surface to best represent the geology seen at the surface.

USING GEOLOGIC MAPS

Map Symbols

A fairly standard set of symbols is used on geologic maps to depict various geologic features. The most common are shown in Figure 14.4 in Chapter 14.

Variations in Outcrop Patterns

Width of Outcrop

The outcrop width of a bed depends on the thickness of the bed, the angle at which that bed intersects the surface, and the slope of the surface. A bed intersecting the surface at a low angle has a wide outcrop; the outcrop width is equal to the thickness of the bed only if the bed intersects the surface at 90°. In *map view,* the apparent width of a bed equals its true thickness (as scaled on the map) only if the bed is vertical.

Because of this relation, horizontal beds have wider outcrop patterns where surface slopes are gentle, and more narrow patterns where slopes are steep, as shown in Figure 15.7A. Similarly, a folded bed of

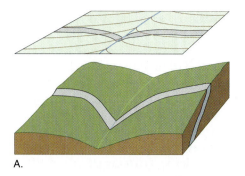

A.

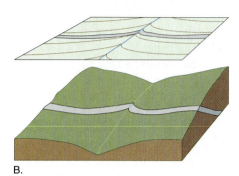

B.

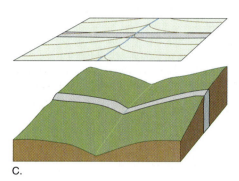

C.

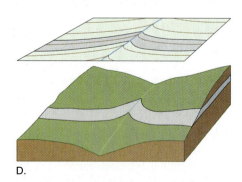

D.

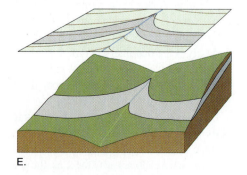

E.

Where dipping beds or contacts between them cross a valley, they generally form a V pattern on a map that points in the direction of dip. A. Beds dip downstream. B. Beds dip upstream. Exceptions are (C.) vertical beds, which are not deflected on crossing a valley; (D.) horizontal beds, which always form Vs that point upstream and are parallel to contours; and (E.) beds that dip downstream but at an angle less than the slope of the valley.

constant thickness but variable dip, such as illustrated in Figure 15.7B, has a variable outcrop width on a flat surface.

Rule of Vs

When contacts between formations cross valleys, the contacts are bent, or deflected, as illustrated in Figure 15.8. The direction and amount of bending depends on the slope of the valley and the dip of the beds. The following rule (often called the **Rule of Vs**) is helpful for determining the general dip of a bed on a geologic map if it is not shown by a symbol:

> *Where a contact crosses a valley, it forms a V, the apex of which points in the direction of dip of the contact (Fig. 15.8A, B).*

Short, wide, open Vs indicate steep dips; longer, more narrow Vs indicate gentle dips. Three exceptions to the rule are:

1. *If beds are vertical, the contact is not deflected on crossing a valley, and no V is formed (Fig. 15.8C).*
2. *If beds are horizontal, the V points up-valley and is parallel to contour lines (Fig. 15.8D).*
3. *If beds dip down the valley, but at an angle less than that of the slope of the valley, the V points up-valley (Fig. 15.8E).*

Reconstructing Geologic History

A geologic map is used to reconstruct the geologic history of an area in the same way that geologic cross sections were used in Chapter 13. Figure 15.9 will serve as an example.

1. First, interpret the map. What is shown? If strike and dip symbols are not given, use the orientation of contacts and the Rule of Vs to determine the strike and dip directions, as has been done in Figure 15.9. The strike is parallel to the contacts, and the dip is in the direction of the Vs where the contacts cross valleys. Put strike and dip symbols at enough places on the map so the structure is clear. Just show strike and direction of dip, not numbers, although you may be able to tell whether the dip is steep or shallow based on the size and shape of the Vs.

2. Remembering the *principle of superposition* (Chapter 13), and that older beds dip under (or toward) younger beds, determine the relative ages of any sedimentary rocks, and locate the oldest.

3. Using the *principle of cross-cutting relations* (Chapter 13), determine the relation between the sedimentary rocks and events such as faulting, folding, erosion, or igneous activity.

4. Finally, list all events, beginning with the oldest, in the sequence in which they occurred. Remember that unconformities usually require uplift and erosion, and that sedimentary rocks originally were deposited as horizontal layers.

Prospecting

Geologic maps are the most important database for prospecting, whether it be for oil or gas, coal, or various mineral commodities. Geologists know from experience that fossil fuels and mineral commodities occur in some geologic settings and not in others. Geologic maps are used to target specific areas for closer examination.

For example, oil and gas form when organic materials that accumulate with marine mud are buried deeply. Once formed, the oil and gas are squeezed out of the mudstone and migrate upward (because they are less dense than water and therefore buoyant) through permeable rock, until they are trapped against an impermeable barrier. Knowing these char-

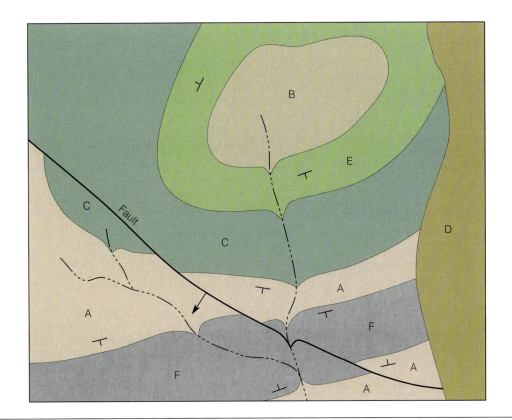

FIGURE 15.9

Reconstructing geologic history. Strike and dip symbols are based on contacts and their "Vs" where they cross valleys. Because older beds generally dip under younger, the oldest bed is *B*, and *E, C, A,* and *F* are successively younger. These beds are folded. The southwest-dipping fault is younger than *F* and the fold, but not *D*. An angular unconformity separates *D* from older units, and the present surface is being eroded. Events occurred as follows, from oldest to youngest: deposition of *B, E, C, A,* and *F;* folding; faulting; erosion; deposition of *D;* and erosion.

acteristics of oil and gas deposits, geologists quickly can eliminate large areas from further exploration. For example, igneous and metamorphic rocks normally would not be very good places to look for oil, because they contain no suitable organic material. Nor would Precambrian rocks, most of which formed before organisms became sufficiently abundant to generate a large amount of oil or gas. Once possible source rocks are identified, the search can be concentrated where impermeable barriers exist that could trap petroleum.

Similar arguments apply to mineral commodities. Certain kinds of ore deposits are found only in certain kinds of rocks, or in association with certain geologic structures. Geologic maps are used to focus exploration projects.

Land Use

Geologic maps, together with topographic maps, are the bases for land-use maps. For example, in one case in Utah (the Sugar House 7½ minute quadrangle, which covers part of Salt Lake City), *all* of the following kinds of maps were developed by the U.S. Geological Survey: surficial geology, age of faults, slopes, landslides, slope stability, construction materials, urban growth, thickness of saturated Quaternary deposits, depth to water in shallow aquifers, depth to top of principal aquifer, concentration of dissolved solids in water from principal aquifer, configuration of potentiometric surface, thickness of loosely packed sediments and depth to bedrock, flood and surface water, and earthquake stability. Other types of maps are developed for areas with different characteristics.

Such maps are used to determine how land can be used most effectively. For example, it would not be effective to build a shopping center in the middle of a landslide area, to route a buried pipeline through an area where slopes are unstable, or to put a landfill on a site where permeability is high.

APPLICATIONS

GEOLOGIC MAPS SUMMARIZE GEOLOGIC FIELD DATA AND PORTRAY THE MAPPING GEOLO-GIST'S INTERPRETATION OF THAT DATA. GEOLOGIC MAPS CONTAIN INFORMATION ON MOST OF THE ASPECTS OF GEOLOGY YOU HAVE ALREADY STUDIED: MINERALS, ROCKS, SURFACE PROCESSES, AND ROCK SEQUENCES AND STRUCTURES. IN THIS LAB, YOU WILL LEARN HOW TO APPLY THE GEOLOG-IC PRINCIPLES YOU HAVE LEARNED THUS FAR TO EXTRACT THAT INFORMATION FROM GEOLOGIC MAPS.

OBJECTIVES

If you complete all the problems, you should be able to:

1. Construct a geologic map from an outcrop map on which rock type and structural information are given.

2. Recognize horizontal, inclined, folded, and faulted strata; unconformities; and igneous intrusions on geologic maps and aerial photographs.

3. Use the Rule of Vs to determine direction of dip for strata.

4. Draw a geologic cross section from a geologic map.

5. Determine the chronologic order of geologic events from a geologic map.

PROBLEMS

1. Figure 15.10 is an outcrop map. The symbols indicating rock types are those given in Figure 13.9.

   a. Using a protractor, plot strike and dip symbols on the map in Figure 15.10 for the outcrops listed below. To plot strike and dip, the boundaries of the map can be used as reference lines for the protractor. The sides of the map are north-south and the top and bottom are east-west.

Outcrop	Attitude	Outcrop	Attitude
A	N43E, 25SE	G	N30W, 18NE
E	N37W, 26SW	H	N53E, 21NW

   b. Make a geologic map by correlating the various rock types from one outcrop to another and by drawing contacts between adjoining units.

   c. Using the principle of superposition, determine the relative age of each sedimentary unit and list them below, with the oldest units at the bottom and the youngest at the top.

   d. Make an east-west geologic cross section of the map along the line from *X* to *Y* in the space provided below the map. Assume the surface is flat. The vertical scale should be the same as the horizontal scale. Use a protractor to plot dips.

   e. What type of structure is present? If it is a fold, indicate whether it is an anticline, syncline, dome, or basin. If it is a fault, indicate whether it is a normal, reverse, thrust, or strike-slip fault; label it with the proper symbol from Figure 14.4; and label footwall and hanging wall, if appropriate, on your cross section.

   f. Using your cross section, determine the thickness of the shale.

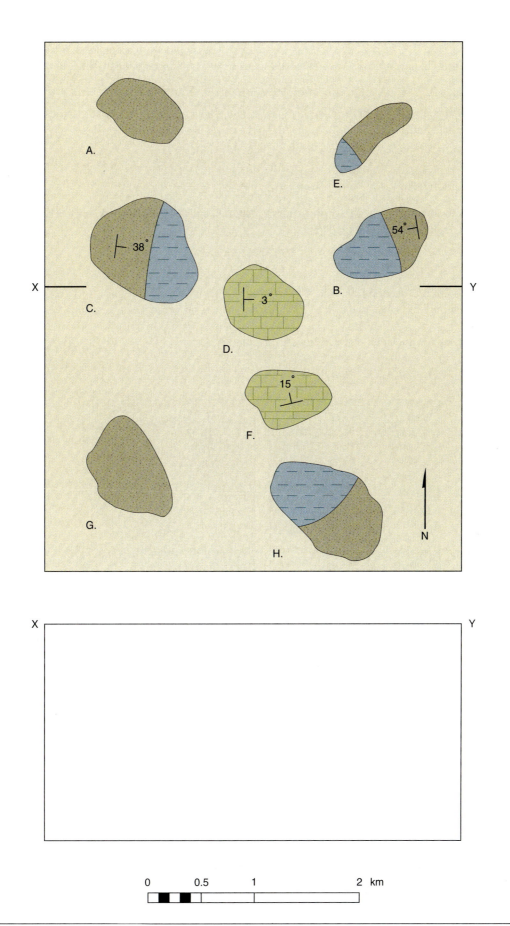

FIGURE **15.10**
Outcrop map for Problem 1.

253

2. The maps in Figure 15.11A, B, C, and D are simple geologic maps with contours superimposed. Place an appropriate strike-and-dip symbol (see Fig. 14.4) in the circles on each map. Use the Rule of Vs and exceptions 1 and 2 to the Rule of Vs, to determine the direction of dip.

3. Figure 15.12 is part of a geologic map of the Gateway Quadrangle, Colorado, and provides the basic geologic data for that area. Answer the following standard geologic questions that might be asked about an area:

   a. What is the oldest rock represented, and where on the map does it occur?

   b. What is the general structure of the Mesozoic (Triassic, Jurassic, and Cretaceous) rock units? That is, are the strata horizontal (or nearly so), inclined, folded, or faulted?

   c. Construct a geologic cross section from A to A′ using the topographic profile in Figure 15.13 as a basis. The vertical scale and horizontal scale are the same, one inch to 2000 feet, so there is no vertical exaggeration.

   d. What is the approximate thickness of the Mesozoic rock units?

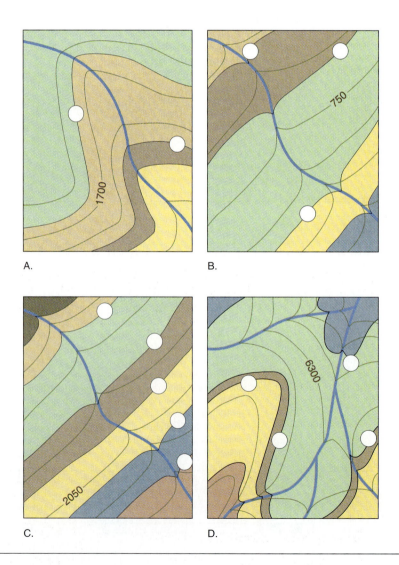

A.

B.

C.

D.

FIGURE 15.11

Geologic maps for Problem 2. Place strike and dip symbols in circles. Brown lines are contour lines, and blue lines represent streams.

e. Notice that one mine and a number of prospects (holes or pits dug to investigate for ore, which are shown with an "X" on the map) are located in the area. The elements of interest are uranium, vanadium, and radium. If you were to explore this area for minerals containing these elements, on what formation(s) would you concentrate your efforts? (Evidence is on the map.)

4. Refer to Figure 15.14, a geologic map of the Black Hills of South Dakota, and to Figure 15.15, a composite Landsat image.

a. Study the outcrop pattern on the geologic map and the satellite image and describe the overall geologic structure.

b. What is the age of the oldest rock in the structure?

c. Why are the areas underlain by the Minnelusa Sandstone and the Minnekahta Limestone–Opeche Formation wider on the west side of the structure than on the east?

d. Are the Cenozoic igneous intrusions older or younger than the surrounding rocks? How could you tell this if you weren't given the age of the rocks?

e. List the *sedimentary* units from the surface downward that would be penetrated by a well drilled at Edgemont in the southwestern part of the map.

f. The brownish red on the Landsat image more or less coincides with the heart of the Black Hills. Why is it brownish red? (See Chapter 7.)

g. Note the similarity in shape or pattern of the brownish red of the satellite photo and some of the contact lines on the geologic map. This is due in part to elevation differences—that is, the elevation is higher within the red area, where rainfall is more plentiful—but that is not the only explanation. Examine the satellite image closely, compare with the geologic map, and suggest another possibility.

h. What are the narrow southeast-trending red streaks on the east side of the satellite image?

Scale 1:24 000
C.I. 20 feet

0   1/2   1 Mile

0   .25   .5   1 Kilometer

FIGURE 15.12
Geologic map of portion of Gateway
Quadrangle, Colorado, for use with Problem 3
(*on following pages*).

EXPLANATION

Alluvium

Light-red wind-deposited sand and silt on benches and mesa tops, reworked in part by water; Recent valley fill and stream deposits.

UNCONFORMITY

Fanglomerate

Poorly sorted, in places rudely bedded, sand and angular fragments and boulders derived from older formations; somewhat indurated.

UNCONFORMITY

Burro Canyon formation

White, gray, and red sandstone and conglomerate with interbedded green and purplish shale.

Morrison formation

Variegated shale and mudstone; white, gray, rusty-red, and buff sandstone; rusty-red conglomerate; local thin limestone beds. At the top the Brushy Basin shale member, Jmb, consisting largely of bentonitic shale but including some sandstone and conglomerate lenses, and at the base the Salt Wash sandstone member, Jms, with more numerous and thicker sandstone beds.

Summerville formation

Thin-bedded red, gray, green, and brown sandy shale and mudstone.

Entrada sandstone and Carmel formation undivided

Orange, buff, and white, fine-grained, massive and cross-bedded Entrada sandstone at the top. Red sandstone and mudstone of the Carmel formation at the base.

UNCONFORMITY

Navajo sandstone

Buff and gray crossbedded, fine-grained sandstone.

Kayenta formation

Irregularly bedded, red, buff, gray, and lavender shale, siltstone, and fine- to coarse-grained sandstone.

Wingate sandstone

Fine-grained reddish-brown, cliff-forming sandstone, thickbedded, massive and crossbedded.

Left margin labels: Pleistocene and Recent — QUATERNARY; Lower Cretaceous — CRETACEOUS; Upper Jurassic, Middle Jurassic, San Rafael group, Glen Canyon group — JURASSIC; JURASSIC(?)

Chinle formation

Red to orange-red siltstone, with interbedded lenses of red sandstone, shale, and limestone-pebble and clay-pellet conglomerate. Lenses of quartz-pebble conglomerate and grit at base.

UNCONFORMITY

Moenkopi formation

Chocolate-brown ripple-bedded shale, brick-red sandy mudstone, reddish-brown and chocolate-brown sandstone, and purple and reddish-brown arkosic conglomerate. Local gypsum beds. The upper member, ℞mu, consisting of thin- and ripple-bedded shale with interbedded sandstone; the middle member, ℞mm, consisting of ledge-forming beds of shale, sandstone, and arkosic conglomerate; and the lower member, ℞ml, consisting of poorly sorted sandy mudstone and local gypsum beds near base.

Cutler formation

Maroon, red, mottled light-red, and purple conglomerate, arkose, and arkosic sandstone. Thin beds of sandy mudstone.

UNCONFORMITY

Gneiss, schist, granite, and pegmatite

Right margin labels: Upper Triassic, Lower Triassic — TRIASSIC; PERMIAN(?); PRE-CAMBRIAN

———————

Contact
Dashed where approximately located.

– – – – – – –

Indefinite contact
Includes inferred contacts and indefinite boundaries of surficial deposits.

Fault
Dashed where approximately located.
U, upthrown side; D, downthrown side.

20

Strike and dip of beds

7500

Structure contours
Drawn on top of Entrada sandstone; dashes indicate projection above surface. Contour interval 100 feet. Datum is mean sea level.

Adit

Prospect

Mine

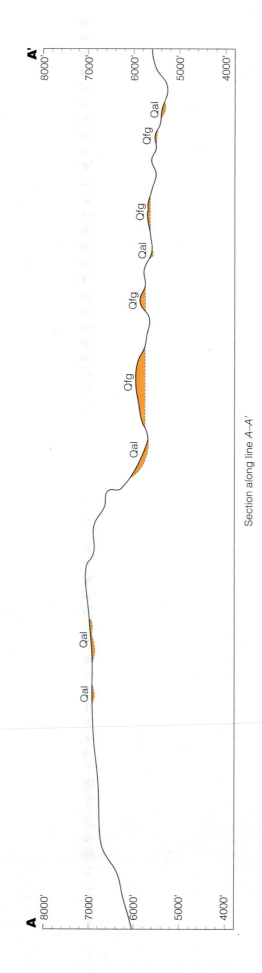

FIGURE 15.13

Cross section along line A–A' of Gateway Quadrangle, Colorado, geologic map, for use with Problem 3c. Qal and Qfg are already located for you.
SOURCE: DATA FROM U.S. GEOLOGICAL SURVEY.

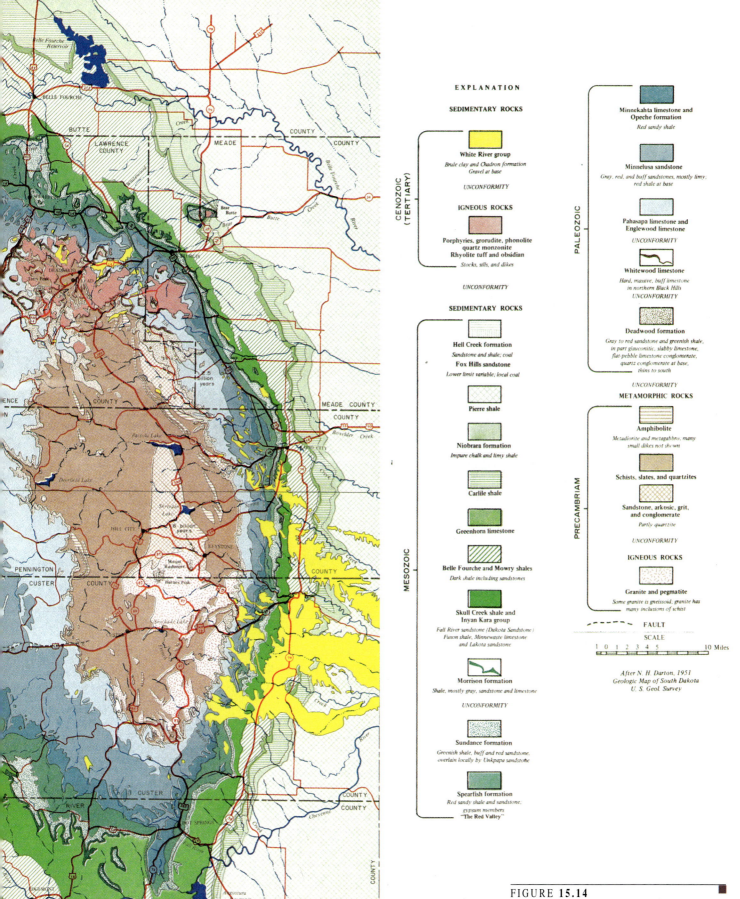

EXPLANATION

SEDIMENTARY ROCKS

CENOZOIC (TERTIARY)

White River group
*Brule clay and Chadron formation
Gravel at base*

UNCONFORMITY

IGNEOUS ROCKS

Porphyries, gorudite, phonolite
quartz monzonite
Rhyolite tuff and obsidian
Stocks, sills, and dikes

UNCONFORMITY

SEDIMENTARY ROCKS

MESOZOIC

Hell Creek formation
Sandstone and shale; coal
Fox Hills sandstone
Lower limit variable; local coal

Pierre shale

Niobrara formation
Impure chalk and limy shale

Carlile shale

Greenhorn limestone

Belle Fourche and Mowry shales
Dark shale including sandstones

Skull Creek shale and
Inyan Kara group
*Fall River sandstone (Dakota Sandstone)
Fuson shale, Minnewaste limestone
and Lakota sandstone*

Morrison formation
Shale, mostly gray, sandstone and limestone

UNCONFORMITY

Sundance formation
*Greenish shale, buff and red sandstone,
overlain locally by Unkpapa sandstone*

Spearfish formation
*Red sandy shale and sandstone,
gypsum members
"The Red Valley"*

PALEOZOIC

Minnekahta limestone and
Opeche formation
Red sandy shale

Minnelusa sandstone
*Gray, red, and buff sandstones, mostly limy,
red shale at base*

Pahasapa limestone and
Englewood limestone

UNCONFORMITY

Whitewood limestone
*Hard, massive, buff limestone
in northern Black Hills*

UNCONFORMITY

Deadwood formation
*Gray to red sandstone and greenish shale,
in part glauconitic, slabby limestone,
flat-pebble limestone conglomerate,
quartz conglomerate at base,
thins to south*

UNCONFORMITY

METAMORPHIC ROCKS

PRECAMBRIAM

Amphibolite
*Metadiorite and metagabbro, many
small dikes not shown*

Schists, slates, and quartzites

Sandstone, arkosic, grit,
and conglomerate
Partly quartzite

UNCONFORMITY

IGNEOUS ROCKS

Granite and pegmatite
*Some granite is gneissoid, granite has
many inclusions of schist*

- - - - FAULT

SCALE

1 0 1 2 3 4 5   10 Miles

*After N. H. Darton, 1951
Geologic Map of South Dakota
U. S. Geol. Survey*

FIGURE 15.14
Geologic map of the Black Hills, South
Dakota, for Problem 4.
SOURCE: SOUTH DAKOTA
GEOLOGICAL SURVEY.

259

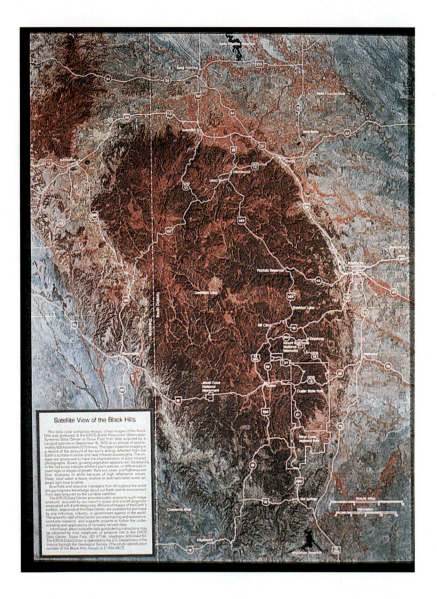

FIGURE **15.15**
Composite false-color Landsat satellite image
of the Black Hills, South Dakota/Wyoming,
for Problem 4.

5.   The map in Figure 15.16 is a geologic map of an area that has undergone faulting, folding, igneous activity, and several periods
    of erosion. Using the rule of Vs, place strike and dip symbols at several places on the map to help you visualize the structures.

    a.   What is the general strike and dip (just directions, not numbers) of the gabbro dike in the northeast corner? How do you know?

    b.   Is the gabbro older or younger than fault *A*? How do you know?

    c.   Is the dip of fault *B* greater or less than fault *A*? How do you know?

    d.   In what direction does fault *A* dip? How do you know?

Which side of fault *A* is the hanging wall and which the footwall? How do you know?

e. Is fault *B* older or younger than fault *A*? How do you know?

Which side of fault *B* is the hanging wall and which the footwall?

f. Is the unconformity between the limestone and conglomerate an angular unconformity, a disconformity, or a nonconformity? How do you know? What type of unconformity is the one between the gneiss and the sandstone?

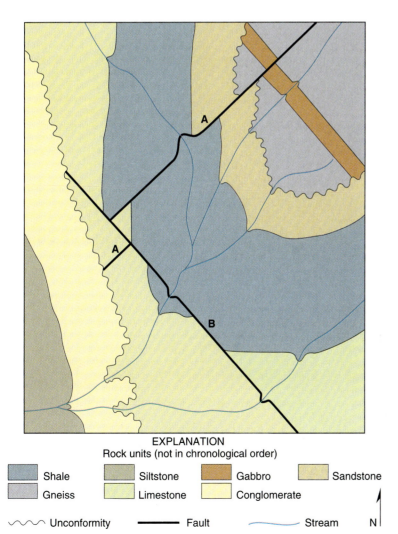

FIGURE **15.16**
Geologic map for use with Problem 5.

g. List the sequence of geologic events that produced the geologic relations shown on the map. Begin with the oldest and number it "1."

EXPLANATION
Rock units (not in chronological order)

Shale   Siltstone   Gabbro   Sandstone

Gneiss   Limestone   Conglomerate

∿∿∿ Unconformity   ▬▬ Fault   ～～ Stream   N

6. Figure 15.17 is part of a geologic map of the Tazewell Quadrangle, Tennessee. The area is part of the Appalachian Mountain system, which here consists of Cambrian and Ordovician sedimentary rocks folded and cut by two prominent faults, the Wallen Valley thrust fault and the Hunter Valley thrust fault (extreme southeast corner).

    a. What is the general structure of the rocks between the Wallen Valley and Hunter Valley thrust faults? That is, are the strata horizontal, inclined, folded?

    b. Why are the contacts between Ccr, Oc, Ol, and On wavy, whereas those farther southeast, such as between Ow and Ohb, are smooth?

    c. In what direction do the two faults dip? Based on the outcrop pattern of the faults, is the dip steep or gentle? Compare the ages of the rocks immediately adjacent to each of the two faults. Do older rocks overlie or underlie younger rocks along the faults?

    d. The outcrop pattern northwest of the Wallen Valley thrust fault is different from that to the southeast. Why? What features are present? Sketch an approximate cross section from the "W" in the word Wallen northwest to the edge of the map. You need not show the topography nor be too accurate in locating contact lines; the purpose is to see whether you understand the general structure.

    e. Depressions, as indicated by depression contours, are abundant in some formations and absent in others. Explain. (Hint: In what kind of rock do the depressions occur?)

7. a. Go to *http://www.uoknor.edu/special/ogs-pttc/geomap.htm* (or link to it through the author's homepage—see *Preface*), and use the geologic map of Oklahoma and the cross sections to answer the following questions.

    (1) What is the age (give the name of the period) of the pre-Quaternary bedrock exposed near Oklahoma City? Near Tulsa?

    (2) Why do the Quaternary sediments (in yellow) appear as streak-line like deposits?

    (3) What is the age of the rocks exposed in the Wichita Mountains and what kinds of rocks are they?

    (4) The Wichita Mountains and the Anadarko Basin are separated by a steep fault. Which side has gone down relatively?

    Is the fault a normal or reverse fault according to the cross section?

    (5) What type of contact (see Ch. 14) separates the Precambrian and Cambrian igneous rocks from the Cambrian sedimentary rocks in the Anadarko Basin?

    (6) Many of the faults in the Ouachita Mountains are thrust faults that originally had low dips. The dips in cross section B-B′ appear steeper than they really are, because the cross sections are vertically exaggerated about 10×. Based on cross section B-B′, in which relative direction were the rocks above the thrust-fault surfaces moved, toward the north or south?

    Rocks of the Ouachitas have also been folded. Based on B-B′, did folding occur before or after thrust faulting?

    b. You may be able to find information about the geology of your state on the Web. Go to *http://ncgmp.usgs.gov/* and link to *State geological surveys* and see what's available, or try some of the universities or colleges in your state.

IN GREATER DEPTH

8. Figure 15.18 is a geologic map of the Wetterhorn Peak Quadrangle, Colorado. This area is within the San Juan Mountains of Colorado, a range composed principally of Tertiary volcanic rocks. Tertiary units present are: (1) San Juan Formation (*Tsj* in gold), which consists of massive tuff, volcanic breccia, and lava flows of rhyolitic to andesitic composition; (2) porphyritic quartz latite, a fine-grained rhyolite-like rock with both plagioclase and K-feldspar (*Tql* in red); and (3) Potosi Volcanic Group, consisting mainly of welded pyroclastic-flow tuff (or ignimbrite) of rhyolitic composition (*Tp* in green).

    a. Is the porphyritic quartz latite (the red bodies) extrusive or intrusive igneous rock? Explain. What geologic term is given to the long, narrow bodies of *Tql*?

    b. List the three Tertiary units in order of age, with youngest first.

    c. Why do the long, narrow bodies of *Tql* follow such straight lines, even though the terrain is mountainous?

    d. Notice that the narrow bodies of *Tql* radiate outward from the irregular-shaped bodies of *Tql*, but the pattern is not random; certain orientations or strikes are preferred. Suggest one or more reasonable hypotheses to explain the preferred orientations.

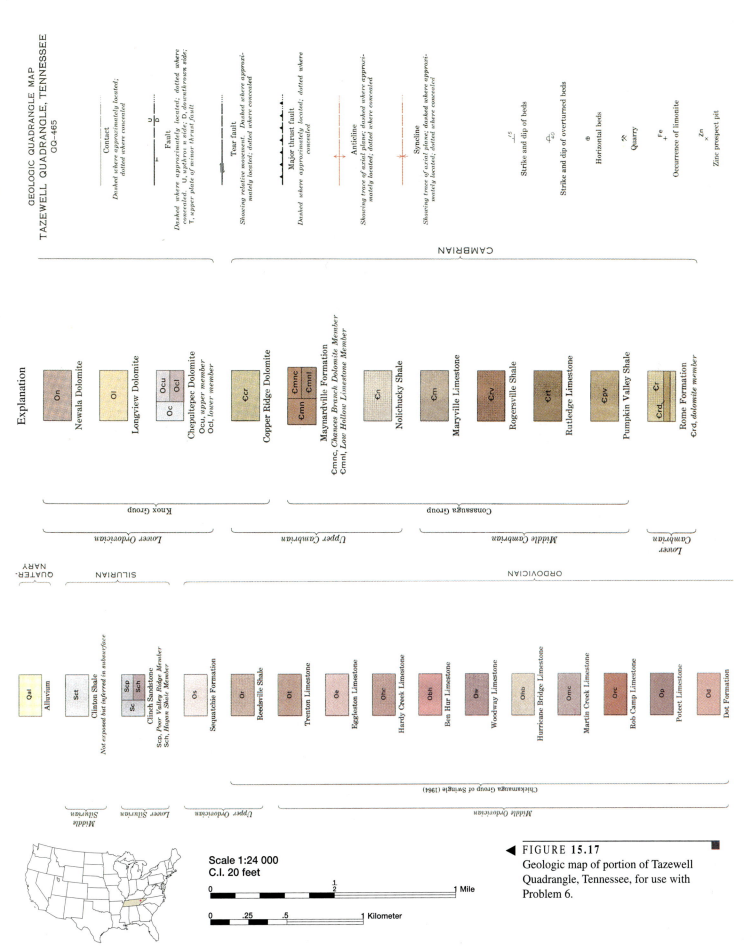

Contact
Dashed where approximately located; dotted where concealed

Fault
Dashed where approximately located; dotted where concealed. U, upthrown side; D, downthrown side; T, upper plate of minor thrust fault

Tear fault
Showing relative movement. Dashed where approximately located; dotted where concealed

Major thrust fault
Dashed where approximately located; dotted where concealed

Anticline
Showing trace of axial plane; dashed where approximately located; dotted where concealed

Syncline
Showing trace of axial plane; dashed where approximately located; dotted where concealed

Strike and dip of beds

Strike and dip of overturned beds

Horizontal beds

Quarry

Occurrence of limonite

Zinc prospect pit

CAMBRIAN

Explanation

On Newala Dolomite

Ol Longview Dolomite

Chepultepec Dolomite
Ocu, upper member
Ocl, lower member
| Ocu | |
| Oc | Ocl |

Ccr Copper Ridge Dolomite

Maynardville Formation
Cmnc, Chances Branch Dolomite Member
Cmnl, Low Hollow Limestone Member
| Cmnc | |
| Cmn | Cmnl |

Cn Nolichucky Shale

Cm Maryville Limestone

Crv Rogersville Shale

Crt Rutledge Limestone

Cpv Pumpkin Valley Shale

Rome Formation
Crd, dolomite member
| Crd | Cr |

Knox Group — Lower Ordovician

Conasauga Group — Upper Cambrian / Middle Cambrian

Lower Cambrian

QUATERNARY

SILURIAN

ORDOVICIAN

CAMBRIAN

Qal Alluvium

Sct Clinton Shale
Not exposed but inferred in subsurface

Clinch Sandstone
Scp, Poor Valley Ridge Member
Sch, Hagen Shale Member
| Scp | |
| Sc | Sch |

Os Sequatchie Formation

Or Reedsville Shale

Ot Trenton Limestone

Oe Eggleston Limestone

Ohc Hardy Creek Limestone

Obh Ben Hur Limestone

Ow Woodway Limestone

Ohb Hurricane Bridge Limestone

Omc Martin Creek Limestone

Orc Rob Camp Limestone

Op Poteet Limestone

Od Dot Formation

Chickamauga Group of Swingle (1964)

Middle Silurian

Lower Silurian

Upper Ordovician

Middle Ordovician

Scale 1:24 000
C.I. 20 feet

0   1/2   1 Mile

0   .25   .5   1 Kilometer

▶ FIGURE **15.17**
Geologic map of portion of Tazewell Quadrangle, Tennessee, for use with Problem 6.

e. Four types of Quaternary surficial deposits are present on the map. All have *Q* as the first letter of their symbol. They originated as (1) stream deposits, (2) landslide deposits, (3) accumulations at the base of a cliff or steep slope, or (4) glacial deposits. Locate the four types on the map, and based on their locations, speculate on the origin for each, using its symbol to identify it.

f. The red-dot pattern on Bighorn Ridge symbolizes a type of alteration, commonly associated with orebodies, that is caused by circulation of hydrothermal (hot water) solutions. Why might the alteration have occurred at that particular locality?

9. Refer to Figure 15.12 and answer the following questions.

a. What kind of contact separates Precambrian rocks from the Cutler Formation? Be specific. Based on the nature of its lower contact, its stratigraphic position, and the description in the *Explanation,* give a hypothesis for the origin of the Cutler Formation.

b. Note that the Navajo Sandstone is present in the southwest part of the map but does not continue to the north. Why? What kind of contact separates the Kayenta and Carmel Formations where the Navajo is absent? Be specific.

c. The red lines running diagonally across the map from northwest to southeast are called *structure contours,* because they are drawn on the top surface of a particular formation and show the shape of that buried surface. On this map, they are drawn on the top of the Entrada Sandstone (notice that they are dashed where the present-day surface is below the top of the Entrada). What is the strike of that surface? What is the gradient of that surface in feet per mile? Use trigonometry to calculate the angle of dip of the surface. (Hint: Convert the numerator and denominator of the gradient to the same units, then divide. The result is the *tangent* of the dip angle. The dip angle is the *arctangent* or *inverse tangent* on most calculators. If necessary, see your lab instructor for help.) Measure the dip on your cross section and compare it with the calculated dip.

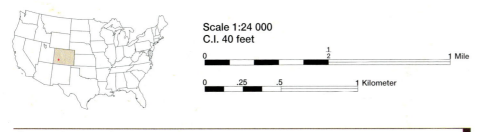

Scale 1:24 000
C.I. 40 feet

0   ½   1 Mile

0   .25   .5   1 Kilometer

◀ FIGURE 15.18
Geologic map of portion of Wetterhorn Peak Quadrangle, Colorado, for use with Problem 8.

chapter

16
Earthquakes

Overview

Earthquakes are both destructive and informative. Energy from them travels through and around the Earth as three types of waves, called P, S, and surface waves. In this lab, you will learn to identify the three types of earthquake waves on a seismogram and to use the seismogram to locate an earthquake and determine its magnitude. You will also learn how the intensity or degree of shaking of an earthquake is measured and mapped and how intensity depends on the characteristics of the ground being shaken. Finally, a simple example will show how earthquakes are used to examine the Earth's interior.

Materials Needed

Pencil and eraser • Drawing compass, or pencil on string,
to draw circles • Calculator

INTRODUCTION

Earthquakes are vibrations resulting from movements within the Earth. They are most common at tectonic plate boundaries, as discussed more fully in Chapter 17. Most earthquakes result from rupture along faults, but some are caused by other kinds of movement. For example, the upward movement of magma into Mount St. Helens in 1980 caused earthquakes that first alerted geoscientists to the probability of an eruption. Huge landslides and collapse of caves can produce earthquakes, and vibrations caused by explosions, such as underground nuclear tests, are very similar to natural earthquakes. We can only presume that those who named the

dinosaur *Seismosaurus* thought that, in its day, the huge dinosaur's footsteps must have generated some excitement.

Seismology is the study of earthquakes, and the branch of seismology known as earthquake seismology focuses specifically on the causes and effects of earthquakes, especially as earthquakes affect people. Because large earthquakes may result in death and destruction, a goal of earthquake seismologists is to be able to predict earthquakes. Other seismologists make use of the energy given off by earthquakes (or explosions) to see within the Earth, just as we process light energy with our eyes or sound energy with our ears to form images of our surroundings.

SEISMIC WAVES

When rocks rupture abruptly, energy radiates outward in all directions (Fig. 16.1). The "point" below the surface at which rupture occurs is called the **hypocenter,** or **focus,** of the earthquake. In fact, it commonly is not just a point, but a narrow zone along a fault that ruptures. Rupture may begin at one point, but then propagates along the fault, in an irregular, jerky fashion. The **epicenter** is the point on the Earth's surface directly above the hypocenter. It is the epicenter that is first reported by the press, because it is of most concern to people and is easily determined.

Seismic energy is released in the form of **seismic waves,** which cause the Earth to

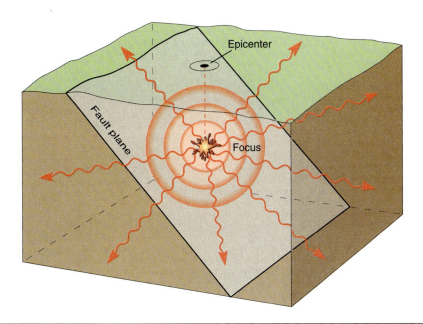

FIGURE 16.1

An earthquake originates at its *focus,* below the surface. The *epicenter* lies directly above the focus, on the surface. Seismic waves radiate outward in all directions from the focus.

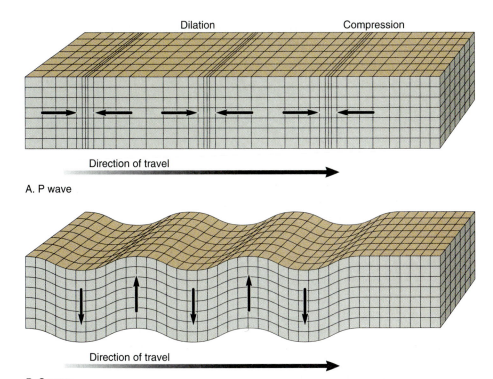

A. P wave

B. S wave

FIGURE 16.2

A. *P* waves vibrate back and forth, parallel to the direction of travel. B. *S* waves vibrate perpendicular to the direction of travel.

vibrate or deform elastically as they pass by. The three principal kinds of seismic waves are as follows:

P **wave** or **primary wave** (Fig. 16.2). The energy of a *P* wave travels through rock as a sequence of back-and-forth vibrations *parallel* to the direction of movement. Its passage causes alternate pushing (compression) and pulling (dilation) on particles in its path. *P* waves can travel in solids or liquids, and are the fastest of the three types of seismic waves.

S **waves, secondary waves,** or **shear waves** (Fig. 16.2). The energy of an *S* wave travels through rock as a sequence of vibrations *perpendicular* to the direction of movement. Its passage causes particles to vibrate up and down or sideways. *S* waves can travel in solids but not in liquids, and their velocity is between that of the other types of seismic waves.

Surface waves. Surface waves travel along the surface of the Earth as two types of waves. One of these has a horizontal shearing motion similar to an *S* wave, and the other a rolling motion in a vertical plane much like water waves. Surface waves are the slowest seismic waves.

SEISMOGRAMS

Earthquakes are very common; hundreds of thousands are detected each year. However, most are too small or too remote to be felt, and relatively few cause damage or loss of life. Earthquakes, even very small ones, are recorded with **seismographs.** A worldwide network of seismographs constantly monitors earthquake activity and provides information about an earthquake within minutes of its occurrence.

The record obtained with a seismograph is a **seismogram.** One type of seismogram is recorded on a rotating drum as a series of wiggly lines (Fig. 16.3). Tick marks or breaks in the lines are precisely spaced and represent time elapsed; actual time is Universal Coordinate (or Greenwich Mean) Time, and for most stations is accurate to a few thousandths of a second. Figure 16.3 shows a simple seismogram in which the only waves recorded are *P, S,* and surface.

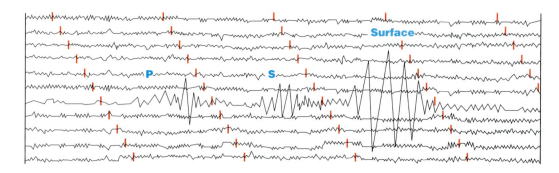

Simple seismogram with *P, S,* and surface waves.

A seismograph station may have three seismographs arranged to record the vertical component and two horizontal components of motion (for example, E-W and N-S). The three together allow one station to give an analysis of an earthquake.

LOCATING EARTHQUAKES

The distance from an earthquake focus can be calculated if the seismic-wave velocity and the time it took to travel to the seismograph are known. Seismograph stations are equipped with voluminous databases regarding wave velocities and can readily produce *travel time curves* from these data. **Travel time curves** show how the time required for various types of waves to travel a given distance varies with distance. In Figure 16.4, curves are shown for a *P* and an *S* wave. If the times at which an earthquake occurred and the *P* wave arrived are known, the distance from the earthquake can be read directly from the graph.

But what if the origin time of the earthquake is unknown, as is commonly the case? Then the difference in arrival times between the *P* and *S* waves can be used.[1] You can see in Figure 16.4 that the vertical distance (travel time) between the *P* and *S* curves increases with distance traveled. The distance from the earthquake can be determined by measuring the *difference* in arrival times of *P* and *S,* and finding the place on the travel time curves where *P* and *S* are that far apart. Its position on the distance scale will correspond to the distance from the earthquake.

For example, let's say the seismo-

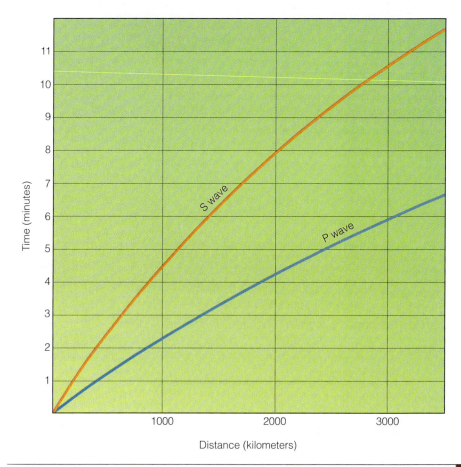

Travel-time curves for *P* and *S* waves. Note that the time interval between curves increases with distance.

[1]*The process is analogous to one you may have used to figure out how far away lightning was during a thunderstorm. You see a lightning flash instantaneously because the velocity of the light waves is very fast (3×10^5 km/sec), but the thunder takes awhile to get to you because the velocity of sound is much slower (about 0.34 km/sec). The farther away the lightning, the longer it takes for the thunder to get to you. If you count the seconds between lightning and thunder and multiply by the velocity of sound—0.34 km/sec—you know how far away the lightning was. A difference of 4 seconds means a distance of 4×0.34 km = 1.36 km, about 0.85 mile.*

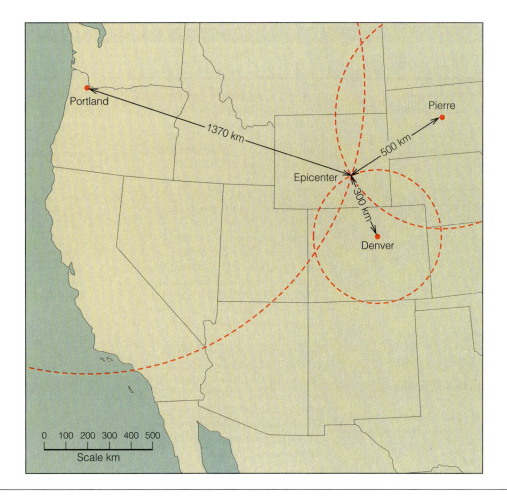

FIGURE **16.5**
Seismograms from seismograph station at Portland, Oregon; Denver, Colorado; and Pierre, South Dakota are used to locate the epicenter near Douglas, Wyoming, with three intersecting circles.

gram in Figure 16.3 was recorded in Pierre, South Dakota, and the difference in arrival times of *P* and *S* waves is about 70 seconds, which on Figure 16.4 corresponds to a distance from the focus of about 500 km. In Figure 16.5, a circle of radius 500 km is drawn around Pierre. The earthquake, if a shallow one, must be located someplace on that circle. But where? One circle is not enough. Three intersecting circles for the same earthquake from three seismograph stations are needed, as illustrated in Figure 16.5. The epicenter lies at the common intersection, near Douglas, Wyoming.

EARTHQUAKE INTENSITY AND MAGNITUDE

The size of an earthquake is described in two ways: intensity and magnitude. **Earthquake intensity** is a measure of the amount of shaking and is a qualitative measurement. Intensity is assessed with the *Modified Mercalli Scale* (Table 16.1), which ranges from I to XII. Questionnaires based on the scale are mailed out after an earthquake, and residents' responses are used to prepare a map of earthquake intensity. Lines of equal intensity, *isoseismals,* are like con-

tours and outline the area where various levels of damage were recorded. Figure 16.6 illustrates the results for the Douglas, Wyoming, earthquake. The intensity felt in a given area depends not only on the size of the earthquake but also on the strength of the underlying ground. An area underlain by loose sediment will be shaken much more than one on solid bedrock. Notice that some people in Pierre felt the earthquake, but it was not felt in the surrounding area.

Earthquake magnitude is a measure of the amount of energy released by the earthquake, so it is a quantitative measurement. The original magnitude scale was the *Richter Magnitude Scale.* It defined

MODIFIED MERCALLI SCALE

I   Not felt by people, except rarely under especially favorable circumstances.

II   Felt indoors only by persons at rest, especially on upper floors. Some hanging objects may swing.

III   Felt indoors by several. Hanging objects may swing slightly. Vibration like passing of light trucks. Duration estimated. May not be recognized as an earthquake.

IV   Felt indoors by many, outdoors by few. Hanging objects swing. Vibration like passing of heavy trucks; or sensation of a jolt like a heavy ball striking the walls. Standing automobiles rock. Windows, dishes, doors rattle. Wooden walls and frame may creak.

V   Felt indoors and outdoors by nearly everyone; direction estimated. Sleepers wakened. Liquids disturbed, some spilled. Small, unstable objects displaced, some dishes and glassware broken. Doors swing; shutters, pictures move. Pendulum clocks stop, start, change rate. Swaying of tall trees and poles sometimes noticed.

VI   Felt by all. Damage slight. Many frightened and run outdoors. Persons walk unsteadily. Windows, dishes, glassware broken. Knickknacks and books fall off shelves; pictures off wall. Furniture moved or overturned. Weak plaster and masonry cracked.

VII   Difficult to stand. Damage negligible in buildings of good design and construction; slight to moderate in well-built ordinary buildings; considerable in badly designed or poorly built buildings. Noticed by drivers of automobiles. Hanging objects quiver. Furniture broken. Weak chimneys broken. Damage to masonry; fall of plaster, loose bricks, stones, tiles, and unbraced parapets. Small slides and caving in along sand or gravel banks. Large bells ring.

VIII   People frightened. Damage slight in specially designed structures; considerable in ordinary substantial buildings, partial collapse; great in poorly built structures. Steering of automobiles affected. Damage or partial collapse to some masonry and stucco. Failure of some chimneys, factory stacks, monuments, towers, elevated tanks. Frame houses moved on foundations if not bolted down; loose panel walls thrown out. Decayed pilings broken off. Branches broken from trees. Changes in flow or temperature of springs and wells. Cracks in wet ground and on steep slopes.

IX   General panic. Damage considerable in specially designed structures; great in substantial buildings, with some collapse. General damage to foundations; frame structures, if not bolted, shifted off foundations and thrown out of plumb. Serious damage to reservoirs. Underground pipes broken. Conspicuous cracks in ground; liquefaction.

X   Most masonry and frame structures destroyed with their foundations. Some well-built wooden structures and bridges destroyed. Serious damage to dams, dikes, embankments. Landslides on riverbanks and steep slopes considerable. Water splashed onto banks of canals, rivers, lakes. Sand and mud shifted horizontally on beaches and flat land. Rails bent slightly.

XI   Few, if any masonry structures remain standing. Bridges destroyed. Broad fissures in ground; earth slumps and landslides widespread. Underground pipelines completely out of service. Rails bent greatly.

XII   Damage nearly total. Waves seen on ground surfaces. Large rock masses displaced. Lines of sight and level distorted. Objects thrown upward into the air.

---

magnitude as the logarithm (base 10)[2] of the amplitude, measured in micrometers (thousandths of a millimeter), of the largest waves recorded on a specified seismograph at a distance of 100 km from the epicenter. The assumption was that more energetic earthquakes would produce bigger wiggles, so that the height—or **amplitude** (top of Fig. 16.7)—of a selected wave would be proportional to the amount of energy released. There are no lower or upper limits to the Richter scale, other than those that may be imposed by the sensitiv-ity of the instruments or by the amount of strain that rocks can accumulate before rupturing; the Richter scale is open ended.

The original Richter scale was designed in 1935 for use with local earthquakes in southern California whose largest waves commonly are surface waves. But what of deeper-focus earthquakes that may not even generate surface waves? What should be measured—*P* waves, *S* waves? Or which seismogram should be used—one measuring vertical motion or one of those recording horizontal motion? These and other considerations have led to the definition of several kinds of magnitude scales that give similar, but not identical, answers for a specific earthquake. However, most have the same general basis—measurement of seismic wave amplitude—so, to keep things simpler, only one of these is illustrated here.

[2]*The logarithm (base 10) of a number is the power to which 10 must be raised to equal that number. A logarithm can be a whole or fractional number. For example, $\log_{10} 100 = 2$; $10^2 = 100$. And $\log_{10} 5 = 0.699$; $10^{0.699} = 5$.*

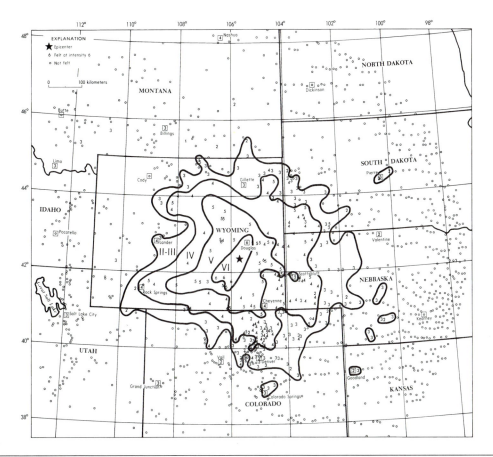

FIGURE 16.6
Intensity map for earthquake near Douglas, Wyoming, with isoseismal lines based on the Modified Mercalli scale.

Figure 16.7 illustrates the measurement of the Richter magnitude of an earthquake modified from the original Richter method (after Bolt, 1988). 1) Measure the amplitude of the largest surface wave in millimeters; in the example, it is 20 mm. 2) Use the graph to determine the Richter magnitude by running a straightedge from the appropriate *Distance to Epicenter* (assumed to be 60 km in the example) to the appropriate *Amplitude;* read the magnitude where this line crosses the *Magnitude* scale (about 5.6 in the example).

USING SEISMIC WAVES TO SEE BENEATH THE SURFACE

Seismic body waves (*P* and *S*) passing through the Earth do not follow straight-line paths but are bent (refracted) and reflected when they encounter different kinds of rocks (Fig. 16.8). Their behavior is just like light in this regard. Light is reflected from some surfaces, like a mirror, but bends as it passes through other sub-

stances, such as water or glass. Figure 16.8 shows a *P* wave being reflected and traveling back to the surface. If the velocity and travel time of the *P* wave are known, the distance traveled can be calculated, and the depth to the reflecting horizon can be determined. Reflection seismology, using energy from human-produced explosions or vibrations, is used to map geologic structures below the surface and is one of the most important methods of exploration used in the oil industry.

273

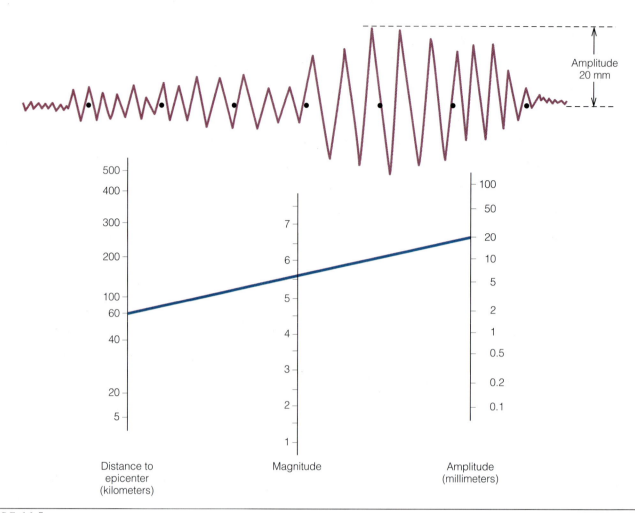

FIGURE 16.7

A line joining the amplitude of the largest wave on the seismogram and the distance to the epicenter crosses the magnitude scale at 5.6.

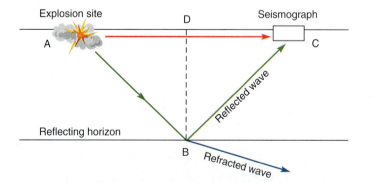

To determine depth to the reflecting horizon

1. Knowing the travel time and distance, ADC, from the explosion site to the seismograph, calculate the *P*-wave velocity (velocity = distance/time).

2. Knowing the travel time of the reflected wave and the velocity found in Step 1, calculate the distance, ABC, travelled by the reflected wave (distance = velocity × time).

3. Use the distances AD and AB to calculate the depth to the reflecting horizon, DB. (Remember the Pythagorean theorem? The square of the hypotenuse of a right triangle equals the sum of the squares of its legs—in this case, the hypotenuse is AB, and AD and DB are the legs, so $AD^2 = AB^2 + DB^2$.)

FIGURE 16.8

Reflection of a P wave and calculation of depth to reflector.

APPLICATIONS

As millions of Americans were tuned in to the beginning of the third game of the World Series (championship baseball) at Candlestick Park, San Francisco, California, on October 17, 1989, a large earthquake occurred. It was the largest to strike the San Francisco Bay region, home to almost 6 million people, since the great earthquake of 1906, when fewer than a million people lived there. The earthquake was felt over 400,000 square miles and resulted in 67 deaths, 12,000 people left homeless, and $6 billion in property damage. In terms of cost, it was one of the most expensive natural disasters ever in the United States, and was the most costly earthquake since 1906.

The following problems illustrate how the epicenter was located, how the magnitude and intensities were determined, how the intensity was influenced by the nature of the surficial material, and how energy from the earthquake could be used to view the interior of the Earth. In each case, the scientific method is used to answer basic problems.

OBJECTIVES

If you complete all the problems, you should be able to:

1. Identify *P, S,* and surface waves on a simple seismogram.

2. Locate the epicenter of an earthquake using seismograms and travel-time curves.

3. Explain the basis for determining the magnitude and intensity of an earthquake, and draw isoseismals from intensity data.

4. Describe how different earth materials (mud, alluvium, bedrock) respond to seismic waves.

5. Use seismic reflection data to determine the depth to a reflector.

PROBLEMS

1. The three seismograms in Figure 16.9 are simplified seismograms of the October 17, 1989, earthquake from three different stations. The stations have the following locations: A. lat 40° 01′N, long 121° 20′W; B. lat 34° 00′N, long 118° 00′W; C. lat 41° 00′N, long 114° 35′W.

    a. Plot the locations of the stations on the map in Figure 16.10.

    b. Where was the epicenter?

        (1) Use the travel-time curves in Figure 16.11 to determine the distance to the epicenter from each of the three stations.

        (2) Locate the epicenter of the earthquake from intersecting circles centered at the three stations.

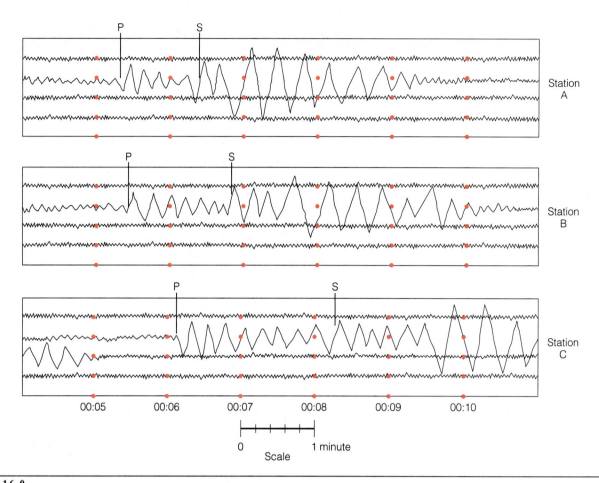

FIGURE 16.9

Three seismograms for Problem 1. Universal Coordinate Time is indicated by the dots on seismograms, which are one minute apart. Time is based on a 24-hour day; 00:05 means 12:05 a.m.

c. When did the earthquake occur?

(1) Use the travel-time curves and the time values indicated on the seismograms to determine the time of occurrence.

(2) The time you determined is Universal Coordinate Time. What time would it have been in the Pacific Daylight time zone if there is a 7-hour difference?

d. How big was the earthquake? Determine the Richter magnitude from seismogram A using the method shown in Figure 16.7.

FIGURE **16.10**
Map for Problem 1.

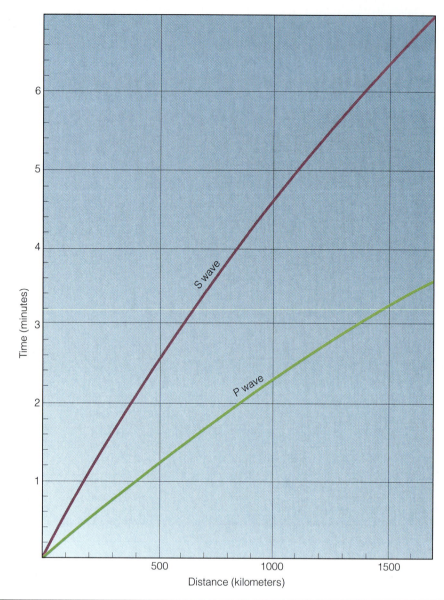

FIGURE **16.11**

Travel time curves for *P* and *S* waves. This enlargement of Figure 16.4 has the same time scale as the seismograms in Figure 16.9 and is to be used for Problem 1.

2. Questionnaires designed for assigning Modified Mercalli scale intensities to the October 17 earthquake were returned from the locations shown on Figure 16.12 A and B. The responses were divided into five groups, as shown in Table 16.2. Using these responses, assign Modified Mercalli scale intensities to the five groups. Next, draw isoseismals on the maps (Fig. 16.12A) based on intensity distribution. How does the area of highest intensity compare with the epicenter of the earthquake?

Table 16.2

RESPONSES TO MODIFIED MERCALLI SCALE QUESTIONNAIRE

A.   Felt by all, and people were extremely frightened. Drivers had great difficulty controlling vehicles. Some buildings collapsed, others very seriously damaged. Concrete bridge collapsed. Smell of gas in some areas. Ground cracked.

B.   Felt by all, most people were frightened. Drivers had some difficulty steering. Some reasonably well-built buildings partially collapsed, poorly built buildings badly damaged. Some water towers damaged. Large trees had tops broken. Some reports of furniture and people being thrown in air. Cracks in ground.

C.   Felt by all, some people knocked down. Noticed by some drivers. Damage to buildings moderate, except for those of poor construction. Windows broken, things knocked off shelves, plaster walls damaged. Trees and bushes shook. Caving noticed on some steep hillsides and along stream banks.

D.   Felt by all, some had trouble walking. Not noticed by most drivers. Damage to structures slight, but windows broken, things knocked off shelves, pictures knocked off walls. Plaster walls cracked.

E.   Felt by most people. Water in swimming pools disturbed. Some reported broken glassware. Pendulum clocks stopped.

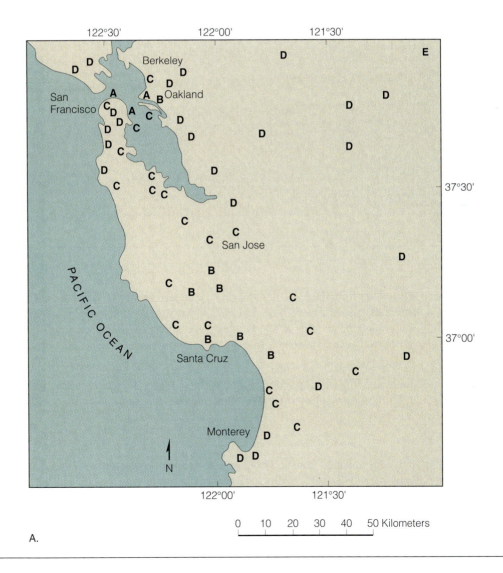

FIGURE 16.12

Maps of San Francisco Bay area showing locations for which intensities have been assigned. Map B is an enlargement to show more detail. Letters A through E correspond to earthquake intensities described in Table 16.2. For use with Problems 2 and 3.

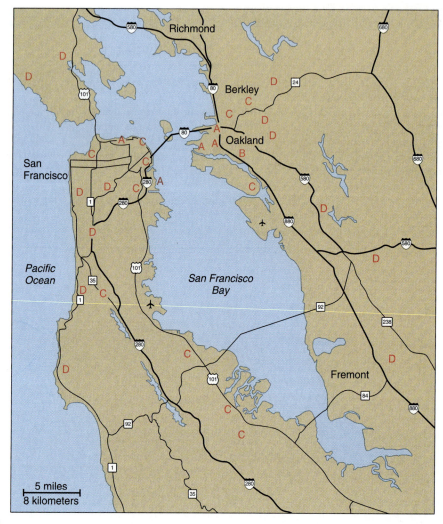

B.

FIGURE 16.12—*Continued.*

3. How can the intensity be so high, so far from the epicenter? Might it be related to the type of sediment or rock at the surface? Figure 16.13 is a Landsat image of the San Francisco Bay to Monterey Bay area. Bedrock appears as reddish brown to dark brown tones. Mud and fill occur around the margins of San Francisco Bay, indicated by either nearly black, green, white, or bluish white tones. Between these two is a light blue area speckled with red, which represents unconsolidated natural alluvium.

    a. Outline the areas in which these three types of materials occur around San Francisco Bay.

    b. Compare your results with Figure 16.14, a map showing the expected effects of ground shaking in the San Francisco Bay area and with the earthquake intensity map you drew. Which of the three types of material shook the most?

    c. Sketch hypothetical seismograms obtained from seismographs, located at Oakland, that sit on (1) mud and fill, (2) unconsolidated soil (mainly alluvium), and (3) stable bedrock. (Because of proximity to the epicenter, *P, S,* and surface waves will not be separated, so just show relative amplitudes of vibrations.)

    d. On what fault was the October 17, 1989, earthquake centered?

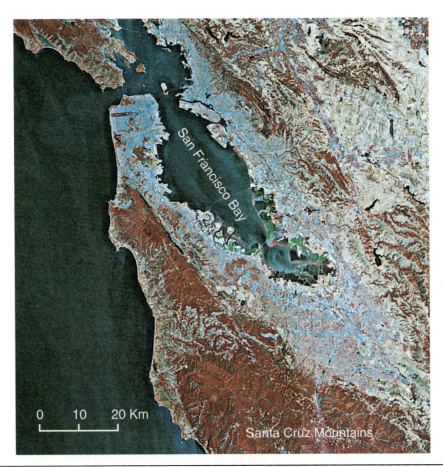

FIGURE 16.13
Landsat image of San Francisco Bay area for use with Problem 3.

4. Go to *http://www.eas.slu.edu/Earthquake_Center/earthquakecenter.html* (or link to it through the author's homepage—see *Preface*), select *Recent Earthquake Maps,* and choose the map of the area in which you live (e.g., *Continental US and Adjacent Canada*).

    a.   When was the map last generated?

    b.   Where was the nearest earthquake in the previous 14 days? Give its approximate latitude and longitude coordinates, as well as its geographic location (e.g., state name).

    c.   What was its magnitude?

    d.   What was the depth of focus?

    e.   If you are not looking at the map *Continental US and Adjacent Canada,* return to *Recent Earthquake Maps,* and select that map. How many earthquakes occurred in California during the previous 14 days, and what was the magnitude of the largest one?

Or go to *http://www.geophys.washington.edu/cnss.cat.html,* and select the epicenter map most appropriate for your area.

    a.   When was the map last generated?

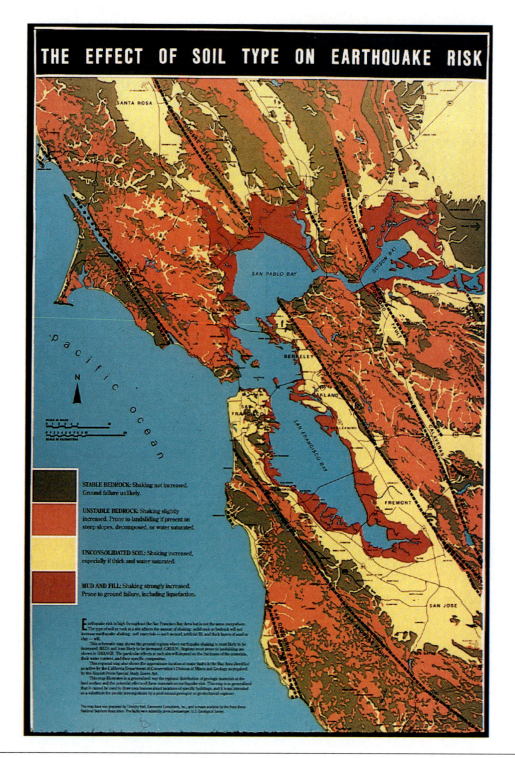

FIGURE **16.14**
Map showing the expected effects of ground shaking in the San Francisco Bay area.

b. Where was the nearest earthquake in the previous 14 days? Give its approximate latitude and longitude coordinates as well as its geographic location (e.g., state name).

c. If you are not looking at the *mainland U.S. epicenter map,* return to the menu of map selections and select that map. How many earthquakes occurred in California during the previous 14 days?

d. Go back to the menu page and scan the list of earthquakes from the past two weeks until you find a description of the earthquake nearest you. When did it occur (date and time—UTC refers to Universal Coordinate Time), what are its latitude and longitude coordinates, and what were the depth of focus and magnitude?

You may find it interesting to look at the Global Earthquake Report by going to the *http://www.geo.ed.ac.uk/quakexe/ quakes* site. There you will find a list of the larger earthquakes that have occurred worldwide in the past week. You can access world and U.S. maps that show the epicenters, and you can zoom in on any area in those maps.

IN GREATER DEPTH

5. While the damage from earthquakes can be devastating, valuable information can be gained by studying them. The energy radiated from earthquakes can be used to look inside the Earth. An aftershock of the October 17 earthquake was recorded at Winnemucca, Nevada, a distance of 660 km away. Some of the P waves arriving in Winnemucca were reflected from the mantle, while others traveled there on a nearly direct path. Figure 16.15 illustrates, in cross section, two paths for the P waves. P_1 is the nearly direct path, and P_2 is reflected from the top of the mantle.

a. If the earthquake occurred at 02:07:00 (Universal Coordinate Time), use the arrival time of P_1 (and the distance of 660 km) to calculate the velocity of P_1; show your work.

b. Assume the velocity of P_2 is the same as P_1, and using its arrival time, calculate the distance traveled by P_2.

c. Half of this distance will be the hypotenuse of a right triangle such as triangle ABD in Figure 16.8. Use the Pythagorean theorem ($c^2 = a^2 + b^2$) to calculate the depth to the mantle. Show your work.

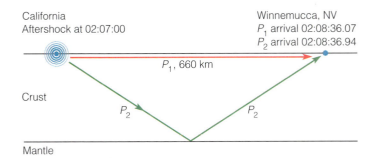

California
Aftershock at 02:07:00

Winnemucca, NV
P_1 arrival 02:08:36.07
P_2 arrival 02:08:36.94

P_1, 660 km

Crust

P_2 P_2

Mantle

FIGURE **16.15**

Cross section illustrating two paths for P waves from aftershock focus in California to Winnemucca, Nevada, 660 km away. Times given in hour:minute:second, Universal Coordinate Time. For use with Problem 5.

chapter

17

Plate Tectonics

Overview

The plate tectonic theory is one of the most important and far-reaching theories in geology. It evolved from detailed observations made over many years and from individual hypotheses that required extensive and imaginative testing. For example, measurement of magnetism of the rocks on the seafloor was used as a test for seafloor spreading; the nature and distribution of earthquakes and volcanoes provided a test for plate convergence; and a variety of geological and geophysical studies proved the existence of transform faults. In this lab, a few predictions based on the plate tectonic hypothesis will be tested by analyzing appropriate data.

Materials Needed

Pencil and eraser • Metric scale • Protractor • Calculator

INTRODUCTION

The theory of plate tectonics was more than 100 years in the making. Beginning in the 19th century, geologists began to compile basic observations about the Earth that could not be explained satisfactorily.

For example:

Fossils of land-dwelling plants and animals were found on opposite sides of oceans; how did these land dwellers cross the ocean?

Fossils of warm-weather plants and animals were found where climates are now cold, and evidence of

glaciation was found in areas that now are warm; were Earth climates radically different in the past?

Earthquakes, volcanoes, and mountains are not randomly distributed about the Earth; why not?

Some interesting hypotheses were developed to explain these observations. To explain the distribution of fossils of land dwellers, it was supposed that continents were connected by land bridges that have since foundered into the oceans. To explain the rock-borne evidence of diverse climates, it was supposed that climates fluctuated widely in the geologic past.

To explain the causes and distribution of earthquakes, volcanoes, and mountains, hypotheses involving contraction or expansion of the Earth were proposed; contraction would cause fold-and-thrust-faulted mountains, and expansion would cause fault-block mountains and provide conduits to the surface for magma. None of these hypotheses withstood testing, however, and eventually they were discarded.

The first detailed attempt at a unifying theory, one that could explain most of the major observations, was the continental-drift hypothesis of Alfred Wegener, proposed in a series of papers and books between 1910 and 1928. Wegener's hypoth-

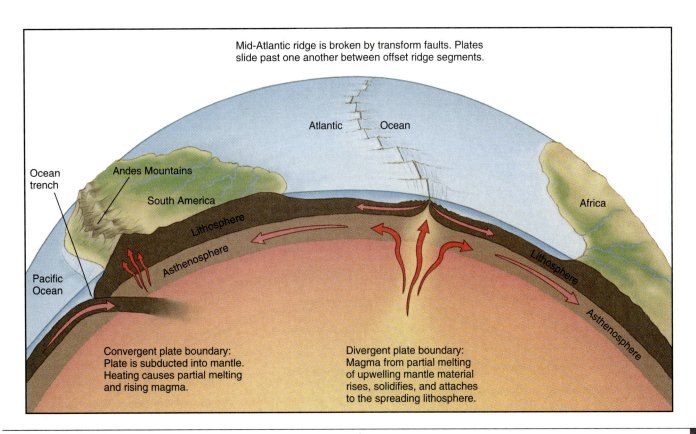

Mid-Atlantic ridge is broken by transform faults. Plates slide past one another between offset ridge segments.

Atlantic Ocean

Ocean trench

Andes Mountains

South America

Lithosphere

Asthenosphere

Pacific Ocean

Africa

Lithosphere

Asthenosphere

Convergent plate boundary: Plate is subducted into mantle. Heating causes partial melting and rising magma.

Divergent plate boundary: Magma from partial melting of upwelling mantle material rises, solidifies, and attaches to the spreading lithosphere.

FIGURE **17.1**
Cut-away view of the Earth, illustrating plates of lithosphere and the three types of plate boundaries. The thickness of lithosphere and asthenosphere is exaggerated.
SOURCE: DATA FROM USGS PROFESSIONAL PAPER 1350.

esis was severely criticized, especially by geologists in the northern hemisphere, on the grounds that it was mechanically impossible.

But more and more evidence seemed to suggest that continents indeed had moved relative to one another. Research that began in the late 1920s, and developed through the thirties and forties, indicated that a mechanism involving convection currents within the solid Earth might be possible. Additional geophysical evidence was especially persuasive, and by the early to mid-1960s, the framework of plate tectonics was in place.

At that time, plate tectonics was still a hypothesis, and there were many doubters, ready with observations that could not be readily explained by the existing model. The mark of a good hypothesis is that it can be tested, and the plate tectonics hypothesis offered many tests. Tests are conceived from predictions. That is, if the hypothesis is correct, then we can predict and test for certain consequent relations or results. In this lab, a series of predictions will be evaluated using real data.

GENERAL THEORY OF PLATE TECTONICS

According to the theory of plate tectonics, the outermost part of the Earth is made of **lithospheric plates,** some 70 to 125 km thick, that move relative to one another (Fig. 17.1). Plates are constructed by successive injections of magma at **divergent plate boundaries.** The plates move away from the divergent boundaries and, at **convergent plate boundaries,** they encounter other plates moving toward them in a relative sense. There, one of the plates is carried beneath the other, and either descends into the mantle or becomes sutured to the overlying plate. A **transform-fault plate boundary** is a type of strike-slip fault; plates slide past one another, and lithosphere is neither created nor destroyed. Let's look at these three types of plate boundaries in a little more detail.

DIVERGENT BOUNDARIES

Plates move apart at divergent plate boundaries as magma from the mantle wells up to fill the gap between adjacent plates, as shown in Figure 17.2. Divergent boundaries in the oceans are expressed by ridges and rises, generally with a narrow rift valley marking the boundary between plates along a central ridge crest.

Perhaps the most critical piece of evidence for the plate tectonic hypothesis, or at least the seafloor-spreading hypothesis, came in the early 1960s from the study of the magnetic properties of rocks on the floor of the North Atlantic Ocean. This story has two parts.

The first has to do with the way in which an igneous rock retains a record (its **remnant magnetism**) of the existing magnetic field as it cools from a molten state. Recall that some minerals, especially magnetite (Fe_3O_4), are strongly magnetic. However, if they are heated above a certain temperature, the **Curie point** (about 580° C

285

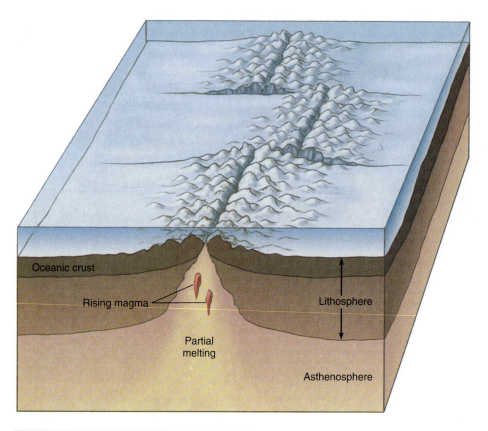

FIGURE 17.2
Block diagram of a mid-ocean, divergent plate boundary.

for magnetite), the magnetism disappears. Conversely, magnetite, which forms at about 1100°C as magma crystallizes, becomes magnetic when the temperature drops below the Curie point. Its magnetism aligns with the existing magnetic field, as shown in Figure 17.3A, and is frozen into the rock as though the rock contained tiny bar magnets. Thus, the rock contains a record of the magnetic lines of force at that time and place; the north-seeking ends of the "magnets" point toward magnetic north, and the axes of their magnetic fields are parallel to the **inclination** of the magnetic lines of force.

The second part of the story has to do with reversals of the magnetic field and **magnetic anomalies** (magnetic values greater or less than expected). In the present-day magnetic field, the north-seeking pole of a magnet points toward the north magnetic pole, which coincides approximately with the north geographic pole. But studies of remnant magnetism show that this has

not always been the case. At many times in the geologic past, the orientation of the Earth's magnetic field has been reversed, as shown in Figure 17.3B. During these periods, the north-seeking poles of newly forming mineral "magnets" actually pointed toward the south geographic pole.

Imagine what could happen along a divergent boundary as magma rose, solidified, and took on the magnetic orientation existing at that time. During times of **normal magnetic polarity,** north-seeking remnant magnetic poles would point toward present-day north; during times of **reversed magnetic polarity,** north-seeking poles would point toward present-day south. As more magma is injected, the previously formed lithosphere (with one magnetic orientation) would be forced to move away from the plate boundary in both directions. Should the magnetic field reverse its polarity, newly injected magma would form rock that has a reversed polarity. In this way,

alternating stripes or bands of rock with either normal or reversed remnant magnetism should be formed parallel to the divergent boundaries. Rocks with normal remnant magnetism reinforce the local magnetic intensity where as those with reversed remnant magnetism decrease the local magnetic intensity. Thus, it was hypothesized in the 1960s that the bands could be recorded by a ship-towed magnetometer as stripes of alternating **positive and negative magnetic anomalies.**

It was an exciting time in geology when parallel bands of positive and negative magnetic anomalies were discovered along the Mid-Atlantic Ridge (Fig. 17.4). Here was a magnetic-tape-like record of past reversals of the magnetic field as recorded by basalt extruded at a divergent plate boundary.

If seafloor spreading occurs as described, and if magnetic anomalies result from magnetic polarity reversals, you should be able to predict the answers to the following questions and propose ways to test your predictions. 1) Will magnetic anomalies on opposite sides of a divergent boundary correlate; that is, will a negative anomaly indicating a specific reversal occur on both sides of the boundary? 2) Will magnetic data from other ocean ridges or rises show the same general pattern of parallel bands as shown in the North Atlantic? 3) Will specific magnetic anomalies correlate on a worldwide basis; that is, will the same anomalies be found worldwide? You also should be able to propose a way to use the magnetic anomalies to determine the rate at which seafloor spreading has occurred in a given area. Problems 1 and 5 deal with magnetic anomalies at divergent plate boundaries.

CONVERGENT BOUNDARIES

Plates come together at convergent boundaries, as illustrated in Figure 17.5. Converging plates contain either oceanic crust or continental crust at their leading edge. When two plates with oceanic crust converge, one dives, or is **subducted,** below the other (Fig. 17.5A). When an oceanic plate and a continental

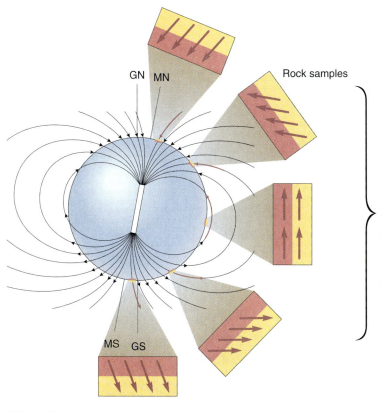

Basalt samples from various latitudes have *normal* remnant magnetism. Arrows in samples are parallel to, and point in the same direction as, magnetic lines of force at each location. Yellow marks upper half of sample.

A. Normal geomagnetic polarity

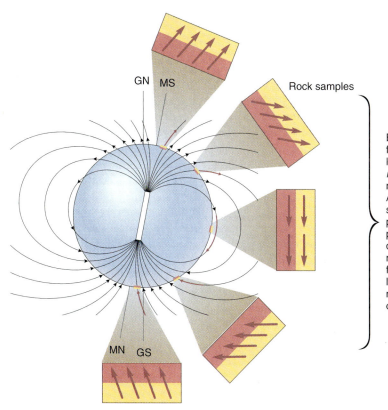

Basalt samples from various latitudes have *reversed* remnant magnetism. Arrows in samples are parallel to, and point in the same directions as, magnetic lines of force at each location. Yellow marks upper half of sample.

B. Reversed geomagnetic polarity

plate converge, the more dense oceanic plate is subducted beneath the continental plate (Fig. 17.5B). When two continental plates collide, partial subduction is accompanied by intense deformation and eventual suturing of segments of the subducted plate to the base of the overriding plate (Fig. 17.5C).

A subducted plate is comparatively cold and rigid when it begins its descent, but it warms and softens as it moves downward. Let's consider two of the many consequences implied by the converging process.

Earthquakes

Most of the world's major earthquakes take place on convergent plate boundaries. The zone along which most occur is named the **Benioff zone,** after seismologist Hugo Benioff. The faulting that generates the earthquakes is both along the boundary between the two lithospheric plates and within the descending slab itself (Fig. 17.5).

If convergence does take place as outlined, you should be able to predict the answers to the following questions, and propose ways to test your predictions. 1) Are earthquakes likely to be shallow (0–70 km), intermediate (70–300 km), or deep focus (300–700 km), or will there be a range of focal depths? 2) Will earthquake foci show the angle at which the

FIGURE **17.3**

Earth's magnetic lines of force during (A) *normal* and (B) *reversed* geomagnetic polarity. During normal polarity, the magnetic north pole (MN) is near the geographic north pole (GN). When the magnetic pole is reversed, MN is near the geographic south (GS). Lavas that cooled through the Curie point during normal polarity have a *normal remnant magnetism* with an orientation appropriate to the latitude at which they formed. This is illustrated in A. by the red and yellow rock samples from various latitudes; samples with *reversed remnant magnetism* are shown in B.

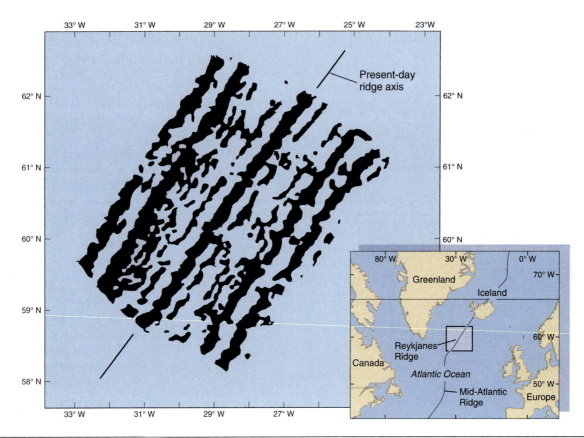

FIGURE 17.4

Symmetrical magnetic anomalies about the Reykjanes Ridge portion of the Mid-Atlantic Ridge, southwest of Iceland. Dark areas are positive magnetic anomalies, light areas between them are negative anomalies.

plate descends? Problem 2 deals with these predictions.

Volcanic Arcs

Volcanic, or magmatic, arcs are associated with convergent plate boundaries. The volcanoes usually occur 100 to 300 km away from the trench that marks the position of the subduction zone on the surface, and are located on the same side of the trench as the earthquake epicenters. The implication is that magma is generated as a result of the subduction process.

Given these general observations and the hypothesis of plate tectonics, 1) suggest a process by which magma forms as a result of subduction; 2) predict how the **arc-trench gap** (the horizontal distance between the arc and the trench) may be related to the dip of the Benioff zone; and 3) predict whether subducted lithosphere must reach a certain depth for magma to form. Problem 2 relates to volcanic arcs.

TRANSFORM FAULTS

Transform faults are so called because they transform motion from one plate boundary to another. For example, mid-ocean divergent boundaries are broken into numerous segments by transform faults, because the entire boundary cannot spread at the same rate (Fig. 17.6). Transform faults are strike-slip faults; the principal movement on them is horizontal, and fault planes are essentially vertical. Like strike-slip faults, the sense of movement is described as right-lateral or left-lateral, depending on whether rocks on one side of the fault appear to have moved to the right or left, when viewed from the other side of the fault. Notice that the *apparent* sense of movement on a transform fault may be different than the *actual* sense of movement. For example, in Figure 17.6, the apparent offset of the ridges is right-lateral, but because of the divergence process, the actual movement is left-lateral, as shown by the arrows. Furthermore, movement is taking place only between the two segments of the offset ridge.

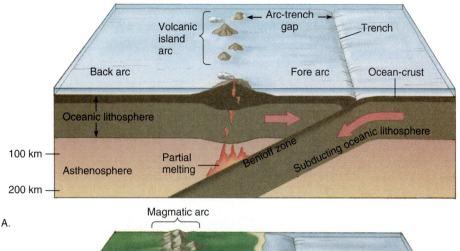

A.

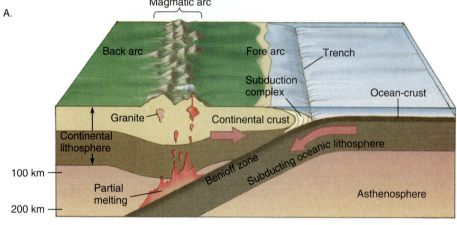

B.

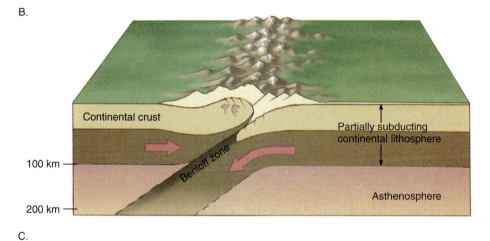

C.

FIGURE **17.5**

Block diagrams of convergent plate boundaries. A. Both plates contain oceanic crust.
B. Subducting plate contains oceanic crust; overriding plate contains continental crust. C. Both
plates contain continental crust.

As with any other kind of fault, movement along a transform fault results in offset of features on opposite sides. Suggest:

1. how the total amount of movement on a transform can be determined from offsets, and

2. how the amount of offset and the ages of the rocks offset can be used to determine the rate of movement and the time at which movement began.

Problem 5 relates to transform faults.

PLATE MOTION AND HOT SPOTS

At more than 100 places on Earth, plumes of hot material rise through the mantle and penetrate the lithosphere. These **hot spots** are sites of persistent volcanic activity. Some, like Iceland, coincide with present-day divergent boundaries, but many do not. Those hot spots within plates may produce strings of volcanoes that form as lithosphere drifts over them (Fig. 17.7). The volcanoes are active when directly over the hot spot, but die out as the lithosphere drifts on.

The position of these plumes within the mantle is fixed, which makes them useful in testing several aspects of plate-tectonic theory. Predict how the ages of hot-spot-produced volcanoes should vary with distance from the hot spot, and suggest a way to test your prediction. Suggest how hot-spot tracks (strings of volcanoes) could be used to monitor the direction of plate movement through time. Problem 3 relates to hot spots.

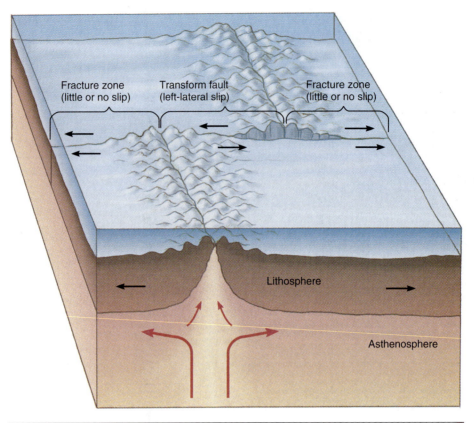

FIGURE **17.6**

Transform fault joining segments of a mid-ocean-ridge, divergent-plate boundary. Movement on fault takes place between ridges but not on fracture-zone extensions.

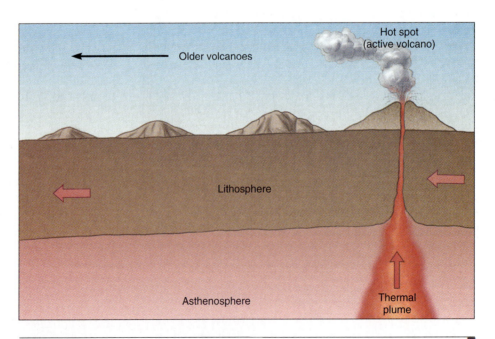

FIGURE **17.7**

A fixed heat and magma source below the lithosphere causes a hot spot on the surface. As the lithosphere moves, a string of volcanoes is formed. Older volcanoes are lower and more rounded, because they have undergone more erosion.

APPLICATIONS

 THE DEVELOPMENT OF THE PLATE TECTONIC THEORY IS AN EXCELLENT EXAMPLE OF HOW SCIENCE WORKS. AS DATA ACCUMULATED AND RELATIONS WERE ESTABLISHED, VARIOUS MODELS OR HYPOTHESES WERE DEVELOPED AND TESTED. THE FOLLOWING PROBLEMS ILLUSTRATE SOME OF THE METHODS USED TO TEST IMPORTANT ASPECTS OF THE THEORY.

OBJECTIVES

If you complete all the problems, you should be able to:

1. Explain how a hypothesis is tested.

2. Explain how magnetic profiles across divergent plate boundaries are interpreted to show divergence and magnetic polarity reversals.

3. Correlate magnetic profiles (a) along a divergent boundary and (b) with the geomagnetic polarity time scale.

4. Calculate divergence rates from magnetic anomalies, given their ages.

5. Draw a cross section of a convergent plate boundary that shows the Benioff zone, volcanic arc, trench, arc-trench gap, forearc, and backarc, and explain their distribution.

6. Use the ages and amount of offset of rocks or surface features to determine the average slip rate along a transform fault.

7. Explain how hot-spot volcanoes form, become extinct, and are used to determine speed and direction of plate movement.

PROBLEMS

1. Figure 17.8 shows magnetic profiles from data gathered in 1965 along four traverses in the South Pacific. The data were collected using a magnetometer (an instrument that measures the magnetic intensity) towed behind a ship. Each profile crosses the Pacific-Antarctic Ridge, the crest of which is shown with a dashed line; water depth is 2 to 4 km. These data were among the first to test the hypothesis of seafloor spreading by comparing magnetic anomalies across and along ocean ridges. The peaks, or positive anomalies, on these profiles correspond to periods of normal magnetic polarity, and the valleys or negative anomalies correspond to times of reversed magnetic polarity.

    a. Four peaks that the authors of the original report found distinctive and useful are labeled on profile A. If the hypothesis of seafloor spreading is correct, similar peaks should be present at roughly equal distances from the ridge axis on the southeastern side of that profile. Identify peaks that are similar to 1, 2, 3, and 4 southeast of the ridge on profile A, and label them 1′, 2′, 3′, and 4′. In searching for matching peaks, use the surrounding peaks and valleys to help identify the numbered ones. To help you get started, 2′ is identified.

    b. A second requirement of the hypothesis of seafloor spreading is that the magnetic anomalies can be connected from profile to profile to form "stripes" parallel to the spreading axis (divergent plate boundary). To test this, first find on profiles B and C the anomalies corresponding to the numbered ones on profile A. To help you, most are already located on profile D. Then connect those anomalies with the same numbers. You may have some difficulty locating the numbered anomalies. That's just the way it is—these are real data—so do your best.

    c. Explain how your results in "a" and "b" do or do not support the seafloor-spreading hypothesis.

2. Figure 17.9 is a map showing a plate boundary in the southwest Pacific Ocean near the Fiji, Samoa, and Tonga Islands. Shown on the map are bathymetry (water depth or depth to seafloor, indicated by contour lines), active volcanoes, and epicenters of earthquakes of varying focal depths.

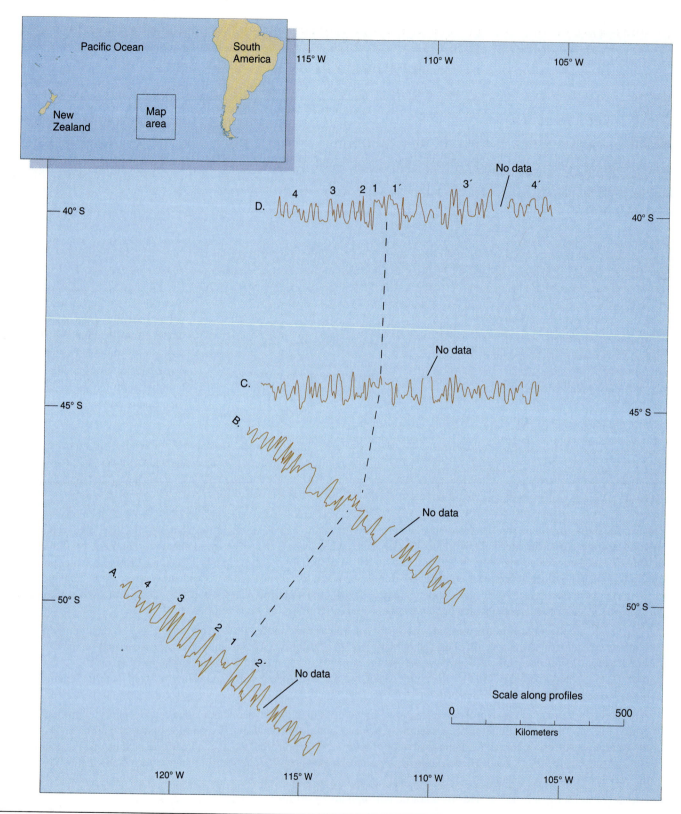

Pacific Ocean

South
America

New
Zealand

Map
area

115° W   110° W   105° W

No data

4   3   2 1   1'   3'   4'

D.

40° S   40° S

No data

C.

45° S   45° S

B.

No data

No data

A.

4

3

2   1

50° S   50° S

2'

No data

Scale along profiles

0   500

Kilometers

120° W   115° W   110° W   105° W

FIGURE **17.8**

Four magnetic profiles across the Pacific-Antarctic Ridge. Dashed line marks the ridge crest. For use with Problem 1.

a. What type of plate boundary is this? Given its location, what type of crust (oceanic or continental) do you think lies to the east? The west?

b. Draw, on the top part of Figure 17.10, a topographic profile of the seafloor along line A-A', Figure 17.9. The topographic profile has a vertical scale of 1 cm to 3,125 m. What is the vertical exaggeration if the horizontal scale is 1 cm to 97.5 km?

c. The bottom part of Figure 17.10 is also a cross section along line A-A', but with no vertical exaggeration. Plot generalized earthquake foci on that cross section by placing circles at the appropriate depths below areas in which earthquakes of various focal groups (different color codes) occur. What is the significance of the focal-depth pattern on your cross section; that is, why are the earthquakes where they are?

d. What is the dip of the earthquake zone? (Measure with a protractor.) Draw arrows on the cross section to illustrate relative motion along the zone.

e. Compare the two profiles. What topographic feature is present where the zone of earthquake foci intersects the surface?

f. Label the trench, forearc, volcanic arc, and backarc on the topographic profile. What is the width of the arc-trench gap? Why are there volcanoes here—how does the magma form? At what depth range does it happen?

3. Figure 17.11 is a map of the Hawaiian Islands and other islands and seamounts (submarine volcanoes) that form the Hawaiian-Emperor chain. All the features along the chain have a volcanic origin, and all are *younger* than the surrounding oceanic crust on which they sit. In 1963, J. Tuzo Wilson proposed that all the volcanoes in the Hawaiian chain had formed above the same hot spot or thermal plume. This idea was later extended to include the Emperor chain. If this hypothesis is correct, then (1) volcanoes should be older farther away from the hot spot, and (2) the distance-age relation can be used to measure the rate of plate movement.

   Table 17.1 lists the ages of these volcanoes and their distance from Kilauea, a currently active volcano on the Island of Hawaii.

   a. Plot the data in Table 17.1 on Figure 17.12, and label each of the points with the name of the volcano. What does the graph indicate about the general relation between age and distance from Kilauea? Draw two best-fit straight-line segments, the first between Kilauea and Laysan, the second between Laysan and Suiko. How do they differ?

   b. Calculate the rate of plate movement in centimeters per year for each of the straight-line segments: divide the distance traveled by the time interval over which travel took place. Remember to convert to the proper units. Are the two values the same? What does this mean? Show your calculations.

   c. The chain bends where the Hawaiian and Emperor chains meet. What could have caused the bend, and when did it happen?

EARTHQUAKE EPICENTERS

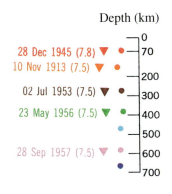

Depth (km)

28 Dec 1945 (7.8) ▼ ●
10 Nov 1913 (7.5) ▼ ●
02 Jul 1953 (7.5) ▼ ●
23 May 1956 (7.5) ▼ ●

28 Sep 1957 (7.5) ▼ ●

0
70
200
300
400
500
600
700

Focal depth Colored dots indicate 1964 through 1977 magnitude 5.0 through 7.4 events; triangles of matching colors labeled by date and magnitude indicate 1899 through 1985 events of magnitude 7.5 or greater (generally surface-wave magnitude)

VOLCANIC CENTERS

Active in historical time Generally within past 1,000 yr. and documented during or soon after eruption; volcano named in color if active 1964 through 1985

Active in Holocene time Within past 10,000 yr; not documented historically

Holocene activity uncertain Includes solfataras and volcano-related thermal springs; no direct evidence for Holocene eruption

FIGURE **17.9**

A plate boundary in the southwest Pacific Ocean near the Fiji, Samoa, and Tonga Islands. Shown on the map are bathymetry (water depth or depth to sea floor), active and recently active volcanoes, and epicenters of earthquakes of varying focal depths. For use with Problem 2.

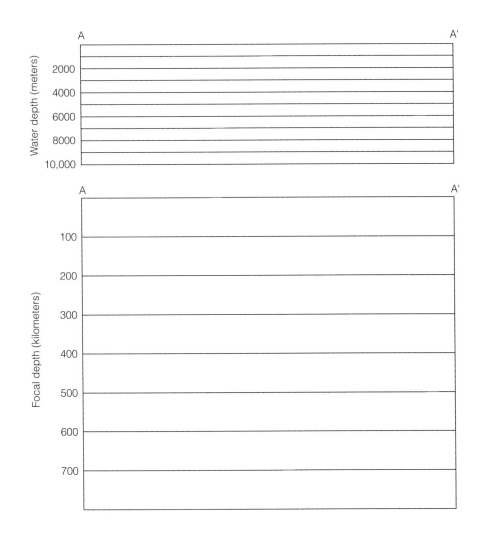

FIGURE **17.10**

Section along line A-A′ of Figure 17.9, for use with Problem 2. Draw a topographic profile on the top part of the diagram, and show earthquake foci on the bottom part. Horizontal scales are the same for top and bottom, but vertical scales are different.

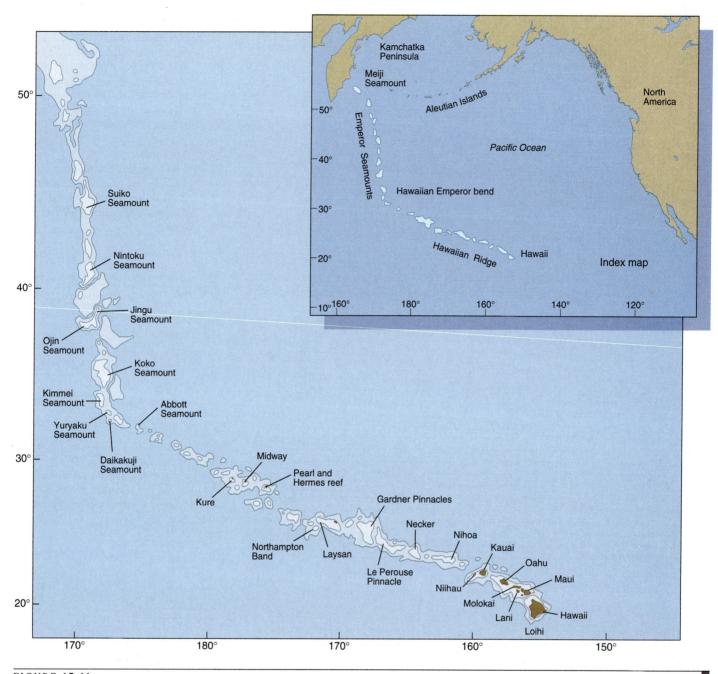

FIGURE 17.11

Hawaiian Islands and other islands and seamounts of the Hawaiian-Emperor chain, shown by 1- and 2-km bathymetric contours. For use with Problem 3.

SOURCE: DATA FROM USGS, PROFESSIONAL PAPER 1350.

Table 17.1

SELECTED VOLCANOES OF THE HAWAIIAN-EMPEROR CHAIN

SOURCE: USGS, *PROFESSIONAL PAPER 1350.*

VOLCANO NAME	DISTANCE FROM KILAUEA (KM)	AGE IN MILLIONS OF YEARS
Mauna Kea (on Hawaii)	54	0.375 ± 0.05
West Maui	221	1.32 ± 0.04
Kauai	519	5.1 ± 0.20
Nihoa	780	7.2 ± 0.3
Necker	1058	10.3 ± 0.4
La Perouse Pinnacle	1209	12.0 ± 0.4
Laysan	1818	19.9 ± 0.3
Midway	2432	27.7 ± 0.6
Abbott	3280	38.7 ± 0.9
Daikakuji	3493	42.4 ± 2.3
Koko (southern)	3758	48.1 ± 0.8
Jingu	4175	55.4 ± 0.9
Nintoku	4452	56.2 ± 0.6
Suiko (southern)	4794	59.6 ± 0.6
Suiko (central)	4860	64.7 ± 1.1

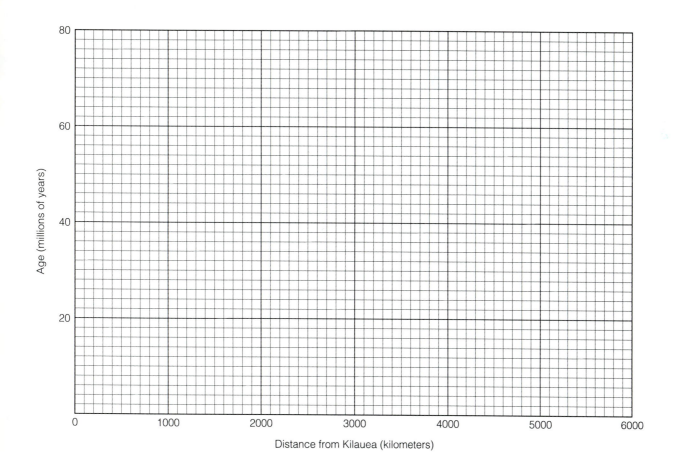

FIGURE **17.12**

Graph of volcano age versus distance from Kilauea volcano along the Hawaiian-Emperor chain, for Problem 3.

4.

a. Go to

ftp://ftp.ngdc.noaa.gov/MGG/images/WorldCrustalAge.gif (or link to it through the author's homepage—see *Preface*), which is a map showing the age of segments of the oceanic crust. Answer the following:

(1) Where is the oldest ocean crust (give a general geographic location and latitude-longitude coordinates)?

(2) How old is the oldest crust? To what geologic period does this correspond?

(3) Where is the youngest oceanic crust, and what is its age?

(4) Explain the relation between the location and age of oceanic crust in terms of plate tectonic theory.

(5) Compare the age of the crust on the east and west sides of South America, and explain the age differences in terms of plate tectonic theory. Hint: look at a map (perhaps in your text, or go or link to *http://www.geo.arizona.edu/saso/eqinfo/plates/index.html*) that shows the tectonic plate boundaries.

b. Go or link to *http://www.ngdc.noaa.gov/mgg/sedthick/sedthick.html,* which is a map showing the total thickness of sediment on the ocean floors. Answer the following:

(1) Where is sediment the thickest, near the continents or in the deep ocean? Why?

(2) The west side of South America is on a convergent plate boundary between a plate with oceanic crust and one with continental crust (e.g., Fig. 17.5B). There is an oceanic trench just offshore, and the volcanic Andes Mountains lie inland. The east side of South America is a trailing continental margin, which was once a divergent plate boundary but now lies well within the South American plate.

(a) On which side of South America is sediment the thickest: the west side, adjacent to the high Andes Mountains, or the east side, where the relief is comparatively low?

(b) Formulate an hypothesis, based on plate tectonic theory, to explain why this is so.

(3) Note the thickness of the sediment in the Bay of Bengal (east of India). This area is also close to a convergent plate boundary, but it is of a type like that in Figure 17.5C.

(a) How does the sediment thickness in the Bay of Bengal compare with that on the west side of South America?

(b) Formulate an hypothesis, based on plate tectonic theory, to explain why this is so.

IN GREATER DEPTH

5. The San Andreas fault is a transform-fault boundary between the North American and Pacific plates in California. Figure 17.13 shows that it is really part of a network of faults collectively called the San Andreas fault *system*. The entire system accommodates movement between the two plates, although at present, most slippage occurs along the San Andreas fault itself. In this problem, you will learn how streams have been offset by the fault and how the study of offset streams and associated sediment can be used to determine the average rate at which movement has occurred in the recent past.

    Figure 17.14 is an aerial view of the San Andreas fault in the arid Carrizo Plain of central California (see Fig. 17.13 for location). A stream, informally named Wallace Creek, drains the Temblor Range to the northeast (look up *temblor* in your dictionary) and crosses the San Andreas fault. The photograph and the map in Figure 17.15 show two prominent stream valleys, both of which bend sharply as they cross the fault. The map also shows three kinds of surficial deposits. Although fan alluvium and slope-wash deposits cover most of the area, it is the two stream-channel alluvium deposits that are of interest here. The older-channel alluvium ranges in age from 10,000 yr BP (years before present, with "present" defined as 1950) to 3680 yr BP. The ages are based on radiocarbon (^{14}C) dates from pieces of charcoal and organic matter found in the deposits.

    a. Why does the present-day stream channel bend sharply at the San Andreas fault? Can you think of two possible explanations? Which is the more likely, and why?

    b. Based on your answer to a, explain the map pattern of the older-channel alluvium.

    c. What evidence is there on the map or photograph for *vertical* movement on the San Andreas fault?

    d. The following statement was made in an article describing Wallace Creek: "Mr. Ray Cavanaugh, who farms at Wallace Creek, reported to us that water actually spilled over the edge [where Wallace Creek bends sharply northwest] in the winter of 1971–1972 or 1972–1973." Based on this statement and your answers to a and b, predict the future of Wallace Creek. (Note the orientation of the north arrow in the *Explanation* in Figure 17.15.)

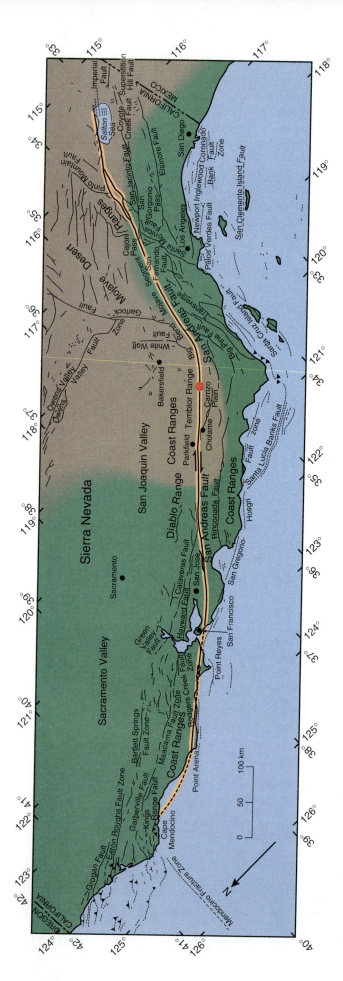

FIGURE **17.13**
The San Andreas fault system. Red dot is location of Wallace Creek.

SOURCE: DATA FROM USGS, PROFESSIONAL PAPER 1515.

FIGURE **17.14**
Oblique aerial photograph of area shown on map in Figure 17.15. The modern and abandoned channels of Wallace Creek are visible on the foreground side of the San Andreas fault.

e. Movement along the San Andreas fault in this area is by "fits and starts." The last large earthquake here, in 1857, was the result of 9.5 m of instantaneous slip. The *average slip rate* of the fault can be determined for two periods in the Wallace Creek area, because two channels of Wallace Creek are displaced by the fault, and the time when the channels first formed is known. Calculate the two slip rates by measuring the offset across the fault for both channels and dividing by the greatest age of the channel deposits. Give your answers in centimeters/year. Show your calculations. Note: the map shows that the channel deposits in both channels just southwest of the fault line "bulge" toward the northwest, where the channels curve abruptly west. The deposits do this because the channels were eroded at the bends before all of the channel alluvium was deposited. Therefore, to get a better idea of the position of the channel when the oldest alluvium was deposited, use points farther downstream for your measurements.

f. The best estimate of the total displacement along the San Andreas fault is about 320 km. This estimate was determined by matching distinctive groups of rocks that are offset by the fault. If the average slip rate you just calculated has been the same for the life of the fault, how old is the San Andreas fault? (Show your work, and remember to keep units consistent.)

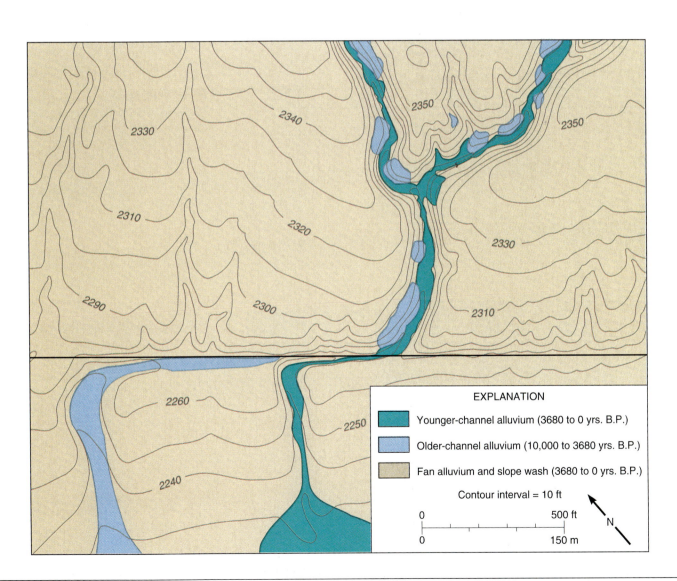

FIGURE **17.15**
Map of Wallace Creek area, for use with Problem 5.

6. Figure 17.16 is an enlargement of profile A of Figure 17.8. If the hypothesis is correct—that peaks on the profile represent normal magnetic polarity, and valleys, negative magnetic polarity—then peaks and valleys should match periods of normal and reversed magnetism known from well-dated rocks from other parts of the world. To test this, fill in the bar above the profile by blackening the areas corresponding to peaks. The dark areas corresponding to large peaks should have widths equal to the width of the peak, as measured about half-way between the top and the bottom. Dark areas for the small peaks should be lines. Several of the trickier ones have been filled in to help you get started, and a few ambiguous parts of the profile have been simplified.

a. Compare your bar diagram with the geomagnetic polarity time scale in Figure 17.17. Start at time zero and see if the patterns match, going back in time. If you are able to match the two, mark points on your diagram (Fig. 17.16) that correspond to time intervals of one million years.

b. Calculate the total *spreading rate* and the *half-spreading rate* in centimeters/year. The spreading rate is the rate at which the two plates are moving away from one another; the half-spreading rate is the rate at which one of the plates moves away from the ridge, and is one-half of the spreading rate, if both plates are moving at the same rate. Use the distance scale given below the profile in Figure 17.16 to determine the distance between correlative anomalies, and the time scale you added to the profile to determine the time it took for them to separate. Spreading rate is distance divided by time, but remember to convert kilometers to centimeters, and millions of years to years, to get the answer in cm/yr. Show your calculations. How does the half-spreading rate compare with the rate of plate movement you calculated in Problem 3?

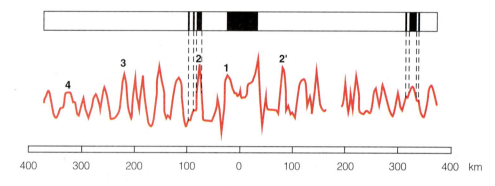

FIGURE **17.16**
Enlargement of Profile A from Figure 17.8, for use with Problem 6.

Magnetic polarity (black = normal, white = reversed)	
Age (millions of years)	1 2 3 4 5 6 7 8 9 10

FIGURE **17.17**
Geomagnetic polarity time scale.

CREDITS

PHOTOGRAPHS

PART OPENER

1: © Doug Sherman/Geofile; **2:** © Doug Sherman/Geofile **3:** Courtesy of Earth Satellite Corporation; **4:** © Doug Sherman/Geofile; **5:** © Doug Sherman/Geofile; **6:** © Doug Sherman

CHAPTER 1

1.1, 1.2A-H, 1.3, 1.4, 1.6B, 1.7, 1.9A, 1.10A (both), **1.11A** (both), **1.12A, 1.13A, 1.14, 1.15, 1.16, 1.17A,B, 1.18A,B, 1.19, 1.20. 1.21:** © Doug Sherman/Geofile

CHAPTER 2

2.1, 2.2A-F, 2.3A,B,C, 2.2, 2.5, 2.6, 2.7, 2.8A,B, 2.9, 2.10, 2.11, 2.12, 2.13, 2.14, 2.15, 2.16: © Doug Sherman/Geofile

CHAPTER 3

3.1A,B, 3.2, 3.3A,B, 3.4, 3.5, 3.6, 3.7, 3.8A,B, 3.9,A,B, 3.10A,B, 3.11A,B, 3.12A,B, 3.13, 3.14A,B, 3.15A,B, 3.16: © Doug Sherman/Geofile

CHAPTER 4

4.3, 4.4, 4.5A,B, 4.6: © Doug Sherman/Geofile; **4.7:** Norris W. Jones; **4.10, 4.11, 4.12, 4.13, 4.14, 4.15A,B, 4.16A-E; 4.17, 4.18, 4.19, 4.20, 4.21:** © Doug Sherman/Geofile

CHAPTER 5

5.3A,B, 5.4, 5.5, 5.6, 5.7, 5.8A,B, 5.9, 5.11, 5.12: © Doug Sherman/Geofile

CHAPTER 6

6.2: U.S. Geological Survey; **6.16:** Massachusetts Office of Travel and Tourism

CHAPTER 7

7.3A,B,C: U.S. Geological Survey, ESIC; **7.3D:** U.S. Geological Survey, EROS Data Center; **7.3E:** Courtesy of Earth Satellite Corporation; **7.3F:** U.S. Geological Survey, EROS Data Center; **7.5:** U.S. Geological Survey, ESIC; **7.61:** U.S. Geological Survey, Denver, CO; **7.8:** U.S. Geological Survey, EROS Data Center; **7.9:** U.S. Geological Survey, ESIC; **7.10A-H:** NOAA

CHAPTER 8

8.3, 8.4, 8.9: U.S. Geological Survey, Rolla, MO; **8.10:** U.S. Geological Survey, Denver, CO; **8.11, 8.12, 8.13:** U.S. Geological Survey, Rolla, MO; **8.16:** U.S. Geological Survey, Denver, CO

CHAPTER 9

9.8: U.S. Geological Survey, Rolla, MO; **9.9:** Courtesy of James F. Quinlan and Joseph A. Ray, 1981, revised 1989; **9.10:** U.S. Geological Survey, Rolla, MO

CHAPTER 10

10.1: Austin Post, U.S. Geological Survey, Tacoma, WA; **10.2:** Robert Felder, U.S. Geological Survey, Menlo Park, CA; **10.4** Norris Jones; **10.7:** U.S. Geological Survey, Denver, CO; **10.8:** Prepared from USC & GS Photography by the University of Illinois Committee on Aerial Photography; **10.9:** U.S. Geological Survey, Denver, CO; **10.10:** Geological Society of America; **10.11:** U.S. Geological Survey, ESIC; **10.12:** U.S. Geological Survey, Rollo, MO; **10.13;** USGS Prof. Paper 1180, 1970, Courtesy U.S. Geological Survey, Denver, CO

CHAPTER 11

11.8: © Doug Sherman/Geofile; **11.9A,B,C:** Prepared from USC and GS Photography by the University of Illinois Committee on Aerial Photography: **11.10:** U.S. Geological Survey, Denver, CO; **11.11:** U.S. Geological Survey, Rollo, MO; **11.12:** U.S. Geological Survey, Denver, CO; **11.14:** U.S. Geological Survey, NAPP

CHAPTER 12

12.1, 12.2, 12.5, 12.6: ©Doug Sherman/Geofile; **12.9:** Prepared from USGS Photography by the University of Illinois Committee on Aerial Photography; **12.10, 12.11, 12.12:** U.S. Geological Survey, Denver, CO; **12.13:** U.S. Geological Survey

CHAPTER 14

14.2: Courtesy of Brian K. McKnight

CHAPTER 15

15.2B: U.S. Geological Survey, Denver, ESIC; **15.3B:** Prepared from USDA-AAA Photography by the University of Illinois Committee on Aerial Photography; **15.4B;** U.S. Geological Survey, ESIC; **15.5B, 15.6B:** U.S. Geological Survey, Denver, ESIC; **15.12:** U.S. Geological Survey; **15.14:** N.H. Darton, 1951. Geologic Map of South Dakota, U.S. Geological Survey; **15.15:** U.S. Geological Survey, EROS Data Center; **15.17, 15.18:** U.S. Geological Survey

CHAPTER 16

16.6: U.S. Geological Survey, Denver, CO; **16.13:** U.S. Geological Survey, EROS Data Center; **16.14:** Timothy Hall, U.S. Geological Survey

CHAPTER 17

17.9: American Association of Petroleum Geologists; **17.14:** U.S. Geological Survey, Menlo Park, CA

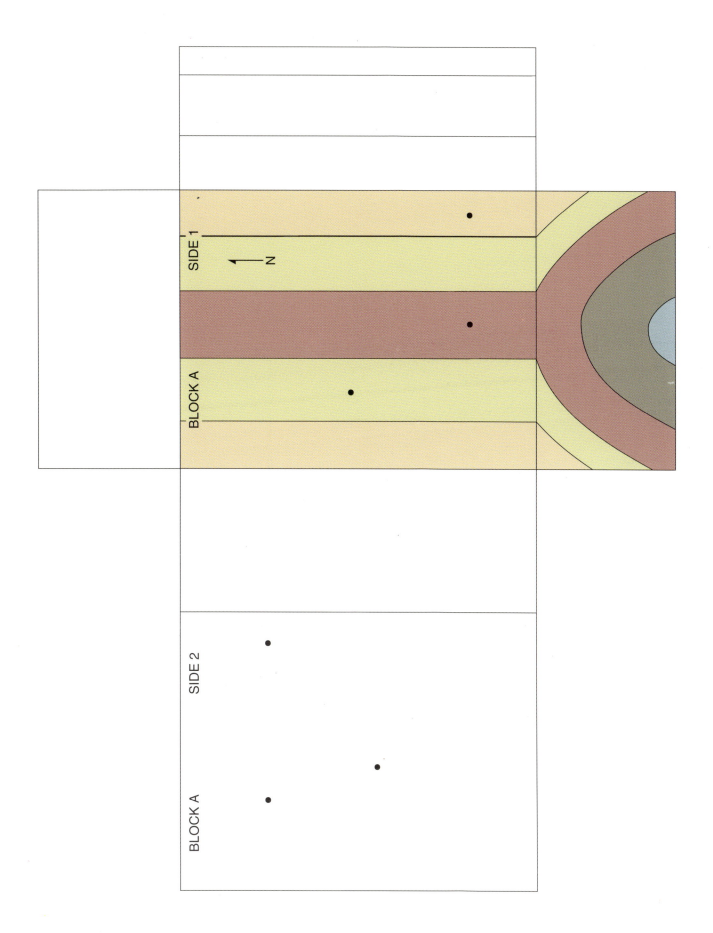

SIDE 1

BLOCK A

N

SIDE 2

BLOCK A

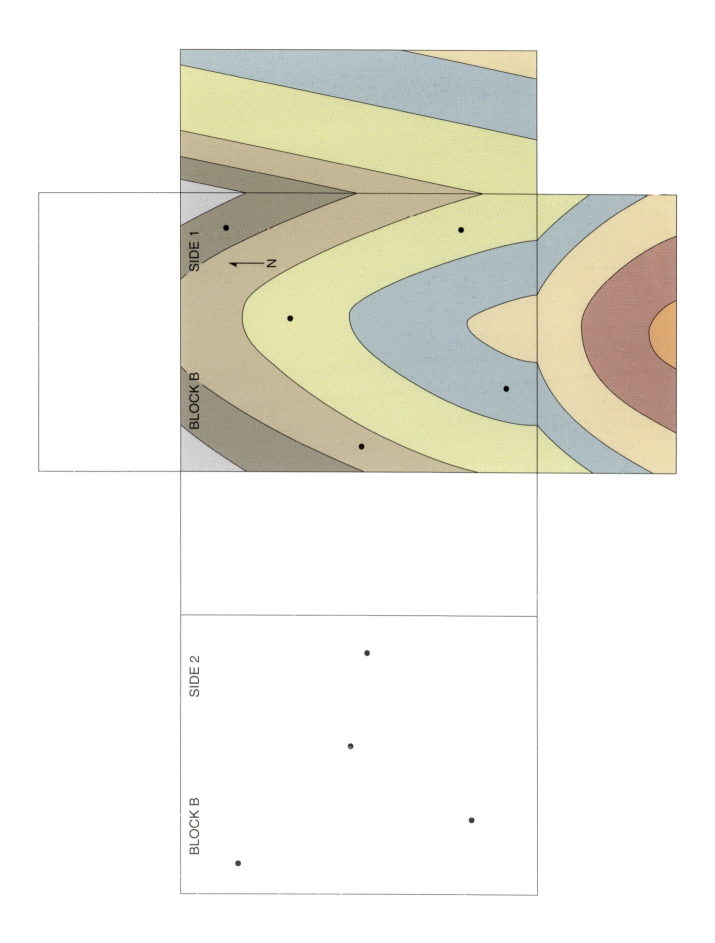

SIDE 1

BLOCK B

N

SIDE 2

BLOCK B

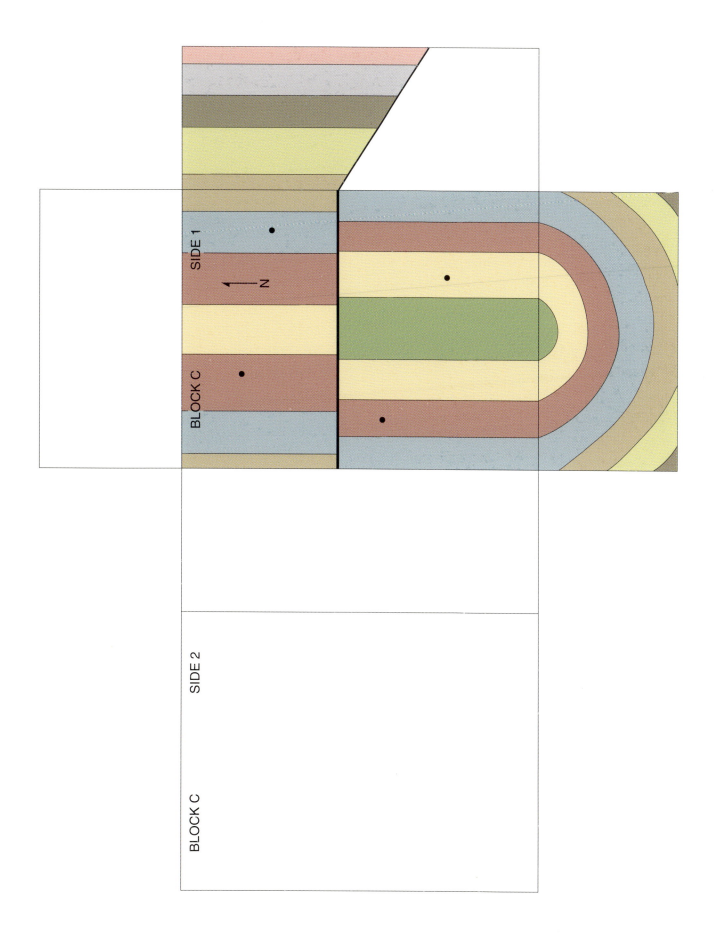

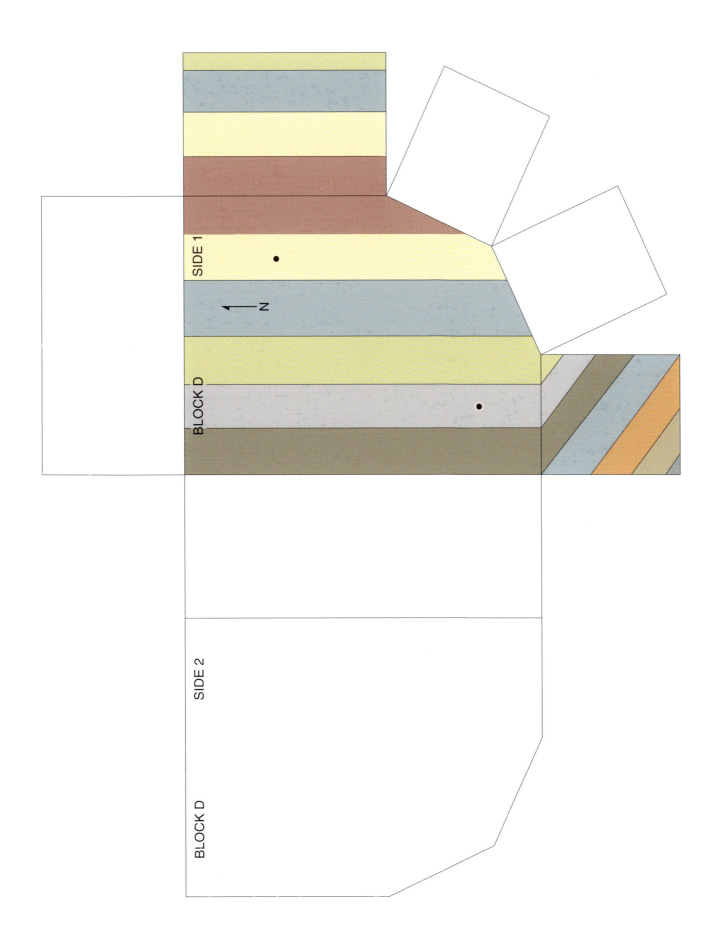